The Muscle Book

Anatomy • Testing • Movement

Klaus-Peter Valerius

Astrid Frank

Bernard C. Kolster

Christine Hamilton

Enrique Alejandre Lafont

Roland Kreutzer

QUINTESSENCE PUBLISHING

Authors' addresses:
Dr. rer. nat. Dr. med. K.-P. Valerius
AG Medizinische Zellbiologie
Institut für Anatomie und Zellbiologie
(Institute of Anatomy and Cell Biology)
Justus-Liebig-Universität Gießen
Aulweg 123, 35385 Gießen, Germany

Astrid Frank
Weiershäuser Str. 19, 35096 Weiershausen, Germany

Dr. med. Bernard C. Kolster
Gabelsberger Str. 24, 35037 Marburg, Germany

Christine Hamilton
Anna-Rosenthal-Weg 41, 91052 Erlangen, Germany

Dr. med. Enrique Alejandre Lafont
Sandfeld 23, 35396 Gießen, Germany

Roland Kreutzer
Unter den Eichen 37, 35041 Marburg, Germany

Quintessence
British Library Cataloguing in Publication Data
The muscle book.
1. Muscles--Atlases. 2. Musculoskeletal system--Atlases.
I. Valerius, Klaus-Peter.
611.7'3'0222-dc22
ISBN-13: 9781850972136

QUINTESSENCE PUBLISHING
© 2011 by Quintessence Publishing Co. Ltd.

Original title "Das Muskelbuch. Anatomie, Untersuchung, Bewegung", 5th edition
ISBN: 978-3-940698-20-9
© 2009 KVM – Der Medizinverlag Dr. Kolster GmbH, Marburg, Germany

Translation: Rosana Jelaska, M. Sc., MITI
Photos: Peter Mertin, Köln
Graphics: David Kühn, Marburg
Production: Quintessenz Verlags-GmbH, Berlin
Printed by: Bosch-Druck, Landshut/Ergolding

ISBN: 978-1-85097-213-6
Printed in Germany

Foreword to the extended 5th edition

The resounding success of The Muscle Book and the impressive feedback it has received have exceeded all expectations, bringing the work close to being a classic. As a result of all this, it is already time for a new edition. This time, since we it would not do to rest on our laurels, we have not only revised the text but also supplemented it considerably. Among other things, the book now includes stretching techniques for certain muscles – a great deal of work which will surely benefit the reader.

However, the basic concept has not changed. It is still quick for the practitioner to find key information on each muscle, in a compact, easily scanned layout. The renewed success of this concept is also due to Mr. Roland Kreutzer, who took on the project at KVM. My particular thanks go to Dr. Wieland Stöckmann, for his tireless work in correcting errors and inaccuracies and for helping me further my functional understanding of the locomotor apparatus.

Gießen, July 2009

Klaus-Peter Valerius
(On behalf of the authors)

Foreword

This representation of the muscles of the human locomotor apparatus is intended for physicians needing a quick overview of a particular muscle, and, most of all, for physiotherapists and members of other professions concerned with the human locomotor apparatus, its training, its disorders and its treatment.

The book's structure is aimed particularly at the requirements of the latter group of readers: rather than being arranged according to systematic groupings, the criteria used in muscle classification are far more functional and relate to the movements of individual joints. However, any muscle required by the reader is easily located using the Table of contents.

Physiotherapy requires not only an accurate knowledge of the anatomy of a muscle but also, above all, an understanding of the great variety of movements in which a muscle is involved. For this reason, muscle functions need to relate to individual joints and movement axes. The synergists and antagonists involved in a muscle's potential movements at various joints are also listed each time. This concept, which may at first seem somewhat redundant, provides the reader with extensive information at a glance, saving him or her from having to consult and decipher complex tables.

Of course, a concept of this sort also has its limits. Thus, muscles generally have many different functions. Some movements will be so weak or have so little impact that including them would go too far, serving only to confuse the reader. A muscle will often have a very different function when the joint it serves is at the extremes of its range of motion than when it is in its neutral position. Consequently, the information shown in this book generally relates to the neutral positions of joints and only considers effects at other starting positions if they are really relevant to the patients' movement sequences. In the majority of cases, functions which are disputed in the literature are not included.

The functional range of the muscles of the neck and thorax has been deliberately left out, as a detailed treatment of these would go far beyond the scope of this book: this applies particularly to a representation of the accessory respiratory functions which come into play in normal breathing.

Particular value has been placed on the superficial anatomy of the active locomotor apparatus, to help the practitioner to find his or her way around the patient and to facilitate the identification and palpation of muscles. However, not every muscle, its belly taut, leaps to the eye immediately through skin and subcutaneous tissue. While some muscle do lie below the skin and are also readily palpable there, they may still not be visible. Other muscles have their origins mainly in the body's fascia and pull it inwards, particularly when contracting isometrically. Where this is the case, the muscle shows up as a hollow in the skin. To minimize this problem, the edges of these muscles or the areas of skin below which they lie have been marked out with arrows. In some cases, dots indicate the places at which contraction can be palpated.

Our heartfelt thanks go to our photo models with their endless patience, who were photographed all day long in strenuous poses. The detailed representation of superficial muscle anatomy, to be found virtually nowhere else in such systematic form, would not have been possible without Peter Düsing and E. Lafont.

We would also like to extend our cordial thanks to Ms. Sabine Rasel and the staff at KVM-Verlag, namely Mr. Bernard Kolster, Ms. Martina Kunze, Ms. Sabine Poppe and Ms. Astrid Waskowiak, for their dedicated, ever-friendly and capable help, as well as to Mr. Peter Mertin for his high-quality photographs.

Gießen, July 2009

Klaus-Peter Valerius
(On behalf of the authors)

Notes on the structure of this book

To ensure clarity and instructive value, every muscle has been given its own page in this book. If the reader needs to look up an individual muscle for any reason, it can be found directly via the index.

Each muscle is described in a brief introductory text detailing its functions in functional context with other muscles. The "Functions" table lists its synergists and antagonists. The muscles included here are differentiated further: according to their functions and also in order of their strength or power compared with the other muscles. First place is given to those muscles which contribute the greatest power to the relevant movement.

The representation of superficial anatomy was an important requirement. With this aim in mind, anatomical drawings are shown in relation to in-vivo photographs wherever possible. In general, the muscles in these pictures are marked out using various symbols.

Key message: The individual entry for each muscle always represents it in interaction with other muscles which also act on the relevant joint.

Should the reader wish to focus on muscles which are involved in a particular direction of movement (e.g. external rotation of the hip), the table "Principal muscles used in individual movements" in the Appendix provides a list of muscles for the joints of the upper and lower extremity.

An assessment of function and power is significant when resolving clinical questions. Thus, muscle function tests are shown for virtually every muscle (the diaphragm and pelvic floor muscles are exceptions). Placing the anatomy of a muscle and its "functional principles" on facing pages provides the user with a basic factual and visual understanding of the functional tests. In this context, it is essential to provide the patient with clear and comprehensible instructions for each test.

For this reason, our suggested wording is given in the form of direct speech. "Lift your shoulder off the couch" is a short, apposite instruction for a specific direction of movement. Please do use these helpful instructions as a guide in your daily practice routine. For simplicity of instructions the patient is referred to as "him".

Another point to note is that a separate function test does not exist for every muscle. If a muscle function test follows the entries for several individual muscles, this means that the test relates to the muscles that precede it.

Functions of the skeletal muscles – classification and stretching capacity

In the muscle function tests, we have used the standard categories of Hislop and Montgomery (2000) and have divided the muscle power being rated into six grades.

Muscle status 5 (normal)

Muscle status 5 corresponds to the full power that can be deployed by a muscle. For this status to be reached in a test, the full range of motion of which the muscle is capable during the contraction needs to be performed against sub-maximal resistance, as applied externally by the therapist. The subject also needs to be able to support the weight of the body part being tested.

Muscle status 4 (good)

Muscle status 4 corresponds to approx. 75% of a muscle's normal power. The full range of motion is required here, as in muscle status 5. Similarly, the test needs to be performed against the weight of the body part being tested, but this time only against moderate resistance by the therapist.

Muscle status 3 (weak)

Muscle status 3 corresponds to approx. 50% of normal muscle function. The full range of motion is tested, once more against the weight of the body part involved, but without the application of additional resistance by the therapist.

Muscle status 2 (very weak)

Muscle status 2 corresponds to approx. 25% of normal muscle power. To reach muscle status 2, the muscle can perform its full range of movement only if the weight of the body part being tested is supported, i.e. the muscle is no longer capable of overcoming the slight resistance exerted by gravity.

Muscle status 1 (contraction virtually undetectable)

Muscle status 1 indicates than only about 10% of normal muscle power is still present. All that can be seen or palpated is the contraction of a muscle, e.g. in the form of a twitch. This means that the muscle is still contracting, but that its power is no longer sufficient to move the body part in question. The test may be performed from the same starting position as for muscle status 2 (without gravity), but it is often more effective to attempt a contraction against gravity (see status 5-3), as this is more likely to lead to contraction of the muscle.

Muscle status 0 (no function)

Muscle status 0 is reached if there is no visible or palpable movement.

The therapist should generally ensure that resistance is applied gradually and slowly, to avoid injuries. The level of resistance, for example when testing for muscle status 5, should be varied according to the size and function of the muscle being tested and adjusted to the patient's general constitution. In many cases, the therapist will be able to draw on his or her own experience when doing this. If a muscle being tested can only achieve part of its full range of motion, it will be given a lower power grading; an additional comment to this effect is a good idea where this is the case.

A further point that will need to be clarified is whether other factors may be restricting mobility, such as reduced stretching capacity of the antagonists or problems in the joint, including the capsules and ligaments. In these cases, it is advisable to perform the movement passively with care, to limit any such factors. If the full range of motion can be performed in this way, the source of the problem is likely to lie in the muscle's innervation.

For small muscles of the hands, fingers, feet and toes, it is no longer possible to have different starting positions for muscle status 2 and 3. The weight of the body part being tested plays only a minor role in these areas.

The muscles of the head (e.g. muscles of facial expression and eye muscles) are not tested for muscle status or stretching but only for muscle activity, as it is very difficult to assess the power of these muscles. Classifying these muscles as showing marked, slight or no contraction is one possible way forward.

Some movements are difficult to perform consciously. For this reason, please give the patient the opportunity to "rehearse" the movement beforehand; only the second or third attempt should be assessed.

Apart from muscle power, the optimum stretching capacity of muscles is another key factor in ensuring harmonious, efficient movement. Single-joint muscles should be able to stretch across the full range of motion of the joint. For two-joint or multiple-joint muscles, stretching capacity depends on the position of the joints spanned by the muscle being tested. Passive insufficiency may occur if a muscle is simultaneously stretched across two or more joints. If this is the case, the muscle stretch test will not assess the potential mobility of the joints in full (e.g. stretching capacity of the ischiocrural muscles, see p. 184). The expected angles of the relevant joints are indicated in these cases.

Table of Contents

Abbreviations and symbols

The following abbreviations and symbols are used in this book:

Joints

DIP	distal interphalangeal joint
PIP	proximal interphalangeal joint
MCP	metacarpophalangeal joint
CMC	carpometacarpal joint
MTP	metatarsophalangeal joint

Spinal segments

C	Symbol for the cervical vertrebrae
T	Symbol for the thoracic vertrebrae
L	Symbol for the lumbar vertrebrae
S	Symbol for the sacral segments

●	Dots indicate zones at which contraction of the relevant muscle is palpable
●	Color indicating the origin of the muscle
●	Color indicating the insertion of the muscle
→	The points of the arrows show the limits of the structures described in the text
▨	Hatched areas represent surfaces which are not palpable or which are not located in the plane of vision

Authors

Dr. rer. nat. Klaus-Peter Valerius, MD

Studied Biology, specializing in anthropology and vertebrate morphology in Berlin and Göttingen, and studied human medicine in Gießen; since 1990, has worked as Research Associate at the Institute of Anatomy and Cell Biology of the University of Gießen; part of the team organizing courses in anatomy and biology for the preclinical degree in human medicine.

Astrid Frank

Trained physiotherapist working in orthopedics, surgery, gynecology, internal medicine and pediatric medicine; since 1981, has taught at the "Rudolf-Klapp-Schule" training institute for physiotherapists in Marburg, specializing in orthopedics; from 1992 to 1998, member of the Instructors' Board of Dr. Brügger in Murnau and St. Peter-Ording (Functional Disorders of the Locomotor Apparatus), where she was responsible for the training and continued professional development of physiotherapists and doctors; since 2002, she has worked freelance at a physiotherapy practice.

Dr. Bernard C. Kolster, MD

After training to become a physiotherapist, studied human biology and human medicine in Marburg; clinical training in physical medicine and other disciplines; specialized in the demonstration and application of reflex therapy methods; author and editor of textbooks in medicine, physiotherapy, natural treatment methods and physical medicine.

Christine Hamilton

Studied physiotherapy, graduating as Bachelor of Physiotherapy, University of Queensland, 1979; worked in Bern (Switzerland) until 1986 and has lived in Germany since 1987. Research stay in Brisbane (Australia) from 1993 to 1995. In 1995, qualified as Master and Research Master in the specialist area "deep muscle function and back problems" in 1995; member of the Joint Stability Research Group of the University of Queensland (Australia) led by Dr. Carolyn Richardson.

Dr. Enrique Alejandre Lafont, MD

Studied human medicine at the University of Gießen. Assistant doctor (resident) since 2004. Further training to qualify as radiology specialist (one clinical year in surgery). Active track and field athlete for 12 years, athletic sports for the past nine years.

Roland Kreutzer

Physiotherapist and Brügger therapy instructor, research associate and lecturer at the Philipps-Universität Marburg, Department of Medicine.

1 Theory

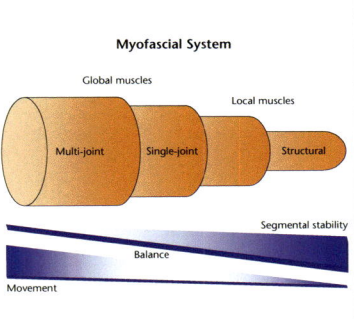

1.1 Functions of the skeletal muscles

The skeletal muscles, together with the skeletal system itself (i.e. the bones, joint capsules and ligaments), fulfill two important roles of the locomotor apparatus, namely movement and protection. In doing so, the muscles exert two seemingly opposing functions, in that they simultaneously initiate and control movement (Twomey and Taylor 1979). This control of movement is referred to as stability (White and Panjabi 1990). The movement of the body takes place at three different levels:

1. movement of the body in space (e.g. jumping)
2. movement between different body parts (e.g. the thorax and pelvis) and
3. intra-articular movement (arthrokinematics).

All movements at these three levels influence each other interactively and need to be controlled in order to achieve overall stability and, consequently, optimum protection (Fig. 1-1). Balance is defined by the control of the movement of the body mass in space and the control of the movement of the body's segments relative to each other. In this process, the various segments of the body adjust relative to one another and also to the force of gravity. The control of intra-articular movements, such as the rolling and sliding of articular surfaces, represents segmental stability. This stability contributes significantly towards the protection of the pain-sensitive structure of the joints and surrounding tissues, for example the nerves and organs.

At all three levels, the muscles are involved in executing movement and maintaining stability.

Thus, the skeletal muscles are responsible for:
- initiating and executing movement
- maintaining balance
- maintaining segmental stability (Fig. 1-1).

Each of these three functions requires certain anatomical, biomechanical and physiological properties from the muscles. In principle, virtually all muscles are capable of fulfilling all three functions. However, various muscles differ from one another in their efficiency in this regard.

Three levels of body movement

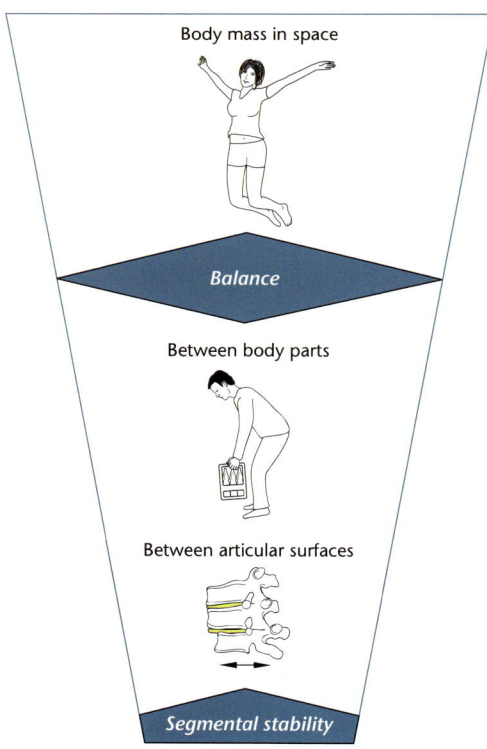

Fig. 1-1: Three levels of body movement and stability (modified from Richardson, Hodges et al. 2004).

1.2 Classification of the skeletal muscles: the myofascial system

The myofascial system (Richardson, Hodges et al., 2004) describes the functions of the skeletal system in a schematic way (Fig. 1-2). In this model, the locomotor apparatus is subdivided into concentric layers (rather like a tree's growth rings). The innermost layer represents the structural skeletal system (the bones, ligaments and joint capsules). It is responsible for the control of intra-articular movement and for segmental protection; i.e. segmental stability.

The deep, local muscle system is closely linked to the skeletal system. With their short, transverse fibers, located close to the joints, the local muscles are highly suited for the direct maintenance of structural segmental stability. In contrast, the outermost layer contains mainly the long, superficial, multi-joint muscles. These muscles are optimally suited for the initiation of rapid movements. The majority of skeletal muscles belong to the middle muscle layer: the single-joint, global system. Depending on requirement, the function of these superficial muscles varies between the initiation of movement and the maintenance of balance (Fig. 1-1).

The classification of a muscle within the myofascial system depends largely on its anatomical properties. To begin with, we differentiate between two systems: the local and the global muscle system. Within the global system, we differentiate between the single-joint and the multi-joint muscles.

The classification of muscles within the myofascial system is a different way of looking at this subject, in addition to the traditional classification of muscle function, movement and power. It may appear somewhat schematic and, furthermore, not all muscles are clearly classifiable within this system. However, the model can be used as plausible tool for describing the functions of the muscles. The model enables individual muscles to be accurately identified, particularly in the area of segmental stability, in which power and movement play a lesser role for the clinical relevance of the myofascial system (see p. 7).

The myofascial system and its functional classification are described in greater detail in the following pages.

Myofascial System

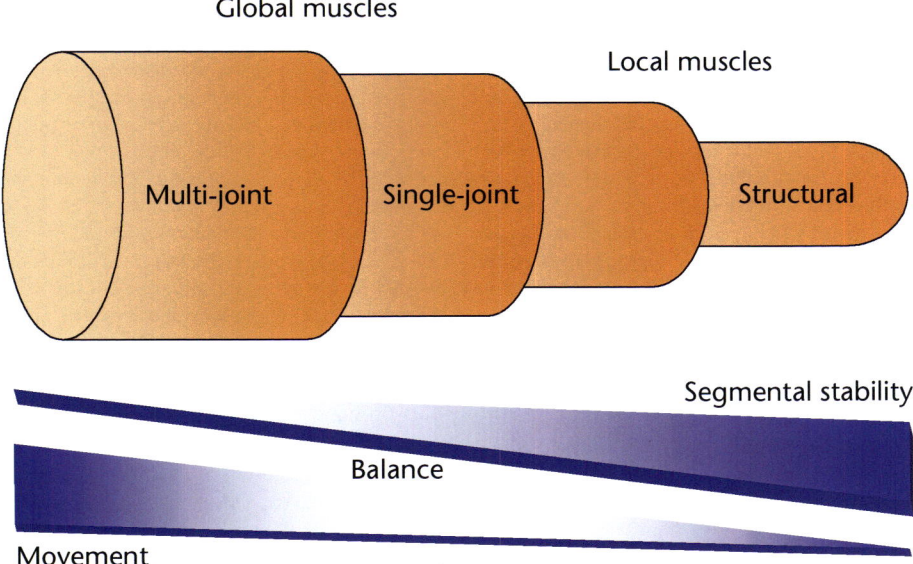

Fig. 1-2: In the myofascial system, the individual skeletal muscles are divided into layers according to their function. The local muscles, which lie deep in the body, are best suited for the maintenance of active segmental stability. Conversely, the superficially located multi-joint muscles produce the most effective movement acceleration. Between these two muscle groups lie the single-joint muscles, which control the stabilization of balance (modified from Richardson, Hodges et al. 2004).

1.2.1 Initiating and executing movement

The skeletal muscles are intrinsically capable of active shortening. This active contraction and the muscles' direct or indirect insertion sites on the skeleton generate torque, movement and force (van den Berg 1999). As a result, muscles are capable of initiating and executing movements, both between various parts of the body and in space. Depending on a muscle's morphological properties, it is assigned one or more directions of movement. The angle of the bundles of muscle fibers (known as fascicles), with the instantaneous physiological axis of movement is measured and divided into its vectors. These vectors give the possible directions of pull of the fascicles. Together with a calculation of the lever arm of the torque, one can then: the direction of movement and the force. Testing the force generated in the muscle's principal direction of movement forms the basis of the traditional muscle function tests.

Factors determining the efficiency of movement

The efficiency with which muscles generate torque depends on a number of biomechanical and physiological factors.
These consist primarily of the physiological cross-sectional size of the muscle, the lever arm and the length and orientation of the fascicles, which influence the movement efficiency. Muscle fibers which lie at a right angle to the principal instantaneous axis of movement produce more effective torque than fibers which are oblique or parallel to that axis. Any change in the position of the joint leads to a change in the length of the muscle and/or its orientation to the axis of movement, thus influencing its power. Shortening or lengthening a muscle by 20% reduces the intrinsic force generated by it; that i, it leads to mechanical insufficiency (Macintosh, Bogduk et al. 1993). For this reason, economical generation of force takes place when a muscle is at its neutral or functional length. At this length, the actin and myosin filaments of the sarcomeres are in their optimum relative positions. For example, the functional length of the biceps muscle of the arm is at c. 90° elbow flexion. Stance and starting position, therefore, play an important role in muscle training. The long, superficial and multi-joint muscles are particularly prone to mechanical insufficiency.
There are also additional extrinsic factors, such as the viscoelastic properties (the fatty and connective tissue fraction) or tendon length of a muscle, which influence force generation and transfer. The distribution of muscle fiber types (I, IIa, IIb) also influences force generation: thus, type I muscle fibers tend to generate a more moderate force which remains constant for a very long time. The type IIa muscle fibers produce a force which is greater, but of shorter duration, while type IIb muscle fibers produce a very powerful force, which can be maintained for only a short time. The type of contraction and the movement speed, together with systematic and physiological factors, also influence the quality and efficiency of the movement.

The initiation of impulses that cause movement is often concentric, rapid and short-term. The muscles give off short, powerful impulses in the desired direction. The long, superficial, multi-joint muscles are particularly suitable in this regard. The continued muscular activity attempts to limit the undesirable consequences of movement initiation, while ensuring that there is as little loss of economy of movement as possible.

1.2.2 Maintaining balance

Each impulse that causes movement has a mechanical effect on balance, on the adjoining joints and on the articular surfaces (pressure and shear force). If a confident, accurate movement is to be executed, its effect on the body as a whole needs to be controlled. The interaction between movement, balance and segmental stability becomes apparent here. In maintaining balance, the muscles perform both braking and straightening actions. These are prerequisites for the orientation, position and posture of the body in space.
Mathematically speaking, balance is a model of equilibrium. For example, when the upper body is inclined forwards, the extensors compensate for the effect of the shear force on the upper body.
    Force acting ventrally = force acting dorsally = equilibrium
Control of the proximal position of a joint is another function involved in the maintenance of balance. In this case, the effect of a peripheral movement and the muscles involved in it need to be compensated for at the proximal joint.
One example of this is flexion of the elbow when standing. In this case, it is necessary to compensate not only for the weight of the forearm but also for the action of the biceps muscle on the shoulder; otherwise, the proximal insertion of the long head of the biceps muscle would pull the shoulder towards the forearm. Here, the job of the muscles of the pectoral girdle is to hold the shoulder firmly in place (i.e. to keep it balanced), in order to allow the intended movement to take place unhindered.
When performing their function of maintaining balance, the muscles often work statically/eccentrically and in a sustained way. Thus, a mixture of power and endurance is often required from the global single-joint muscles. The ratio between power and endurance appears to be activity - and training -specific (Mandell, Weitz et al. 1993; Mannion, Junge et al. 2001). Typical measures of muscle function, such as power and endurance, therefore show high inter- and intraindividual variability. For example, the trunk of a computer programmer will have a different power to endurance ratio to that of a bricklayer. And similarly, a footballer's supporting leg will show different quadriceps femoris muscle activity to his swinging leg. The efficiency with which the skeletal muscles maintain balance is subject to similar biomechanical and physiological factors as in the initiation of movement (see above).

1.2.3. Maintaining segmental stability

The term segmental stability refers to the control of arthrokinematics. The structural, skeletal system on its own is rarely capable of providing sufficient segmental protection for the joints and the surrounding tissues (Cholewicki and McGill 1996). Consequently, the muscles need to be called in to help protect the joints. The deep local muscles located close to the joints cushion impacts and stresses and also limit any rolling, sliding and shear movements which deviate from the normal axis. At the same time, they enable primary joint movements to be executed as economically as possible.

In this role, the local muscles function as an additional, active joint capsule. With their intrinsic elastic rigidity (i.e. their elastic spring force), the muscles provide immediate resistance to undesirable and abnormal movements between the articular surfaces. Nevertheless, the muscle – in contrast to the joint capsule – is capable of changing its elasticity. An increase in muscle activity increases the muscle's elastic rigidity and thus also its cushioning effect. However, this effect is more or less limited to 25% of the muscle's maximum activity. A co-contraction (i.e. the simultaneous activation of an antagonistic muscle) is therefore the only other remaining possibility of increasing active segmental stability to any appreciable extent (Hogan 1990; Johansson, Sjolander et al. 1991; Lloyd 2001). A co-contraction of two local muscles with 25% activity is more effective and more efficient than 100% activity of a single muscle (Hoffer and Andreassen 1981; Hodges, Kaigle Holm et al. 2003). Depending on its position, a muscle may give more or less elastic rigidity to a joint. Short, deep muscles lying close to a joint can give more effective protection than long, superficial ones. Accordingly, the biomechanical, anatomical and physiological prerequisites for effective segmental stability are almost opposite to those involved in the efficiency of movement. Therefore, local muscles provide outstanding segmental stabilization but are rarely capable of initiating a movement or maintaining balance. By comparison, global muscles are virtually incapable of segmental stabilization.

The different properties of the local and global systems confer a considerable advantage to the locomotor apparatus. Because of their transverse, deep fascicles, a co-contraction of the local muscles provides hardly any resistance to the agonists of primary joint movement. Furthermore, the mechanical efficiency of segment-stabilizing function does not depend either on the initial position or on posture. This provides long-lasting, constant protection throughout the range of movement, with minimal loss of economy of movement.

The arrangement of the muscles in the myofascial system is highly effective and efficient with regard to their differing roles, namely movement and stability. The global muscles initiate movement and maintain balance, whereas the local muscles preserve segmental stability.

1.3 Characteristics of the individual muscle systems

Whether a muscle belongs to the local or the global system depends primarily on its anatomical and physiological properties (Bergmark 1989; Janda 1996; Richardson, Hodges et al. 2004). Global muscles are superficial and long, and their fascicle force vectors run at right angles to the main axis of movement. Local muscles are deep, small, close to joints, frequently single-joint and their fascicle force vectors run parallel to the main axis of movement. Furthermore, local muscles have a high proportion of type I muscle fibers and a high muscle spindle density (Bajek, Bobinac et al. 2000). The two systems differ with respect to the motor control of their function. During alternating arm movements (flexion–extension) while standing upright, the local muscles (e.g. the transversus abdominis muscle) demonstrate low-level, but continual co-contraction (i.e. tonic activity). In contrast, a typical global trunk muscle such as the iliocostalis exhibits variable activity in the direction and rhythm of the movement (i.e. phasic activity) (Fig. 1-3).

The preprogramming of the two muscle systems is also different. The preactivation of the transversus abdominis muscle prior to elevation of the arm is constant and independent of the direction of the elevation. The purpose of this preprogramming is more to stabilize and protect the spinal segments in good time by means of a co-contraction, and less to maintain balance. In contrast, the preprogramming of a global muscle, such as the iliocostalis lumborum, is oriented according to the direction of movement of the arm elevation. It is guided by the balancing requirements of the trunk. If the arm is elevated forwards, the iliocostalis muscle is activated prior to the arm movement. When the arm is elevated backwards, however, this muscle is activated considerably later, and only after the movement of the arm. At this point, the global muscle control is dependent on the direction of movement, with the objective of controlling the movement of the thorax and pelvis and of maintaining balance (Hodges and Richardson 1997; Hodges and Gandevia 2000; Gandevia, Butler et al. 2002).

A number of motor controls provide support for these differently allocated muscle functions (Richardson and Bullock 1986; Hodges, Sapsford et al. 2007; Belavy, Richardson et al. 2007).

Coordination of muscle synergy

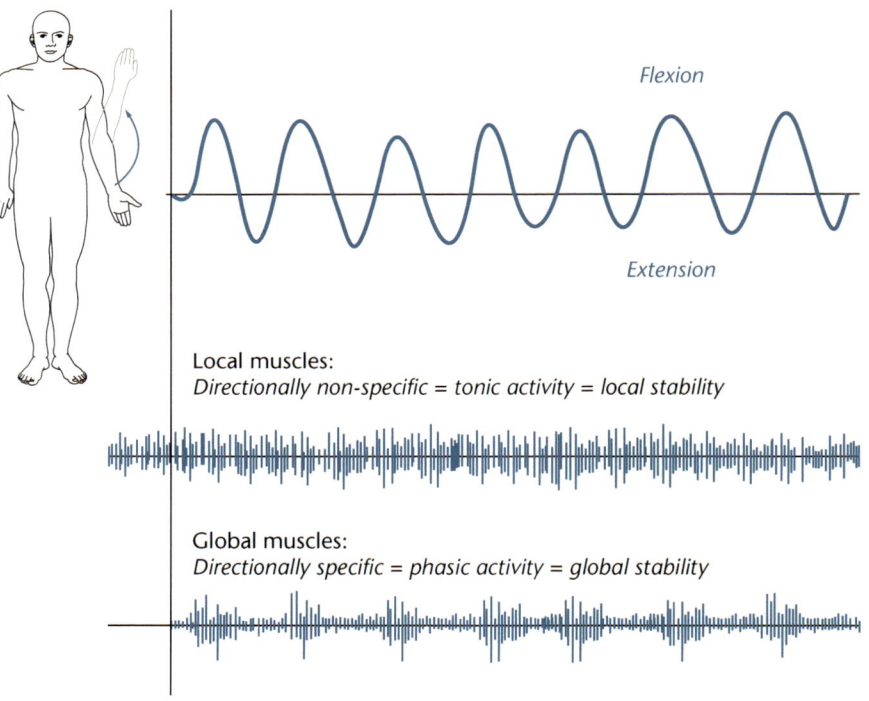

Flexion

Extension

Local muscles:
Directionally non-specific = tonic activity = local stability

Global muscles:
Directionally specific = phasic activity = global stability

Fig. 1-3: Schematic representation of muscle synergy during alternating movements (modified from Richardson, Hodges et al. 2004.)

1.4 Clinical relevance

The classification of the muscles within the myofascial system is of particular value in muscle dysfunction or during rehabilitation. The individual muscle systems show different deficits in injury or pain, so that they need correspondingly different therapeutic exercise strategies for rehabilitation.

Local muscles

Dysfunction of the local muscles in musculoskeletal pain has been investigated in numerous studies. The main cause of recurrent pain is impaired muscle coordination. In lower back pain, for example, the normal preprogramming of the local muscles prior to arm elevation is delayed. When alternating arm movements are performed, there is only "phasic" activity on alternate sides, instead of constant "tonic" co-contraction. As a result, the usual timely and constant protection of the segments is absent (Hodges and Richardson 1996; Hodges, Heijnen et al. 2001; Ng, Parnianpour et al. 2001; Belavy, Richardson et al. 2007; Belavy, Ng et al. 2010). The main objective of therapeutic exercises for chronic pain is to remedy this coordination impairment (Tsao and Hodges 2007).

Other typical forms of local muscle dysfunction associated with chronic pain are atrophy and morphological changes, such as in increase in the fatty and connective tissue fraction (Hides, Stokes et al. 1994; Zhao, Kawaguchi et al. 2000). These changes often manifest specifically in the symptomatic joint or spinal segment (Kjaer, Bendix et al. 2007). At the first acute onset of pain, reflex inhibition of a local muscle already leads to a rapid reduction in the muscle's cross-section. These deficits show a marked association with the pain and are independent of the diagnosis, the activity status and posture. Moreover, the muscles do not recover spontaneously following resolution of the symptoms, despite the resumption of daily activities and sport by the patient (Hides, Jull et al. 2001; Sterling, Jull et al. 2004). This absence of constant protection for the segments is probably the underlying cause of the recurrent lower back pain. Therefore, targeted treatment of the local system is essential for rehabilitation and the prevention of recurrent pain. Several randomized, controlled studies confirm the efficacy of targeted exercise therapy for local muscles (Goldby, Moore et al. 2006; O´Sullivan, Phyty et al. 1997; Jull, Trott et al. 2002; Moseley, Nicholas et al. 2004; Stuge, Holm et al. 2006).

Global muscles

Compared with local muscle dysfunction, deficits in the global muscle system associated with pain are more variable and more individual in nature. It is more common for them to recover spontaneously following resolution of the symptoms and following the resumption of daily activities and sport.

The characteristics of muscle dysfunction in the long, multi-joint muscles are almost completely opposite to those of the local muscles. They frequently show a tendency towards increased stretch sensitivity (shortening), premature activity and overactivity (hypertonicity). They are also closely linked to mechanosensitive neural structures. In contrast, the deficits of the single-joint global muscles are often more variable. This muscle group can show not only changes in power, endurance, coordination and muscle morphology, but also stretch sensitivity and overactivity.

Dysfunctions of the global muscle system are often referred to as "muscle imbalance." The objective of the various methods used in assessing muscle imbalance is to identify disproportion between the various muscles. This involves measuring not only power ratios between agonists and antagonists but also ratios in muscle activity between single- and multi-joint muscles (Janda 1996; Ng and Richardson et al. 2002) or their effects on the body's posture and on balance (Sahrmann 1990; Klein-Vogelbach 1991).

Yet the various forms of dysfunction of the global system suggest only an indirect link with the pain. Therefore, they are often closely associated with the nature and level of an individual's activity, such as sport (Mandell, Weitz et al. 1993; Wang and Cochrane 2001; Nadler, Malanga et al. 2002). Weakness of the trunk muscles is often seen in chronic back pain and is, therefore, thought to be the cause of the back pain. However, this apparent muscle weakness is more a consequence of muscle deconditioning (due to an inactive lifestyle) than a consequence of the pain. Consequently, athletic, active individuals with or without lower back pain show hardly any weakness of the trunk muscles.

There exists a great variety of methods for assessing muscle power endurance, stretch sensitivity, balance and posture. The challenge is not so much to document muscle deficits than to identify the significance of the dysfunction for the patients and for the pain they suffer. Many forms of muscle dysfunction (e.g. muscle weakness) are self-limiting. They normalize spontaneously, as soon as the pain subsides and general physical activity is resumed. Other forms of dysfunction normalize less spontaneously and/or prevent the individual from resuming his or her usual activities. For example, a loss of balance could cause a fear of movement because of the attendant unsteady gait. This fear leads to a loss of quality of life and favors. This positive effect of movement on the pain process probably underlies the efficacy of active versus passive treatment measures in pain that has become chronic (Mannion, Junge et al. 2001).

The close association between the function/dysfunction of the global system and the nature and level of an individual's daily activities, sport and physical exercise explains the high intra- and interindividual variability. Training to achieve a muscle power and endurance at a level greater than the patient requires for his or her daily life is, therefore, futile. The general aim of exercise therapy for the global system is to guide the patient back into his or her daily activities and/or sport. Consequently, the content and structure of a training program for the global system should be guided by this individual goal (Mannion, Junge et al. 2001).

The following table summarizes the characteristics of the local and global muscle systems with regard to their function and dysfunction associated with musculoskeletal disorders, and the most important diagnostic measures used.

Examples of the classification of muscles into individual categories in the myofascial system may be found in the Appendix from p. 416.

Characterization of the muscle groups of the myofascial system			
Characteristic	Local muscles	Global muscles, single-joint	Global muscles, multi-joint
Anatomy	Close to joint, segmental	Bridges a joint	Bridges several joints
Muscle size	Small	Large	Very long
Course	Deep, short	Less deep, longer	Superficial, long
Position relative to the direction of movement	Oblique and at right angles	Parallel	Usually parallel, but highly variable (due to complex biomechanics resulting from the multi-joint nature)
Fiber type	Predominantly type I	Mixed type I and II, high variability	Predominantly type II, long, fusiform
Receptor type	Mostly muscle spindles	Mixed, variable	Mostly sensory endings
Link to neighboring structures	Closely linked to joint capsule and fascia	Middle layer	Closely linked to neural structures
Function	Segmental stability	Balance	Initiation of movement
Susceptibility to mechanical insufficiency	Not susceptible	Susceptible	Highly susceptible
Force generation	Constant, 30% maximum contraction	Variable mixture of power and endurance, 30–80% maximum contraction	Short acceleration force, 80% maximum contraction
Typical contraction types	Static, tonic	Static and eccentric, closed chain	Concentric, open chain
Control	Very early pre-programming at all times, independent of direction of movement	Early pre-programming, dependent on direction of movement	Early pre-programming, dependent on direction of movement
Dysfunction	Atrophy/inhibition	Atrophy/inhibition	"Spasm"
Clinical signs of dysfunction	Fatigue	Weakness, fatigue	Stretch sensitivity, "shortening"
Coordination	Always delayed coordination impairment control depending on direction of movement	Occasionally delayed	Premature activity
Histopathological correlates	Increased fatty and connective tissue fraction, reduced capillary and fiber circumference (mostly type I > type II)	Increased fatty and connective tissue fraction, reduced capillary and fiber circumference	Type I atrophy, extremely reduced muscle circumference
Association with symptoms	Close association with symptoms	Variable, indirect association with symptoms	Associated with pain sensitivity of the neural structures
Clinical investigations	Test of voluntary selective submaximal tension	Muscle function tests: power and duration, muscle imbalance	Stretch sensitivity tests, provocation tests, muscle imbalance, neural structures
Classification of selected muscles according to regions within the myofascial system			
Region	Local muscles	Global muscles, single-joint	Global muscles, multi-joint
Spine (cervical spine, lumbar spine)	Transversus abdominis, deep multifidus muscles (short), longus colli, longus capitis, rectus capitis, rotatores muscles	Obliquus externus, superficial multifidus muscles	Rectus abdominis, erector spinae (thoracic part), sternocleido-mastoideus, scalenus muscles, trapezius (descending part)
Upper extremity	Rotator cuff: supra- and infraspinatus muscles, subscapularis, teres minor	Deltoid muscle	Latissimus dorsi, Biceps brachii (long head)
Lower extremity	Vastus medialis, obliquus, popliteus	Vastus lateralis, intermedius, medialis muscles	Rectus femoris, biceps femoris

2 Upper extremity

Muscles of the pectoral girdle

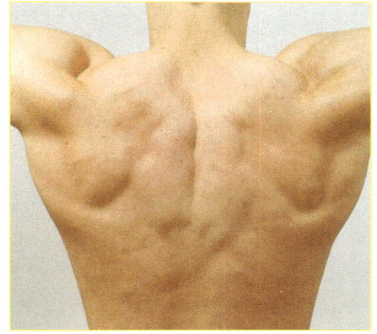

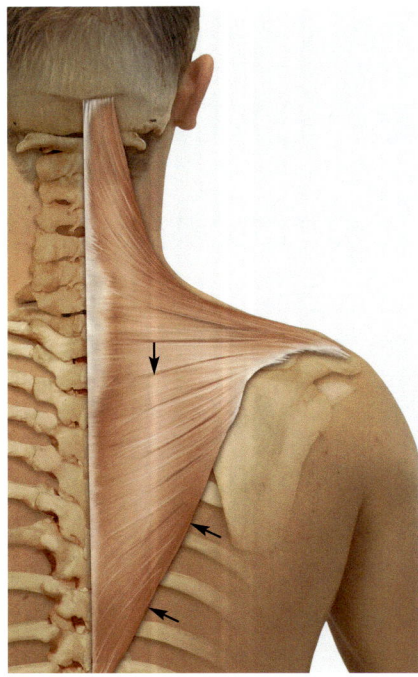

Trapezius muscle, ascending part

The ascending part of the trapezius muscle displaces the scapula in the caudal direction and, if contracting together with the descending part, rotates the shoulder blade so that its glenoid surface points in the cranial direction and the inferior angle of the shoulder blade deviates in the lateral direction (elevated position).

Origin	Spinous processes of the thoracic vertebrae 4–12 Supraspinal ligament
Insertion	At the medial scapular spine via an aponeurosis
Innervation	Accessory nerve (cranial nerve XI)

Functions

 Synergists   Antagonists

Acromioclavicular and sternoclavicular joints

Displacement of the scapula in the caudal direction

Synergists	Antagonists
Serratus anterior muscle (caudal part)	Trapezius muscle (descending part)
Pectoralis minor muscle	Levator scapulae muscle
Indirectly via insertion at the humerus via adduction	Rhomboid muscles
Latissimus dorsi muscle	Serratus anterior muscle (cranial part)
Pectoralis major muscle	

Displacement of the scapula in the medial direction

Synergists	Antagonists
Trapezius muscle (descending and transverse parts)	Serratus anterior muscle
Rhomboid muscles	
Levator scapulae muscle	
Indirectly via insertion at the humerus via adduction	
Latissimus dorsi muscle	
Pectoralis major muscle	

Rotation of the scapula into the elevated position

Synergists	Antagonists
Serratus anterior muscle (caudal part)	Rhomboid muscles
Trapezius muscle (descending part)	Serratus anterior muscle (cranial part)
	Pectoralis minor muscle
	Indirectly via insertion at the humerus via adduction
	Latissimus dorsi muscle
	Pectoralis major muscle

Muscle function testing

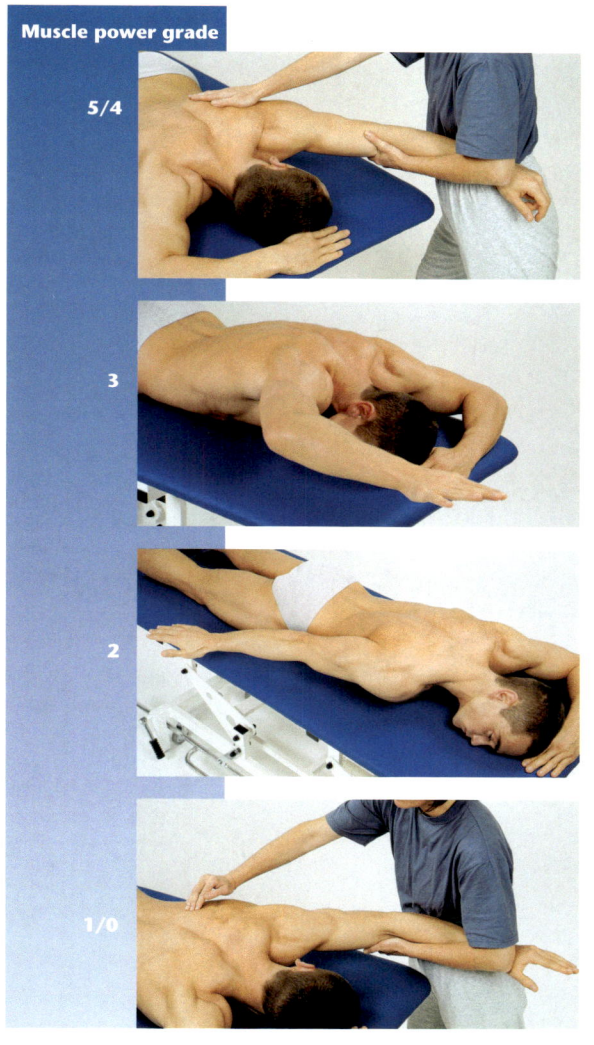

Muscle power grade

5/4

3

2

1/0

Starting position: The patient lies on his front, with the arm being tested extended alongside the head.

Test procedure: The examiner supports the patient's elevated arm with one hand and, with the other, applies pressure to the inferior angle of the scapula in the direction of elevation of the scapula.

Instruction: "Holding your arm outstretched, pull your shoulder blade towards your lower back against my resistance and hold this position."

Starting position: The patient lies on his front, with the arm being tested extended alongside the head.

Test procedure: The examiner looks at the movement of the shoulder blade.

Instruction: "Raise your arm off the couch and pull your shoulder blade towards your lower back."

Starting position: The patient lies on his front, with the arm being tested lying alongside the body in external rotation.

Test procedure: The examiner looks at the patient.

Instruction: "Raise your arm off the couch and pull your shoulder blade towards your lower back."

Starting position: The patient lies on his front.

Test procedure: The examiner palpates the ascending part of the trapezius muscle.

Instruction: "Try to pull your shoulder blades towards your lower back."

🔱 Clinical relevance

- Weakness of the trapezius muscle resulting from an accessory nerve lesion often manifests in the form of characteristic, wing-like protrusion of the scapula (winged scapula). This winging is particularly apparent when the arm is abducted at the shoulder.
- Unilateral contracture of the trapezius muscle is frequently seen in torticollis.
- Weakness of the trapezius muscle makes it more difficult to abduct and elevate the upper arm above shoulder height.
- Active trigger points can often be found in this muscle.

⚠ Problems/comments

- Restricted shoulder mobility can also cause the arm to hang laterally over the edge of the examination couch.

Trapezius muscle, transverse part

The transverse part of the trapezius muscle displaces the scapula in the medial direction and fixes it to the trunk.

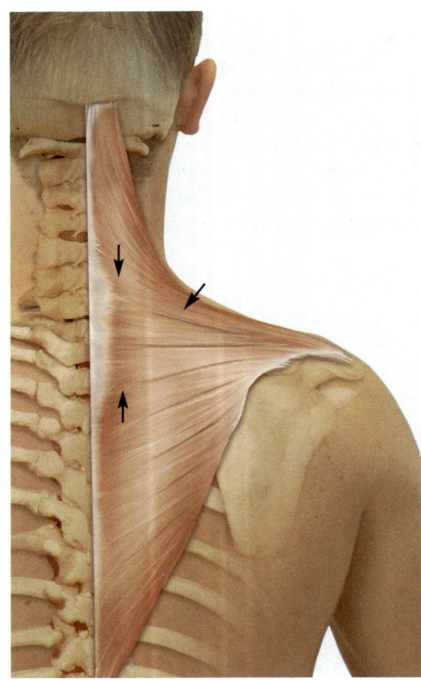

Origin	Nuchal ligament Spinous processes of the thoracic vertebrae C5–T3
Insertion	Scapular spine Acromion
Innervation	Accessory nerve (cranial nerve XI)
Special features	The transverse part originates from the spinous processes with a rhomboid aponeurosis

Functions

 Synergists

 Antagonists

Acromioclavicular and sternoclavicular joints

Displacement of the scapula in the medial direction
Trapezius muscle (descending and ascending parts)
Rhomboid muscles
Levator scapulae muscle (weak)
Indirectly via insertion at the humerus via adduction
Latissimus dorsi muscle
Pectoralis major muscle

Serratus anterior muscle

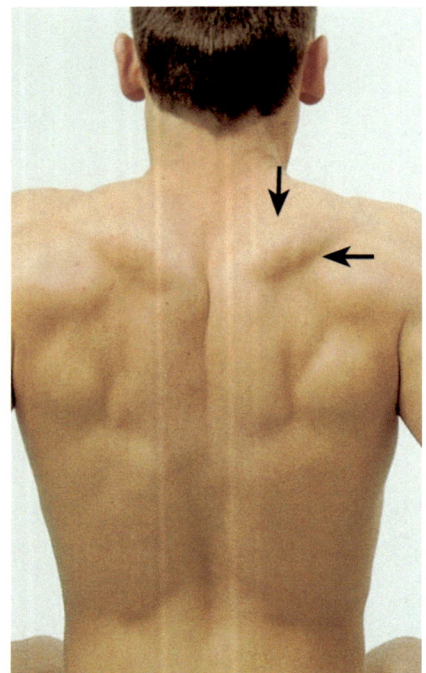

Muscle function testing

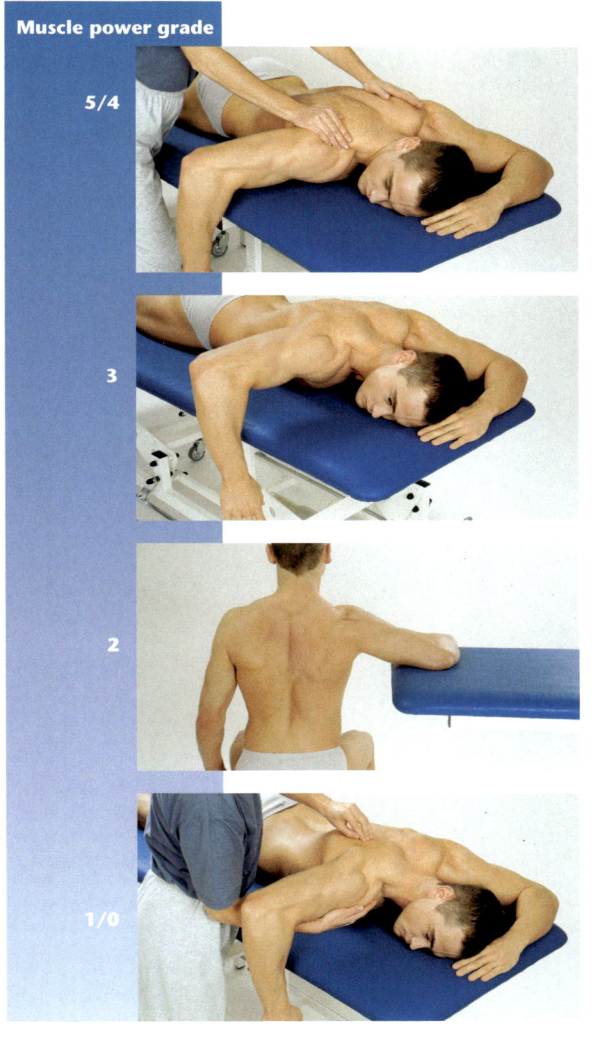

Muscle power grade

5/4

3

2

1/0

Starting position: The patient lies on his front. The arm is abducted 90° at the shoulder and flexed 90° at the elbow.

Test procedure: The examiner holds the thorax in place with one hand and with the other applies downward pressure (towards the couch) to the shoulder joint.

Instruction: "Raise your arm and shoulder off the couch against my resistance and hold this position."

Starting position: The patient lies on his front. The arm is abducted 90° at the shoulder and flexed 90° at the elbow.

Test procedure: The examiner looks at the movement of the shoulder.

Instruction: "Raise your arm and shoulder off the couch."

Starting position: The patient sits, resting the arm sideways on the couch with the shoulder abducted 90° and the elbow flexed 90°.

Test procedure: The examiner looks at the movement of the shoulder.

Instruction: "Push your arm backwards along the couch."

Starting position: The patient lies on his front.

Test procedure: The examiner palpates the transverse part of the trapezius muscle.

Instruction: "Try to raise your arm and shoulder off the couch."

⚕ Clinical relevance

- Weakness of the trapezius muscle resulting from an accessory nerve lesion often manifests in the form of characteristic, wing-like protrusion of the scapula (winged scapula). This winging is particularly apparent when the arm is abducted at the shoulder.
- Unilateral contracture of the trapezius muscle is frequently seen in torticollis.
- Weakness of the trapezius muscle makes it more difficult to abduct and elevate the upper arm above shoulder height.
- Active trigger points can often be found in this muscle.

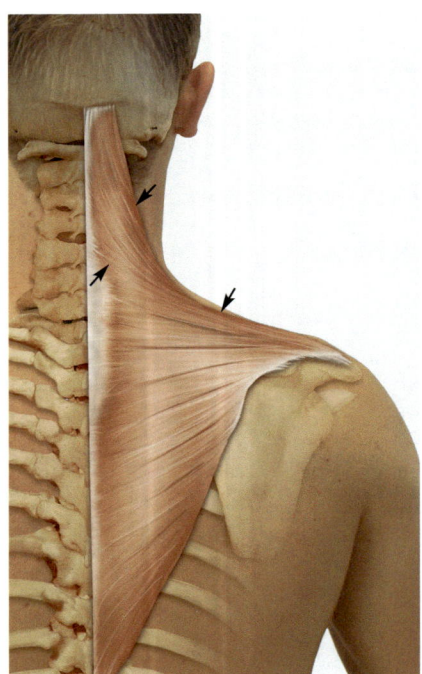

Trapezius muscle, descending part

The descending part of the trapezius muscle displaces the scapula cranially. Together with the ascending part, it can rotate the shoulder blade so that its glenoid surface points cranial and the lower angle of the shoulder blade deviates in the lateral direction (elevated position). This muscle also extends the cervical spine and inclines it to the same side.

Origin	External occipital protuberance, medial section of the superior nuchal line, nuchal ligament (cranial part) Spinous processes of cervical vertebrae 1–4
Insertion	Lateral third of the clavicle Acromion
Innervation	Accessory nerve (cranial nerve XI) Also ventral branches, C2–C4

Functions

🏃 Synergists   🏃 Antagonists

Intervertebral joints

Displacement of the scapula in the caudal direction
Sternocleidomastoideus muscle (in the ipsilateral direction)
Levator scapulae muscle
Iliocostalis muscle
Longissimus muscle
Intertransversarii muscles
Spinalis muscle
Multifidus muscle
Semispinalis muscle

Contralateral counterparts of the muscles listed as synergists

Extension of the cervical spine
Deep muscles of the back of the neck (ipsilateral parts)
Sternocleidomastoideus muscle (both sides)
Levator scapulae muscle (both sides

Longus colli muscle
Longus capitis muscle
Sternocleidomastoideus muscle (both sides, with the head already inclined forwards)

Displacement of the scapula in the cranial direction
Levator scapulae muscle
Rhomboid muscles
Serratus anterior muscle (cranial part)

Trapezius muscle (ascending part)
Serratus anterior muscle (caudal part)
Pectoralis minor
Indirectly via insertion at the humerus via adduction
Latissimus dorsi muscle,
pectoralis major muscle

Displacement of the scapula in the medial direction
Trapezius muscle (descending and transverse part)
Rhomboid muscles
Levator scapulae muscle
Indirectly via insertion at the humerus via adduction
Latissimus dorsi muscle,
pectoralis major muscle

Serratus anterior muscle

Rotation of the scapula into the elevated position
Serratus anterior muscle (caudal part)
Trapezius muscle (descending part)

Rhomboid muscles
Serratus anterior muscle (cranial part)
Pectoralis minor muscle
Indirectly via insertion at the humerus via adduction
Latissimus dorsi muscle
Pectoralis major muscle

Muscle function testing

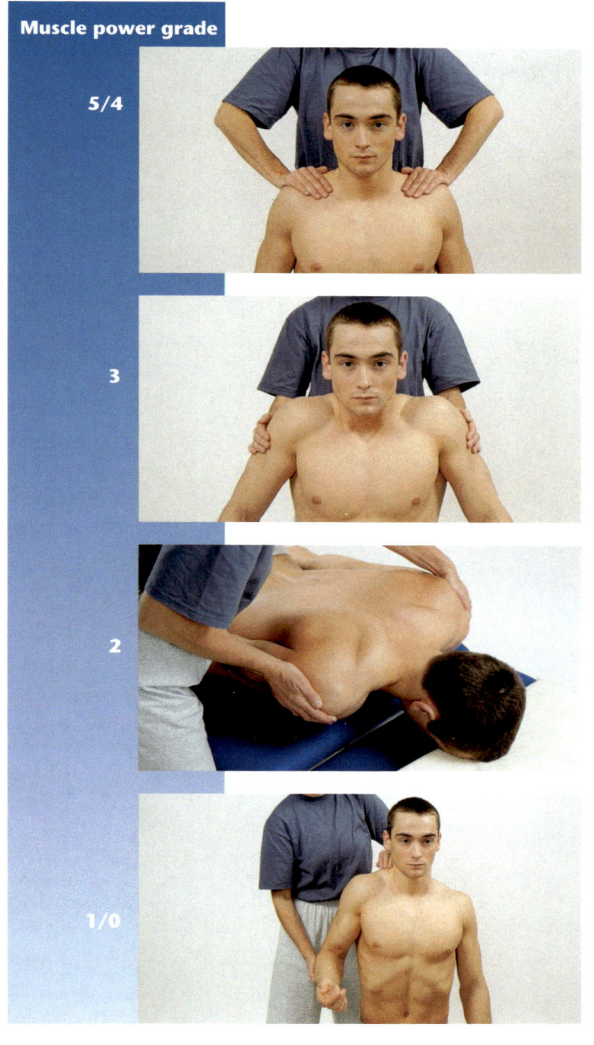

Muscle power grade

5/4

3

2

1/0

Starting position: The patient sits with his arms hanging down loosely.

Test procedure: The examiner presses down on the patient's shoulders.

Instruction: "Raise your shoulders towards your ears against my resistance and hold your final position."

Starting position: The patient sits with his arms hanging down loosely.

Test procedure: The examiner looks at the patient's shoulder movements.

Instruction: "Raise your shoulders as far as possible."

Starting position: The patient lies on his front with his arms placed alongside the body. His forehead lies on the examination couch.

Test procedure: The examiner supports the shoulders from the ventral side, if necessary.

Instruction: "Move your shoulder towards your ear as far as possible."

Starting position: The patient sits.

Test procedure: The examiner palpates the descending part of the trapezius muscle.

Instruction: "Try to raise your shoulders as far as possible."

⚕ Clinical relevance

- Unilateral contracture of the trapezius muscle is frequently seen in torticollis.
- Weakness of the trapezius muscle makes it more difficult to abduct and elevate the upper arm above shoulder height.
- A shortened costoclavicular ligament obstructs elevation of the scapula.
- Active trigger points can often be found in this muscle.

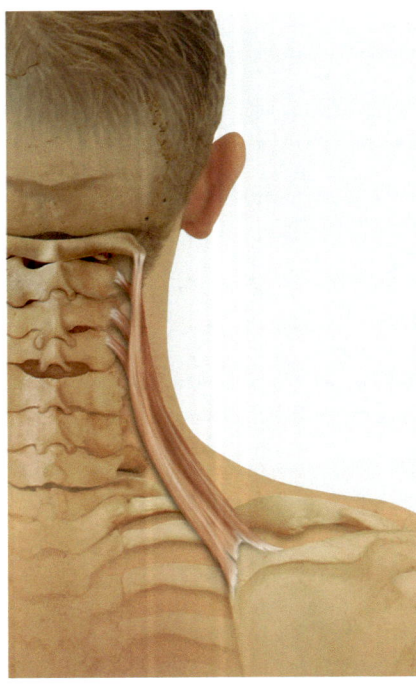

Levator scapulae muscle

The levator scapulae muscle (levator muscle of the scapula) either lifts the shoulder blade or prevents it from sinking (e.g., when heavy loads are carried), depending on the location of its mobile or fixed point. It also pulls the shoulder blade in the medial direction. If the levator scapulae muscles contract on both sides, they can extend the cervical spine; if only one side contracts, it can incline the cervical spine to the same side.

Origin	Posterior tubercles of the transverse processes of the cervical vertebrae 1–4
Insertion	Superior angle of the scapula and the adjoining medial margin of the scapula
Innervation	Dorsal nerve of the scapula, C3–C5 Ventral branches of C3–C5

Functions

 Synergists

 Antagonists

Acromioclavicular and sternoclavicular joints

Displacement of the scapula in the cranial direction
Trapezius muscle (descending part)
Rhomboid muscles
Serratus anterior muscle (cranial part)

Trapezius muscle (ascending part)
Serratus anterior muscle (caudal part)
Pectoralis minor muscle
Indirectly via insertion at the humerus via adduction
Latissimus dorsi muscle
Pectoralis major muscle

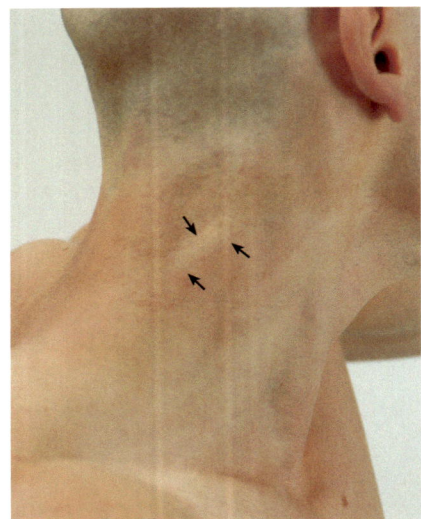

Displacement of the scapula in the medial direction
Trapezius muscle
Rhomboid muscles
Indirectly via insertion at the humerus via adduction
Latissimus dorsi muscle
Pectoralis major muscle

Serratus anterior muscle

Muscle function testing

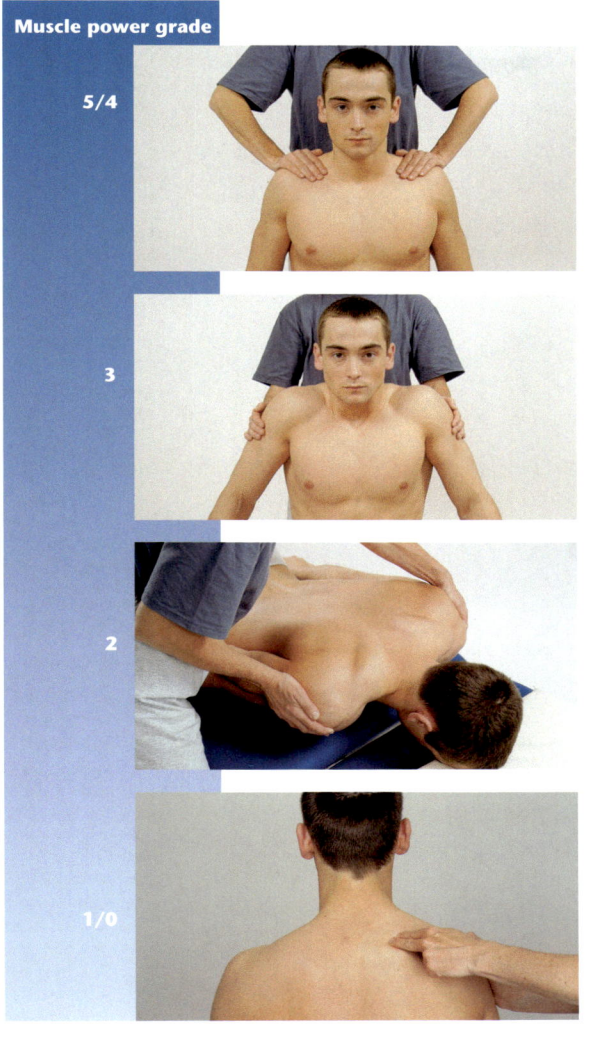

Muscle power grade

5/4

Starting position: The patient sits with his arms hanging down loosely.

Test procedure: The examiner presses down on the patient's shoulders.

Instruction: "Raise your shoulders towards your ears against my resistance and hold your final position."

3

Starting position: The patient sits with his arms hanging down loosely.

Test procedure: The examiner looks at the patient's shoulder movements.

Instruction: "Raise your shoulders as far as possible."

2

Starting position: The patient lies on his front with his arms placed alongside the body. His forehead lies on the examination couch.

Test procedure: The examiner supports the shoulders from the ventral side, if necessary.

Instruction: "Move your shoulder towards your ear as far as possible."

1/0

Starting position: The patient sits with his arms hanging down loosely.

Test procedure: The examiner palpates the levator scapulae muscle at the superior angle.

Instruction: "Try to raise your shoulders towards your ears as far as possible."

⚕ Clinical relevance

- If there is weakness of the trapezius muscle, predominance of the levator scapulae muscle can cause the superior angle of the scapula to be pulled upwards.
- An active trigger point is often found at the insertion of the levator scapulae muscle.

⚠ Problems/comments

- It is difficult to distinguish between the functions of the levator scapulae muscle and of the descending part of the trapezius muscle.

Rhomboideus major muscle

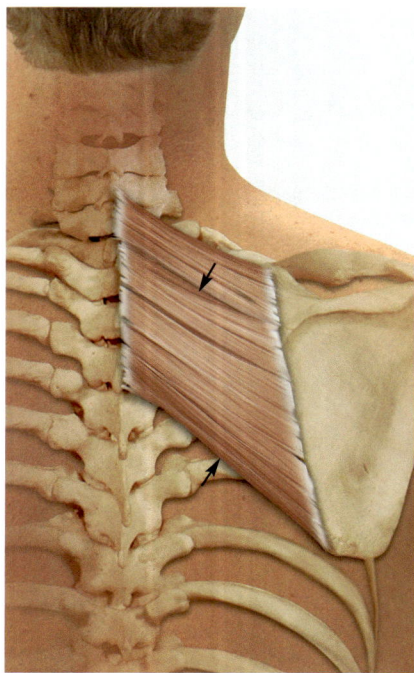

The rhomboideus major and minor (greater and lesser rhomboid) muscles lift the shoulder blade and move it towards the spine. Together with the antagonistic serratus anterior, they press the medial margin of the scapula against the thorax, thus fixing it into place.

Origin	Spinous processes of thoracic vertebrae 1–5
Insertion	Medial margin of the scapula between the scapular spine and the inferior angle
Innervation	Dorsal nerve of the scapula, C4–C5

Functions

Synergists

Antagonists

Acromioclavicular and sternoclavicular joints

Displacement of the scapula in the cranial direction

Trapezius muscle (descending part)
Levator scapulae muscle
Rhomboideus minor muscle
Serratus anterior muscle (cranial part)

Trapezius muscle (ascending part)
Serratus anterior muscle (caudal part)
Pectoralis minor muscle
Indirectly via insertion at the humerus
Latissimus dorsi muscle
Pectoralis major muscle

Displacement of the scapula in the medial direction

Trapezius muscle
Rhomboideus minor muscle
Levator scapulae (weak)
Indirectly via insertion at the humerus via adduction
Latissimus dorsi muscle
Pectoralis major muscle

Serratus anterior muscle

Muscle function testing

5/4

Starting position: The patient lies on his front. The arm being tested lies on the couch with the shoulder internally rotated.

Test procedure: The examiner holds the opposite scapula in place with one hand and applies pressure in the lateral and caudal direction with the other.

Instruction: "Raise your arm and shoulder against my resistance and hold your final position."

3

Starting position: The patient lies on his front. The arm being tested lies on the couch with the shoulder internally rotated.

Test procedure: The examiner looks at the movement of the scapula.

Instruction: "Raise your arm and shoulder off the couch."

2

Starting position: The patient sits. The arms are internally rotated at the shoulders.

Test procedure: The examiner looks at the movement of the scapula.

Instruction: "Bring your shoulder blades together as far as possible."

1/0

Starting position: The patient lies on his front.

Test procedure: The examiner palpates the rhomboideus major muscle.

Instruction: "Try to raise your arm and shoulder off the couch."

⚕ Clinical relevance

- If the scapula is not fixed by the rhomboid muscles, the power of adduction and extension of the upper arm at the shoulder is diminished.
- Normal arm function is reduced less by any deficit in the rhomboid muscles than by a deficit in the trapezius and serratus anterior muscles.
- Weakness of this muscle can cause the scapula to stick out medially (winged scapula)

ⓘ Problems/comments

- The rhomboideus major is tested together with the rhomboideus minor.
- Ensure that the patient does not push the head of the humerus against the examination couch and attempt to raise the arm. The arm and shoulder blade should move together.

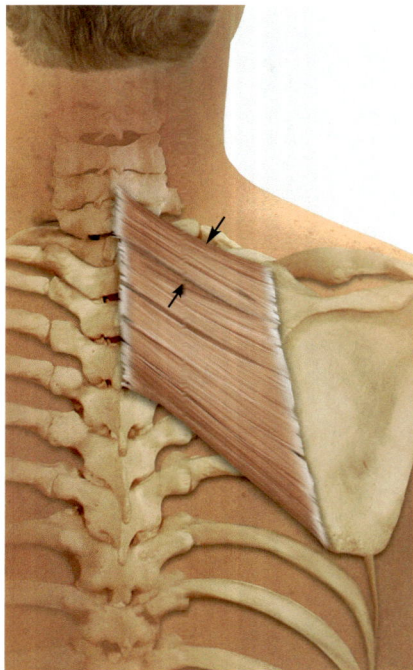

Rhomboideus minor muscle

The rhomboideus major and minor (greater and lesser rhomboid) muscles lift the shoulder blade and move it towards the spine. Together with the antagonistic serratus anterior, they press the medial margin of the scapula against the thorax, thus fixing it into place.

Origin	Spinous processes of cervical vertebrae 6 and 7
Insertion	Medial margin of the scapula in the region of the scapular spine
Innervation	Dorsal nerve of the scapula, C4

Functions

🏃 Synergists   🏃 Antagonists

Acromioclavicular and sternoclavicular joints

Displacement of the scapula in the cranial direction
Trapezius muscle (descending part)
Levator scapulae muscle
Rhomboideus major muscle
Serratus anterior muscle (cranial part)

Trapezius muscle (ascending part)
Serratus anterior muscle (caudal part)
Pectoralis minor muscle
Indirectly via insertion at the humerus via adduction
Latissimus dorsi muscle
Pectoralis major muscle

Displacement of the scapula in the medial direction
Trapezius muscle
Rhomboideus major muscle
Levator scapulae

Serratus anterior muscle

Muscle function testing

Muscle power grade

5/4

Starting position: The patient lies on his front. The arm being tested lies on the couch with the shoulder internally rotated.

Test procedure: The examiner holds the opposite scapula in place with one hand and applies pressure in the lateral and caudal direction with the other.

Instruction: "Raise your arm and shoulder against my resistance and hold your final position."

3

Starting position: The patient lies on his front. The arm being tested lies on the couch with the shoulder internally rotated.

Test procedure: The examiner looks at the movement of the scapula.

Instruction: "Raise your arm and shoulder off the couch."

2

Starting position: The patient sits. The arms are internally rotated at the shoulders.

Test procedure: The examiner looks at the movement of the scapula.

Instruction: "Bring your shoulder blades together as far as possible."

1

Starting position: The patient lies on his front.

Test procedure: The examiner palpates the rhomboideus major muscle.

Instruction: "Try to raise your arm and shoulder off the couch."

⚕ Clinical relevance

- If the scapula is not fixed by the rhomboid muscles, the power of adduction and extension of the upper arm at the shoulder is diminished.
- Normal arm function is reduced less by any deficit in the rhomboid muscles than by a deficit in the trapezius and serratus anterior muscles.
- Weakness of this muscle can cause the scapula to stick out medially (winged scapula)

! Problems/comments

- The rhomboideus minor is tested together with the rhomboideus major.
- Ensure that the patient does not push the head of the humerus against the examination couch and attempt to raise the arm. The arm and shoulder blade should move together.

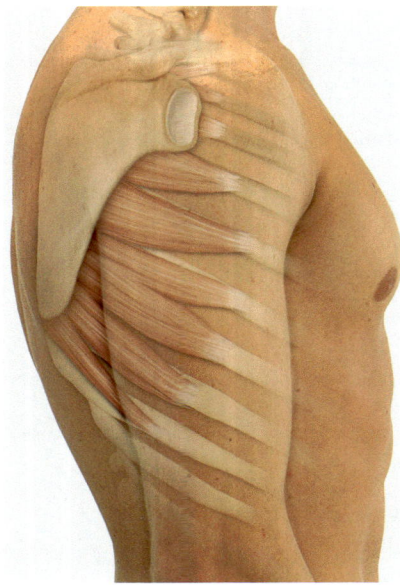

Serratus anterior muscle

The serratus anterior (costoscapularis) muscle can displace the shoulder blade in the lateral and caudal directions and, significantly when acting in conjunction with the antagonistic trapezius muscle, rotate the scapula into the elevated position. Together with the antagonistic rhomboid muscles, it presses the medial margin of the scapula against the thorax, thus fixing it into place.

Origin	Ribs 1–9, originating in an arch below the axilla
Insertion	Ventral surface of the medial margin of the scapula between the superior and inferior angle
Innervation	Long thoracic nerve, C5–C7
Special features	Together with the ribs, the serratus anterior muscle forms the medial wall of the axilla

Functions

 Synergists

Antagonists

Acromioclavicular and sternoclavicular joints

Displacement of the scapula in the lateral direction
Indirectly via insertion at the fixed humerus
Pectoralis major muscle

Trapezius muscle (all parts, especially the transverse part)
Rhomboid muscles
Levator scapulae muscle
Indirectly via insertion at the humerus via adduction
Latissimus dorsi muscle

Displacement of the scapula into the elevated position
Trapezius muscle (descending and ascending parts)

Rhomboid muscles
Pectoralis minor muscle
Indirectly via insertion at the humerus via adduction
Latissimus dorsi muscle
Pectoralis major muscle

Muscle function testing

Muscle power grade

5/4

Starting position: The patient lies on his back, his arm anteverted 90° and abducted slightly at the shoulder.

Test procedure: The examiner holds the distal forearm in place with one hand and the elbow with the other. He applies pressure longitudinally along the upper arm and downwards (towards the couch).

Instruction: "Push your arm upwards against my resistance."

3

Starting position: The patient lies on his back, his arm anteverted 90° and abducted slightly at the shoulder.

Test procedure: The examiner looks at the movement of the scapula.

Instruction: "Push your arm up towards the ceiling."

2

Starting position: The patient sits. His shoulder is anteverted 90°, with the arm resting on the examination couch.

Test procedure: The examiner looks at the movement of the scapula.

Instruction: "Push your arm forwards along the couch."

1/0

Starting position: The patient lies on his back, his arm anteverted 90° and abducted slightly at the shoulder.

Test procedure: The examiner palpates the serratus anterior muscle.

Instruction: "Try to push your arm up towards the ceiling."

℞ Clinical relevance

- Weakness of the serratus anterior muscle results in a winged scapula (particularly the inferior angle).
- Weakness of the serratus anterior muscle makes it more difficult to flex and abduct the arm at the shoulder

⚠ Problems/comments

- The movement of the shoulder needs to be seen and the inferior angle of the scapula palpated to ensure that the movement is not being aided in some other way.
- The term "flexion" is often used instead of "anteversion" for the ventral movement of the arm in the shoulder, analogously to similar movements in other joints.

Pectoralis minor muscle

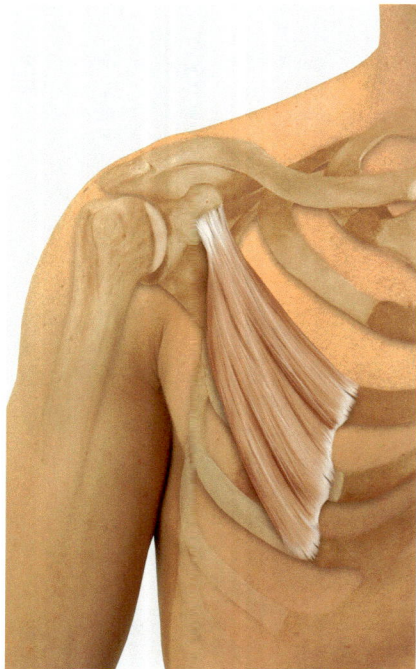

The pectoralis minor (lesser pectoral) muscle fixes the shoulder blade powerfully to the trunk and counteracts its displacement in the dorsal direction (e.g. during push-ups) or in the cranial direction (e.g. during pull-ups). Thus, it can pull the shoulder blade in the caudal and medial directions.

Origin costal cartilage

Upper margin and ventral surface ribs 3–5, close to the

Fascia of the relevant intercostal muscles

Insertion

Coracoid process

Innervation

Medial and lateral pectoral nerves, C6–C8

Special features

The pectoralis minor muscle is one of the structures forming the anterior wall of the axilla

Functions

 Synergists

Antagonists

Acromioclavicular and sternoclavicular joints

Displacement of the scapula in the caudal direction
Trapezius muscle (ascending part)
Serratus anterior muscle (caudal part)
Indirectly via insertion at the humerus via adduction
Latissimus dorsi muscle
Pectoralis major muscle

Trapezius muscle (descending part)
Levator scapulae muscle
Rhomboid muscles
Serratus anterior muscle (cranial part)

Displacement of the scapula in the medial direction
Rhomboid muscles
Levator scapulae muscle
Trapezius muscle
Indirectly via insertion at the humerus via adduction
Latissimus dorsi muscle
Pectoralis major muscle

Serratus anterior muscle

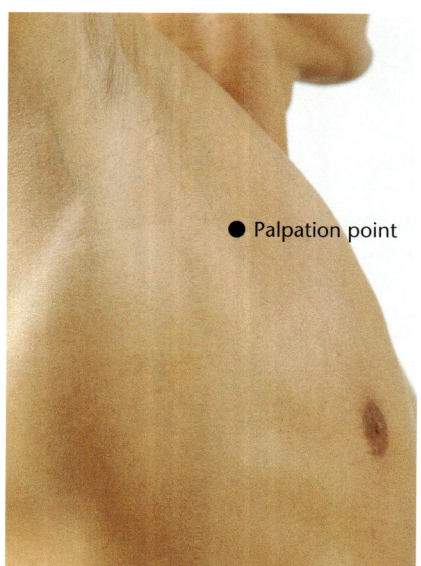

● Palpation point

Muscle function testing

Muscle power grade

5/4

Starting position: The patient lies on his back. The arms lie alongside the body.

Test procedure: The examiner holds the thorax in place with one hand and with the other applies pressure to the shoulder in the direction of elevation and downwards towards the couch.

Instruction: "Raise your shoulder off the couch against my resistance."

3

Starting position: The patient lies on his back. The arms lie alongside the body.

Test procedure: The examiner looks at the movement of the shoulder.

Instruction: "Raise your shoulder off the couch."

2

Starting position: The patient lies on his side.

Test procedure: The examiner supports the arm and looks at the movement of the shoulder.

Instruction: "Pull your shoulder in the direction of your navel."

1/0

Starting position: The patient lies on his back.

Test procedure: The examiner palpates the pectoralis minor muscle caudally of the coracoid process

Instruction: "Try to raise your shoulder off the couch."

⚕ Clinical relevance

- Retroversion of the arm at the shoulder may be weakened if weakness of the pectoralis minor muscle is present, as a result of the reduced stabilization of the scapula.

- If the muscle is shortened, it may trap the brachial plexus or the axillary blood vessels, causing pain in the arm (thoracic outlet syndrome).

- Contracture of the pectoralis minor muscle restricts anteversion of the arm at the shoulder.

⚠ Problems/comments

- The patient should not be in a position to press the hand downwards in order to force the shoulder forwards. It is important to ensure that neither hand nor elbow is pressing against the couch during the test.

- The pectoralis minor muscle is an accessory muscle of the pectoralis major muscle.

Subclavius muscle

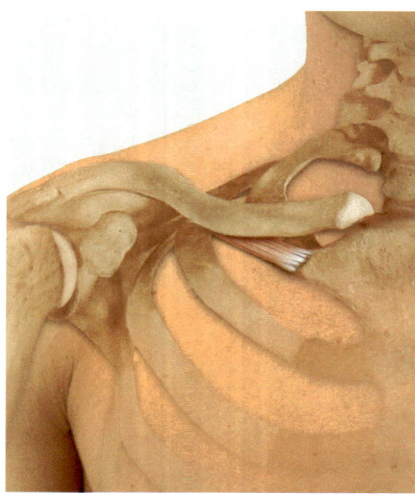

The subclavius muscle depresses the acromial end of the clavicle and presses it against the sternum. In doing so, it also indirectly fixes the scapula via the acromioclavicular joint. Furthermore, it forms a muscle pad between the first rib and the clavicle, ensuring that there is always sufficient distance between them to maintain the blood supply to the subclavicular vessels.

Origin	Cranial surface of the first rib, close to the costal cartilage
Insertion	Acromial extremity of the clavicle
Innervation	Subclavian nerve, C5–C6
Special features	This muscle becomes hypertrophic in the presence of a pectoralis minor muscle deficit

Functions

 Synergists

 Antagonists

Acromioclavicular and sternoclavicular joints

Depressing the clavicle
Indirectly via the scapula
Pectoralis minor muscle
Trapezius muscle (ascending part)
Indirectly via insertion at the humerus via adduction
Pectoralis major muscle
Latissimus dorsi muscle

Sternocleidomastoideus muscle
Indirectly via the scapula
Trapezius muscle (descending part)
Rhomboid muscles
Levator scapulae muscle

Stretch tests

Pectoralis minor muscle

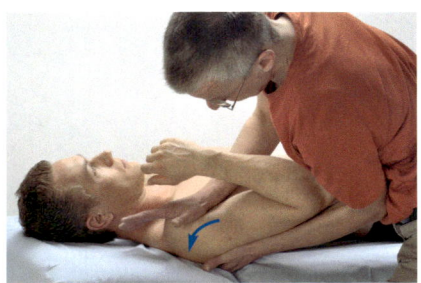

Method
The therapist moves the patient's pectoral girdle into maximum retraction and elevation. The elbow is flexed.

Finding
Muscle shortening is present if the movement cannot be performed to its full extent and if the endpoint feels soft and elastic.
The patient reports a stretching sensation along the course of the muscle.

Comment
Symptoms radiating into the arm suggest compression of the brachial plexus below the pectoralis minor muscle.

Trapezius muscle (descending part)

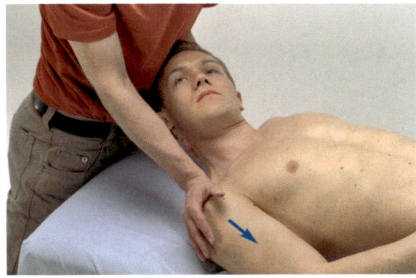

Method
The therapist applies gentle traction and moves the patient's cervical spine and head into flexion, lateral flexion to the opposite side and rotation to the same side. The therapist moves the patient's pectoral girdle over the scapula in the caudal and dorsal directions into maximum depression and retraction.

Finding
Shortening of the descending part of the trapezius muscle is present if the movement cannot be performed to its full extent and if the endpoint feels soft and elastic.
The patient reports a stretching sensation along the course of the muscle.

Levator scapulae muscle

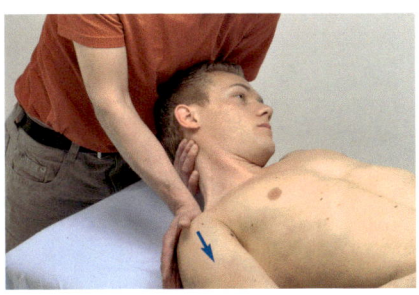

Method
The therapist applies gentle traction and moves the patient's cervical spine and head into flexion, lateral flexion to the opposite side and rotation to the same side. The therapist moves the patient's pectoral girdle over the scapula in the caudal direction into maximum depression and retraction.

Finding
Muscle shortening is present if the movement cannot be performed to its full extent and if the endpoint feels soft and elastic.
The patient reports a stretching sensation along the course of the muscle.

2 Upper extremity

Muscles of the shoulder

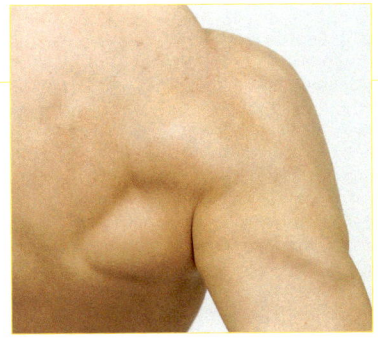

Deltoideus muscle, clavicular part

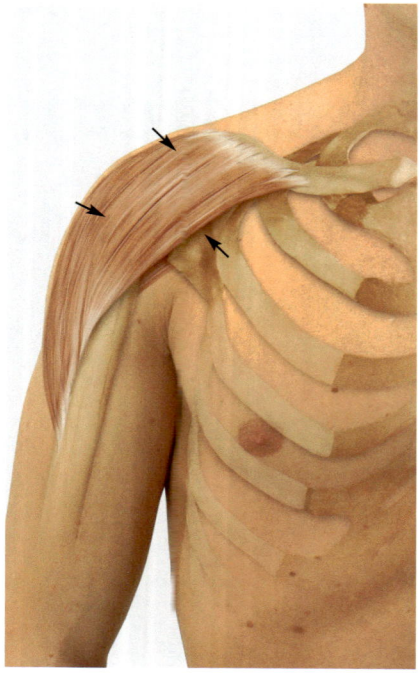

The deltoideus (deltoid) muscle is the most powerful abductor of the shoulder. Furthermore, with its various parts, it is also involved in ante- and retroversion, as well as internal and external rotation. When heavy loads are carried, this muscle counteracts caudal luxation of the humerus in the shoulder joint.

The clavicular part of the deltoideus muscle, when contracting in isolation, brings about anteversion and internal rotation of the humerus at the shoulder. When it contracts in conjunction with the spinal part, its function depends on the position of the shoulder joint: when the arm is adducted, both these parts act jointly as antagonists to the acromial part, thus acting as powerful adductors. If the arm is already abducted, they take over to continue the abduction, once the acromial part becomes insufficient as an abductor.

Origin	Lateral third of the clavicle
Insertion	Deltoid tuberosity of the humerus
Innervation	Axillary nerve, C5–C6
Special features	The clavicular part of this muscle demarcates the infraclavicular fossa; the deltoideus muscle is an indicator muscle for spinal segment C5

Functions

 Synergists   Antagonists

Glenohumeral joint

Anteversion
Pectoralis major muscle
Biceps brachii muscle (long head)
Coracobrachialis muscle
Infraspinatus muscle (cranial part)

Latissimus dorsi muscle
Triceps brachii muscle (long head)
Teres major muscle
Deltoideus muscle (spinal part)

Internal rotation
Subscapularis muscle
Pectoralis major muscle
Latissimus dorsi muscle
Teres major muscle

Infraspinatus muscle
Teres minor muscle
Deltoideus muscle (spinal part)

Adduction (with arm adducted)
Pectoralis major muscle
Latissimus dorsi muscle
Teres major muscle
Teres minor muscle
Coracobrachialis muscle
Deltoideus muscle (spinal part)
Biceps brachii muscle (long head)
Infraspinatus muscle (caudal part)
Triceps brachii muscle (long head)

Deltoideus muscle (acromial part)
Biceps brachii muscle (long head)
Infraspinatus muscle (cranial part)
Subscapularis muscle (cranial part)

Adduction (with arm abducted)
Deltoideus muscle (spinal part)
Biceps brachii muscle (long head)
Infraspinatus muscle (cranial part)
Subscapularis muscle (cranial part)

Pectoralis major muscle
Latissimus dorsi muscle
Teres major muscle
Teres minor muscle
Coracobrachialis muscle
Biceps brachii muscle (long head)
Infraspinatus muscle (caudal part)

Muscle function testing

Muscle power grade

5/4

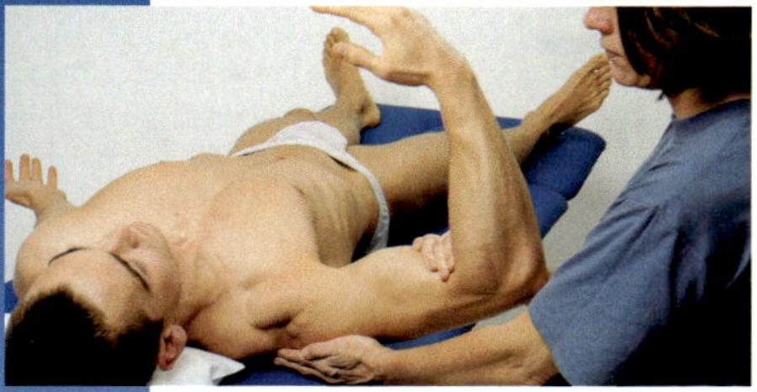

Starting position: The patient lies on his back. The arm is abducted 90° at the shoulder and the elbow is flexed.

Test procedure: The examiner holds the shoulder in place and applies downward resistance (towards the couch) to the distal part of the upper arm.

Instruction: "Raise your arm off the couch and towards your other shoulder against my resistance and hold this position."

3

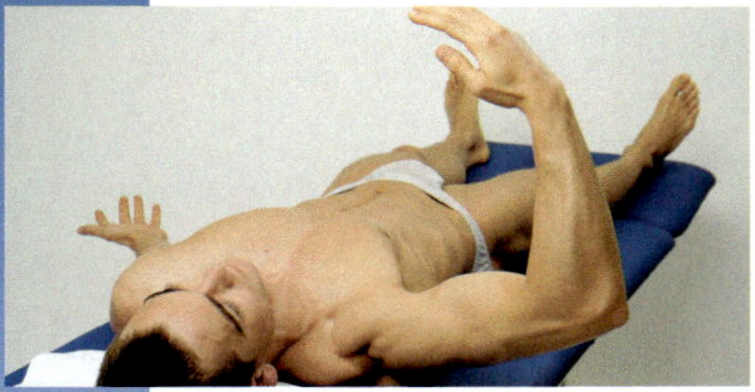

Starting position: The patient lies on his back. The arm is abducted 90° at the shoulder and the elbow is flexed.

Test procedure: The examiner looks at the movement of the shoulder joint.

Instruction: "Raise your arm off the couch until it is vertical."

2

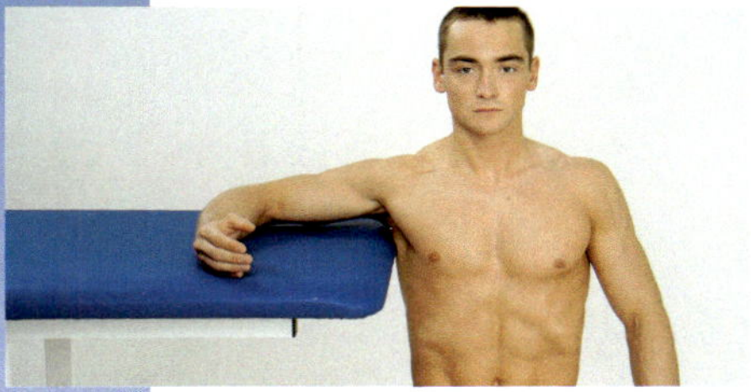

Starting position: The patient sits. The arm rests on the couch or other flat surface with the shoulder abducted 90°.

Test procedure: The examiner looks at the movement of the shoulder joint.

Instruction: "Move your arm forwards along the couch."

1/0

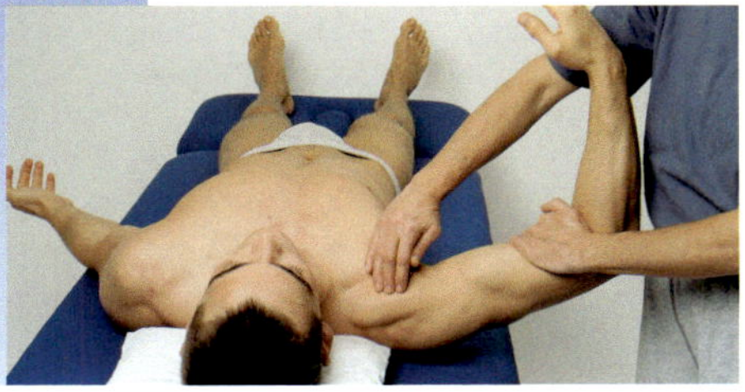

Starting position: The patient lies on his back. The arm is abducted 90° at the shoulder and the elbow is flexed.

Test procedure: The examiner palpates the clavicular part of the deltoideus muscle.

Instruction: "Try to raise your arm off the couch."

 Problems/comments

- The function of the clavicular part of the deltoideus muscle cannot be distinguished from that of the pectoralis major muscle.

Deltoideus muscle, spinal part

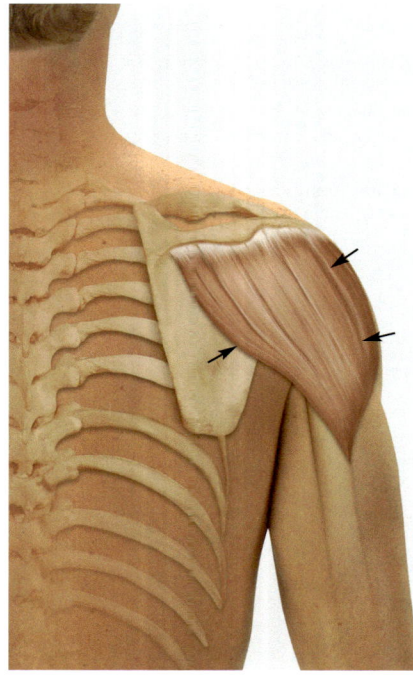

The spinal part of the deltoideus (deltoid) muscle, when contracting in isolation, causes retroversion and external rotation of the humerus at the shoulder. When it contracts in conjunction with the clavicular part, its function depends on the position of the shoulder joint: when the arm is adducted, both these parts act jointly as antagonists to the acromial part, thus acting as powerful adductors. Otherwise, if the arm is already abducted, they take over to continue the abduction.

Origin	Scapular spine
Insertion	Deltoid tuberosity of the humerus
Innervation	Axillary nerve, C5–C6
Special features	The deltoideus muscle is an indicator muscle for spinal segment C5

Functions

 Synergists

 Antagonists

Glenohumeral joint

Retroversion

Synergists	Antagonists
Latissimus dorsi muscle	Pectoralis major muscle
Triceps brachii muscle (long head)	Deltoideus muscle (clavicular part)
Teres major muscle	Biceps brachii muscle (long head)
	Coracobrachialis muscle
	Infraspinatus muscle (cranial part)

External rotation

Synergists	Antagonists
Infraspinatus muscle	Subscapularis muscle
Teres minor muscle	Pectoralis major muscle
Biceps brachii (long head)	Latissimus dorsi muscle
	Teres major muscle

Adduction (with arm adducted)

Synergists	Antagonists
Pectoralis major muscle	Deltoideus muscle (acromial part)
Latissimus dorsi muscle	Biceps brachii muscle (long head)
Teres major muscle	Infraspinatus muscle (cranial part)
Teres minor muscle	Subscapularis muscle (cranial part)
Coracobrachialis muscle	
Biceps brachii muscle (short head)	
Deltoideus muscle (clavicular part)	
Infraspinatus muscle (caudal part)	
Triceps brachii muscle (long head)	

Adduction (with arm abducted)

Synergists	Antagonists
Deltoideus muscle (clavicular part)	Pectoralis major muscle
Biceps brachii muscle (long head)	Latissimus dorsi muscle
Infraspinatus muscle (cranial part)	Teres major muscle
Subscapularis muscle (cranial part)	Teres minor muscle
	Coracobrachialis muscle
	Biceps brachii muscle (short head)
	Infraspinatus muscle (caudal part)
	Triceps brachii muscle (long head)

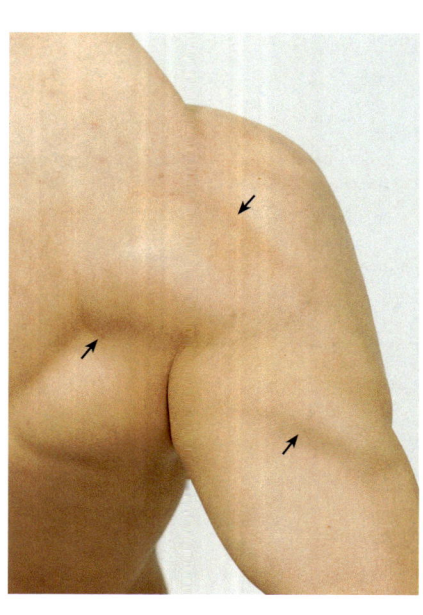

Muscle function testing

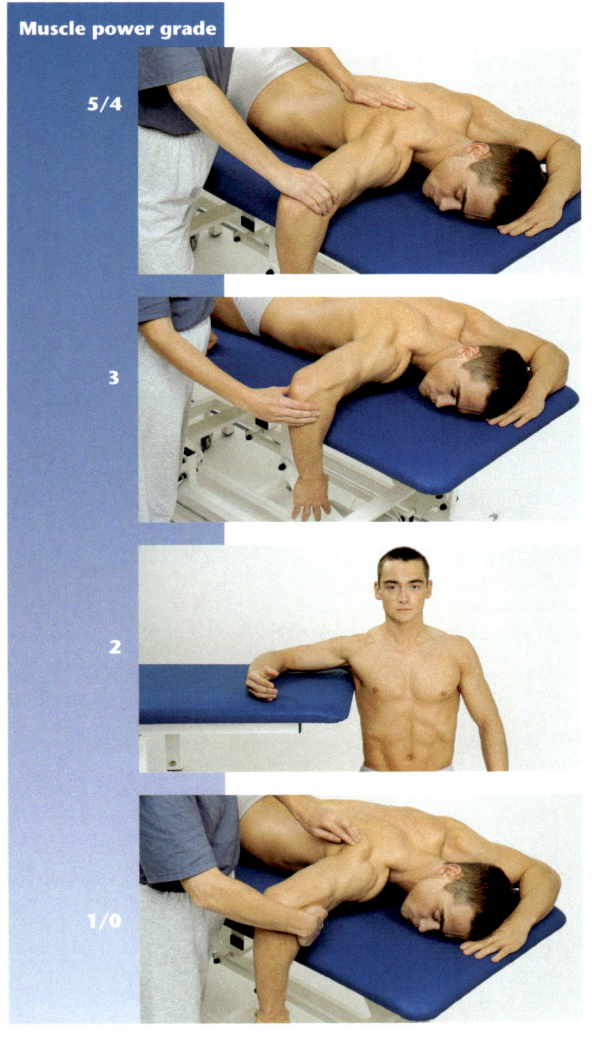

Muscle power grade

5/4

3

2

1/0

Starting position: The patient lies on his front with the arm abducted 90° at the shoulder. The upper arm lies on the couch, the forearm hangs vertically over the edge.

Test procedure: The examiner holds the scapula in place with one hand and applies downward pressure to the elbow with the other.

Instruction: "Raise your arm against my resistance and hold this position."

Starting position: The patient lies on his front with the arm abducted 90° at the shoulder. The upper arm lies on the couch, the forearm hangs vertically over the edge.

Test procedure: The examiner looks at the movement of the shoulder joint.

Instruction: "Raise your arm off the couch."

Starting position: The patient sits. The arm is abducted 90° at the shoulder.

Test procedure: The examiner looks at the movement of the shoulder joint.

Instruction: "Move your arm backwards along the surface it is resting on."

Starting position: The patient lies on his front with the arm abducted 90° at the shoulder. The upper arm lies on the couch, the forearm hangs vertically over the edge.

Test procedure: The examiner palpates the spinal part of the deltoideus muscle.

Instruction: "Try to raise your arm off the couch."

 Problems/comments

- To reduce the involvement of the long head of the triceps brachii in this movement, the test may be performed with the elbow extended.

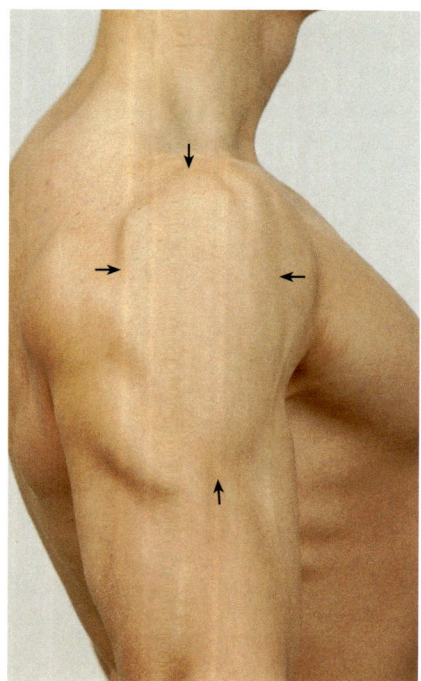

Deltoideus muscle, acromial part

The acromial part of the deltoideus (deltoid) muscle brings about abduction of the arm, whereby the supraspinatus muscle centers the head of the humerus in its socket. When the acromial part of the deltoideus muscle becomes insufficient during abduction, this task is taken over by the spinal and clavicular parts of the deltoideus.

Origin	Acromion
Insertion	Deltoid tuberosity of the humerus
Innervation	Axillary nerve, C5–C6
Special features	The deltoideus muscle is an indicator muscle for spinal segment C5

Functions

 Synergists Antagonists

Glenohumeral joint

Abduction

Synergists	Antagonists
Deltoideus muscle (clavicular and spinal parts, with the arm already abducted)	Pectoralis major muscle
Infraspinatus muscle (cranial part)	Latissimus dorsi muscle
Biceps brachii muscle (long head)	Teres major muscle
Subscapularis muscle (cranial part)	Teres minor muscle
	Coracobrachialis muscle
	Biceps brachii muscle (short head)
	Deltoideus muscle (clavicular and spinal parts, with the arm adducted)
	Infraspinatus muscle (caudal part)

Muscle function testing

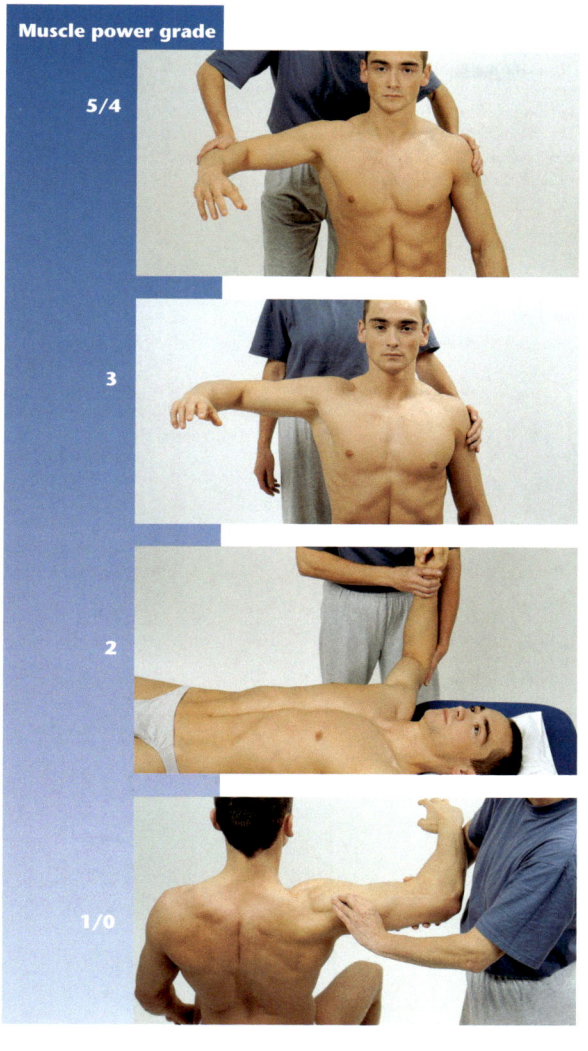

Muscle power grade

5/4

3

2

1/0

Starting position: The patient sits, his arms hanging down loosely at his side. The arm being tested is flexed 90° at the elbow.

Test procedure: The examiner applies pressure to the distal upper arm in the direction of adduction.

Instruction: "Extend your arm away from your body against my resistance and hold this position."

Starting position: The patient sits, his arms hanging down loosely at his side. The arm being tested is flexed 90° at the elbow.

Test procedure: The examiner holds the pectoral girdle in place on the opposite side and looks at the movement of the arm.

Instruction: "Extend your arm away from your body."

Starting position: The patient lies on his back. The arm being tested is flexed 90° at the elbow.

Test procedure: The examiner supports the upper arm and forearm and takes their weight during the movement.

Instruction: "Extend your arm away from your body."

Starting position: The patient sits, his arms hanging down loosely at his side. The arm being tested is flexed 90° at the elbow.

Test procedure: The examiner palpates the acromial part of the deltoideus muscle.

Instruction: "Try to extend your arm away from your body."

⚠ Problems/comments

- Do not allow the patient to rotate the arm externally to any extent, as the biceps brachii muscle is used for this movement.
- When testing the acromial part of the deltoideus muscle, ensure that the patient does not raise his shoulder and tilt the upper body to the other side, as this may look like abduction of the shoulder.
- The supraspinatus muscle aids the function of the acromial part of the deltoideus.

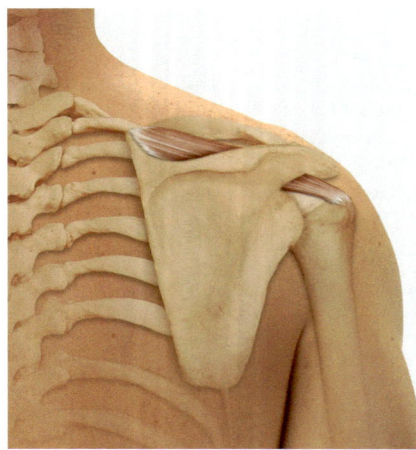

Supraspinatus muscle

The supraspinatus (supraspinous) muscle has no significant rotatory component; instead, as part of the rotator cuff, it fixes the head of the humerus within the glenoid cavity, particularly when the deltoideus muscle tries to pull the head of the humerus out of its socket during the initial phase of abduction. If supraspinatus muscle palsy is present, the deltoideus muscle pulls the greater tuberosity of the humerus against the roof of the shoulder, particularly during the middle phase of abduction.

Origin	Supraspinous fossa
	Supraspinous fascia
Insertion	Superior facet of the greater tuberosity of the humerus
Innervation	Suprascapular nerve, C4–C6
Special features	The innervation of the supraspinatus muscle is supplied by the supraclavicular part of the brachial plexus
	The muscle forms part of the rotator cuff of the shoulder

Functions

 Synergists

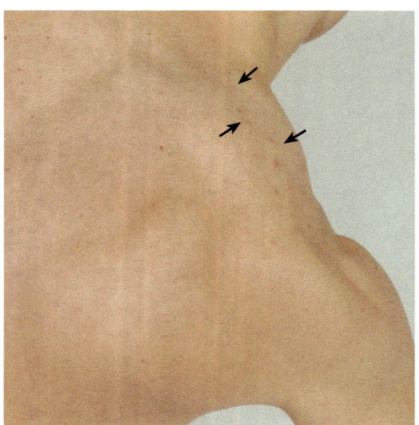

 Antagonists

Glenohumeral joint

Fixing the head of the humerus to prevent sliding in the cranial direction
Infraspinatus muscle (cranial part)
Teres minor muscle
Teres major muscle
Subscapularis muscle
Latissimus dorsi muscle
Pectoralis major muscle

Deltoideus muscle

Muscle function testing

Muscle power grade

5/4

3

2

1/0

Starting position: The patient sits, his arms hanging down loosely at his side. The arm being tested is flexed 90° at the elbow.

Test procedure: The examiner applies pressure to the distal upper arm in the direction of adduction.

Instruction: "Extend your arm away from your body against my resistance and hold this position."

Starting position: The patient sits, his arms hanging down loosely at his side. The arm being tested is flexed 90° at the elbow.

Test procedure: The examiner holds the pectoral girdle in place on the opposite side and looks at the movement of the arm.

Instruction: "Extend your arm away from your body."

Starting position: The patient lies on his back. The arm being tested is flexed 90° at the elbow.

Test procedure: The examiner supports the upper arm and forearm and takes their weight during the movement.

Instruction: "Extend your arm away from your body."

Starting position: The patient sits, his arms hanging down loosely at his side. The arm being tested is flexed 90° at the elbow.

Test procedure: The examiner holds the pectoral girdle in place on the opposite side and looks at the movement of the arm.

Instruction: "Try to extend your arm away from your body."

 Clinical relevance

- In supraspinatus muscle palsy, the head of the humerus on the affected side is higher and the risk of luxation is greater.
- Rupture of the supraspinatus tendon reduces the stability of the shoulder joint.
- Supraspinatus tendon syndrome is a condition involving chronic irritation of the supraspinatus tendon with severe pain. The resultant protective posture assumed by the patient can lead to considerable shrinkage of the joint capsule within a few weeks, restricting the mobility of the joint.

 Problems/comments

- It is difficult to distinguish the supraspinatus muscle from the acromial part of the deltoideus muscle, since both execute the same movement simultaneously. The supraspinatus is the more active muscle at the start of abduction, pulling the humerus below the acromion. As abduction progresses, the deltoideus gains more favorable leverage and can deploy greater force.
- The patient must not rotate the arm externally to any extent during this muscle function test, as the movement may otherwise be aided too much by the biceps brachii muscle. No shoulder elevation and no rotation or lateral flexion of the trunk should be permitted either, as these movements may also lead to apparent abduction.

Infraspinatus muscle

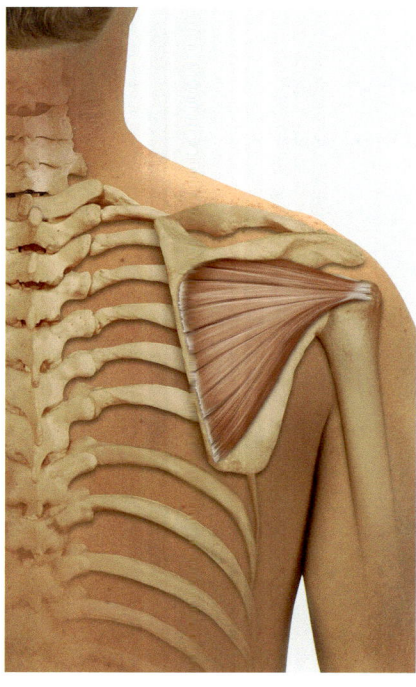

The infraspinatus (infraspinous) muscle produces powerful external rotation, particularly during the final phase of abduction, when the greater tubercle needs to be externally rotated in such a way as to avoid impairing continued abduction by impacting on the roof of the shoulder. It abducts the upper arm with its cranial part and adducts it with its caudal fibers.

Origin	Infraspinous fossa
	Caudal margin of the scapular spine
	Infraspinal fascia
Insertion	Middle facet of the greater tuberosity of the humerus
Innervation	Suprascapular nerve, C5–C6
Special features	The infraspinatus muscle forms part of the rotator cuff

Functions

 Synergists    Antagonists

Articulatio humeri

External rotation

Synergists	Antagonists
Teres minor muscle	Subscapularis muscle
Deltoideus muscle (spinal part)	Pectoralis major muscle
Biceps brachii (long head)	Deltoideus muscle (clavicular part)
	Latissimus dorsi muscle
	Teres major muscle

Adduction (caudal part)

Synergists	Antagonists
Pectoralis major muscle	Deltoideus muscle (acromial part)
Latissimus dorsi muscle	Deltoideus muscle (spinal and clavicular parts,
Teres major muscle	with arm abducted)
Teres minor muscle	Biceps brachii muscle (long head)
Coracobrachialis muscle	Subscapularis muscle (cranial part)
Biceps brachii muscle (short head)	
Deltoideus muscle (spinal and clavicular parts,	
with arm adducted)	
Triceps brachii muscle (long head)	

Adduction (cranial part)

Synergists	Antagonists
Deltoideus muscle (acromial part)	Pectoralis major muscle
Deltoideus muscle (spinal and clavicular parts,	Latissimus dorsi muscle
with arm abducted)	Teres major muscle
Biceps brachii muscle (long head)	Teres minor muscle
Subscapularis muscle (cranial part)	Coracobrachialis muscle
	Biceps brachii muscle (short head)
	Deltoideus muscle (spinal and clavicular parts,
	with arm adducted)
	Triceps brachii muscle (long head)

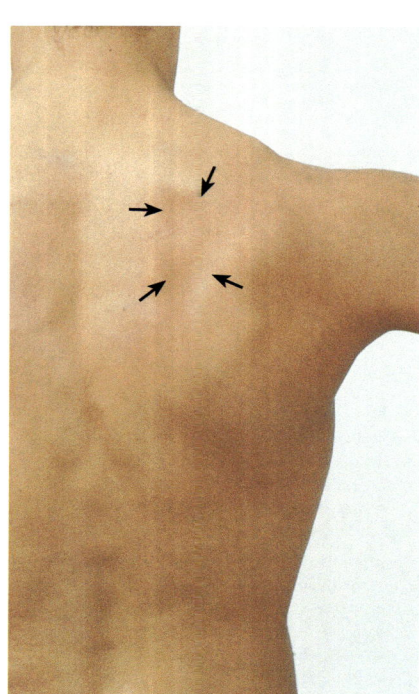

Muscle function testing

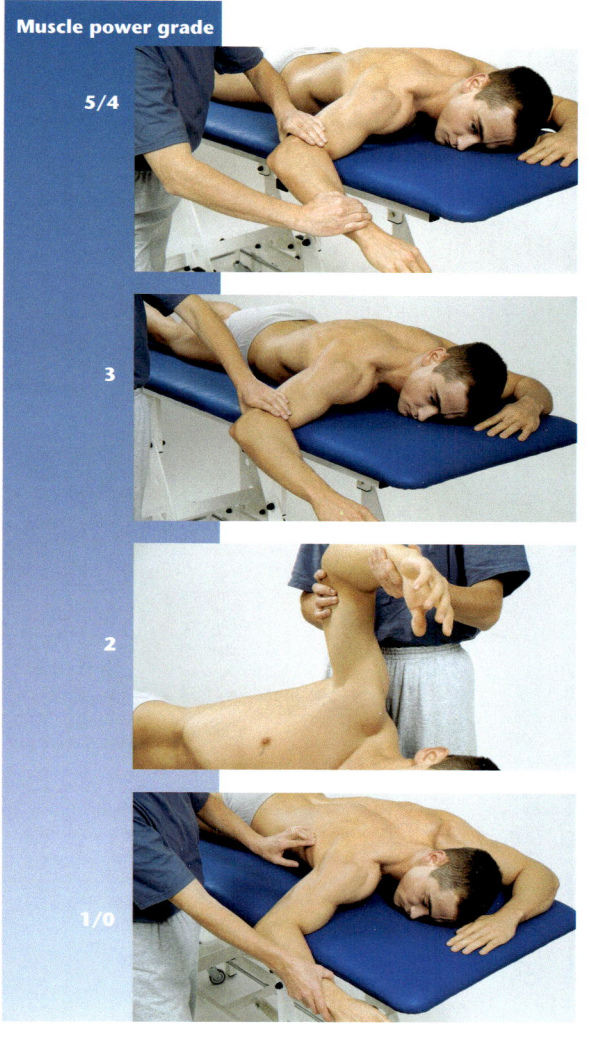

Muscle power grade

5/4

3

2

1/0

Starting position: The patient lies on his front. The arm is abducted 90° at the shoulder and the elbow is flexed 90°. The forearm hangs over the edge of the treatment couch.

Test procedure: The examiner holds the upper arm in place with one hand and with the other applies pressure to the forearm in the direction of internal rotation of the shoulder.

Instruction: "Rotate your forearm forwards and upwards against my resistance and hold this position."

Starting position: The patient lies on his front. The arm is abducted 90° at the shoulder and the elbow is flexed 90°. The forearm hangs over the edge of the treatment couch.

Test procedure: The examiner holds the upper arm in place.

Instruction: "Rotate your forearm forwards and upwards."

Starting position: The patient lies on his side. The arm is abducted 90° at the shoulder and the elbow is flexed 90°. The forearm hangs over the edge of the treatment couch.

Test procedure: The examiner supports the forearm.

Instruction: "Rotate your forearm towards your head as far as possible."

Starting position: The patient lies on his front.

Test procedure: The examiner palpates the infraspinatus muscle.

Instruction: "Try to rotate your forearm forwards and upwards."

 Problems/comments

- The teres minor muscle, the infraspinatus muscle and the spinal part of the deltoideus muscle are virtually indistinguishable in this test.
- If the arm is extended, supination in the elbow joint can give the appearance of external rotation.
- It is not always possible to palpate the infraspinatus muscle due to the overlying latissimus dorsi muscle.

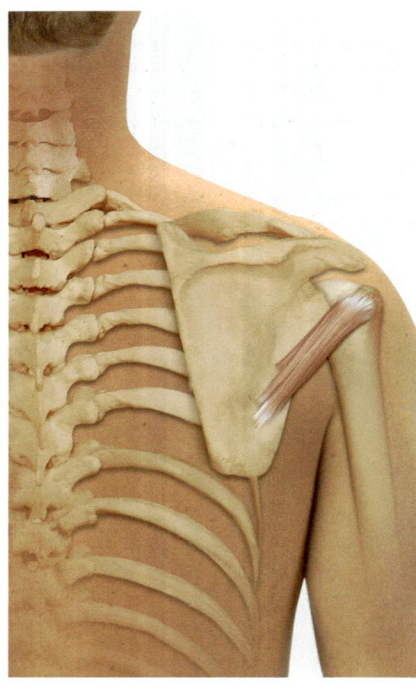

Teres minor muscle

The teres minor muscle produces external rotation of the shoulder and, if the arm is abducted, it also causes adduction of the shoulder. As part of the rotator cuff, it stabilizes the shoulder joint.

Origin	Upper two thirds of the lateral margin of the scapula Fascia which separates the teres minor muscle from the teres major and infraspinatus muscles
Insertion	Inferior facet of the greater tuberosity of the humerus, caudally of the insertion of the infraspinatus muscle
Innervation	Axillary nerve, C5–C6
Special features	The teres minor muscle forms the superior border of the lateral axillary hiatus and is part of the rotator cuff

Functions

 Synergists   Antagonists

Glenohumeral joint

External rotation

Infraspinatus muscle
Deltoideus muscle (spinal part)

Subscapularis muscle
Pectoralis major muscle
Deltoideus muscle (clavicular part)
Latissimus dorsi muscle
Teres major muscle

Adduction

Pectoralis major muscle
Latissimus dorsi muscle
Teres major muscle
Coracobrachialis muscle
Biceps brachii muscle (short head)
Deltoideus muscle (spinal and clavicular parts, with arm adducted)
Infraspinatus muscle (caudal part)
Triceps brachii muscle (long head)

Deltoideus muscle (acromial part)
Deltoideus muscle (spinal and clavicular parts, with arm abducted)
Infraspinatus muscle (cranial part)
Biceps brachii muscle (long head)
Subscapularis muscle (cranial part)

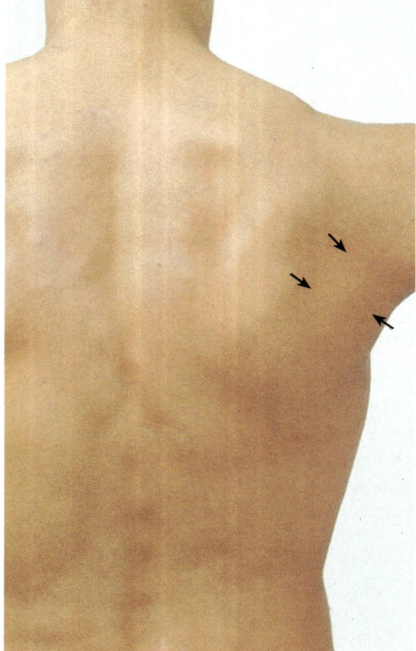

Muscle function testing

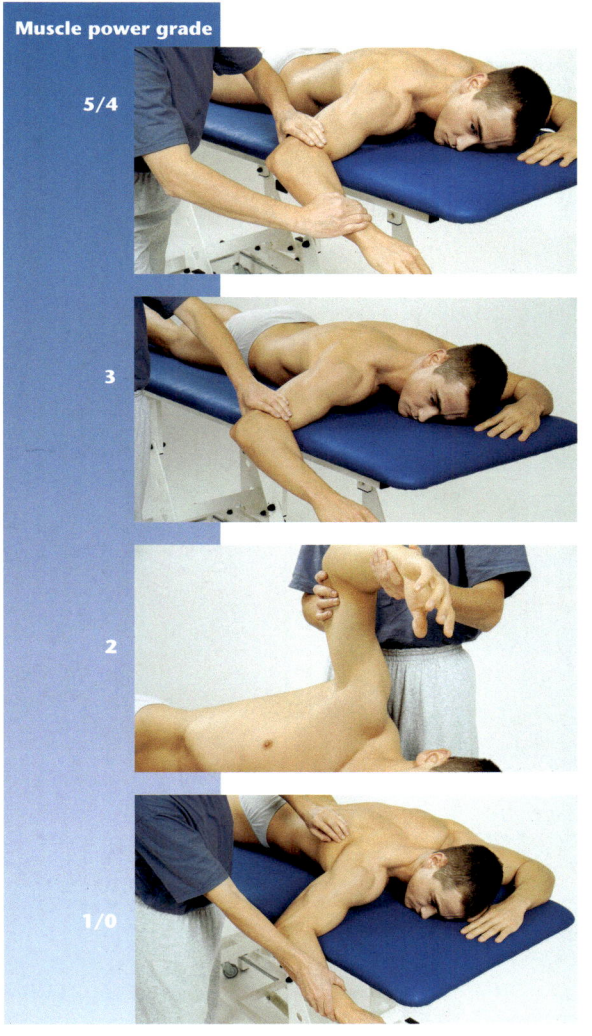

Muscle power grade

5/4

3

2

1/0

Starting position: The patient lies on his front. The arm is abducted 90° at the shoulder and the elbow is flexed 90°. The forearm hangs over the edge of the treatment couch.

Test procedure: The examiner holds the upper arm in place with one hand and with the other applies pressure to the forearm in the direction of internal rotation of the shoulder.

Instruction: "Rotate your forearm forwards and upwards against my resistance and hold this position."

Starting position: The patient lies on his front. The arm is abducted 90° at the shoulder and the elbow is flexed 90°. The forearm hangs over the edge of the treatment couch.

Test procedure: The examiner holds the upper arm in place.

Instruction: "Rotate your forearm forwards and upwards."

Starting position: The patient lies on his side. The arm is abducted 90° at the shoulder and the elbow is flexed 90°. The forearm hangs over the edge of the treatment couch.

Test procedure: The examiner supports the forearm.

Instruction: "Rotate your forearm towards your head as far as possible."

Starting position: The patient lies on his front.

Test procedure: The examiner palpates the teres minor muscle.

Instruction: "Try to rotate your forearm forwards and upwards."

⚠ Problems/comments

- The teres minor muscle, the infraspinatus muscle and the spinal part of the deltoideus muscle are virtually indistinguishable in this test.
- If the arm is extended, supination in the elbow joint can give the appearance of external rotation.

Subscapularis muscle

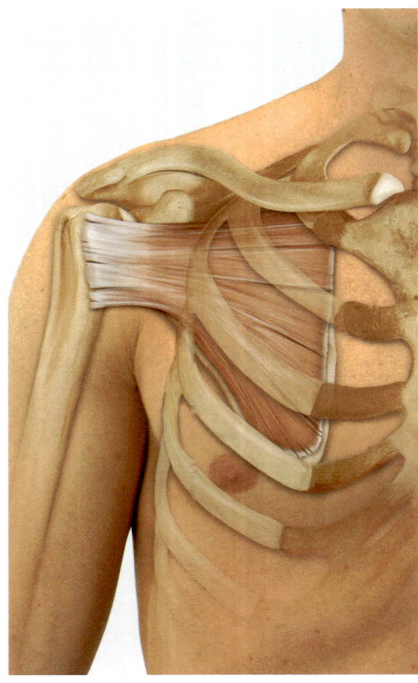

The subscapularis (subscapular) muscle is a powerful internal rotator of the upper arm and pulls the abducted arm back to the trunk. As part of the rotator cuff, it stabilizes the shoulder joint.

Origin   Subscapular fossa

Insertion   Lesser tuberosity of the humerus

Innervation   Subscapular nerve, C5–C6

Special features Together with the scapula, the subscapularis muscle forms the posterior wall of the axilla. It is part of the rotator cuff

Functions

 Synergists

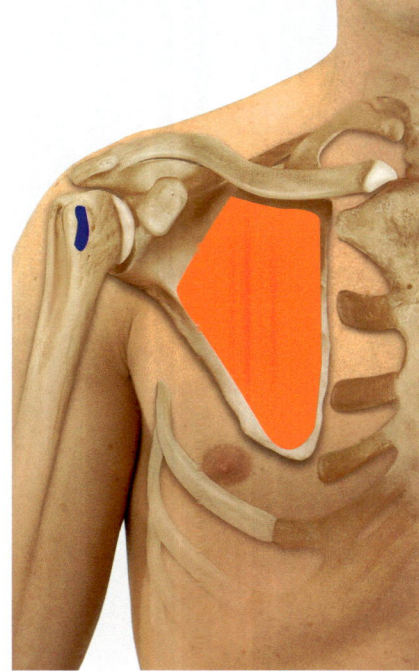

Synergists	Antagonists

Glenohumeral joint

Internal rotation

Synergists	Antagonists
Pectoralis major muscle	Infraspinatus muscle
Deltoideus muscle (clavicular part)	Teres minor muscle
Latissimus dorsi muscle	Deltoideus muscle (spinal part)
Teres major muscle	

Abduction (caudal part)

Synergists	Antagonists
Deltoideus muscle (acromial part)	Pectoralis major muscle
Deltoideus muscle (spinal and clavicular parts, with arm abducted)	Latissimus dorsi muscle
Infraspinatus muscle (cranial part)	Teres major muscle
Biceps brachii muscle (long head)	Teres minor muscle
	Coracobrachialis muscle
	Biceps brachii muscle (short head)
	Deltoideus muscle (spinal and clavicular parts, with arm adducted)
	Infraspinatus muscle (caudal part)
	Triceps brachii muscle (long head)
	Deltoideus muscle (acromial part)

Anteversion (from marked abduction)

Synergists	Antagonists
Pectoralis major muscle	Latissimus dorsi muscle
Deltoideus muscle (clavicular part)	Triceps brachii muscle (long head)
Biceps brachii muscle (long head)	Teres major muscle
Coracobrachialis muscle	Deltoideus muscle (spinal part)

Retroversion (from marked anteversion)

Synergists	Antagonists
Latissimus dorsi muscle	Pectoralis major muscle
Triceps brachii muscle (long head)	Deltoideus muscle (clavicular part)
Teres major muscle	Biceps brachii muscle
Deltoideus muscle (spinal part)	Coracobrachialis muscle

Muscle function testing

Muscle power grade

5/4

Starting position: The patient lies on his front. The arm is abducted 90° at the shoulder and the elbow is flexed 90°. The forearm hangs over the edge of the treatment couch.

Test procedure: The examiner holds the upper arm in place with one hand and with the other applies pressure to the forearm in the direction of external rotation of the shoulder.

Instruction: "Rotate your forearm backwards and upwards against the pressure of my hand and hold this position."

3

Starting position: The patient lies on his front. The arm is abducted 90° at the shoulder and the elbow is flexed 90°. The forearm hangs over the edge of the treatment couch.

Test procedure: The examiner holds the upper arm in place with one hand.

Instruction: "Rotate your forearm backwards and upwards as far as possible."

2

Starting position: The patient lies on his side. The arm is abducted 90° at the shoulder and the elbow is flexed 90°.

Test procedure: The examiner supports the forearm.

Instruction: "Rotate your forearm towards your pelvis as far as possible."

1/0

Starting position: The patient lies on his front.

Test procedure: The examiner palpates the subscapularis muscle in the axilla.

Instruction: "Try to rotate your forearm backwards and upwards as far as possible."

 Problems/comments

- The subscapularis muscle is tested together with the teres major muscle.
- It is difficult to distinguish between the function of the subscapularis muscle and that of the other powerful internal rotators.
- If the arm is extended from the elbow, pronation of the forearm can give the appearance of internal rotation of the upper arm from the shoulder.

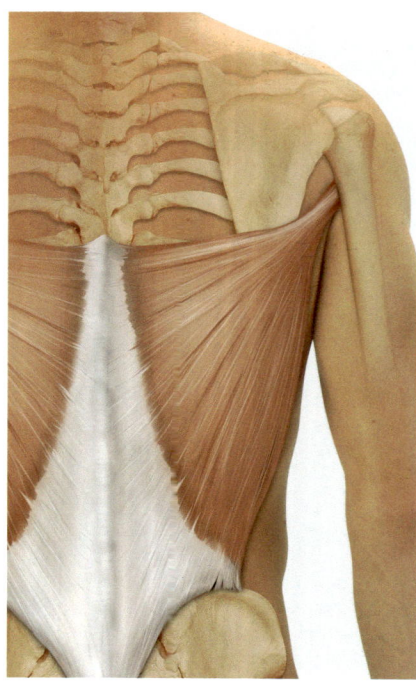

Latissimus dorsi muscle

The latissimus dorsi muscle brings about the retroversion, adduction and internal rotation of the humerus at the shoulder. It displaces the scapula caudally over the thorax, both indirectly via the shoulder and directly via its small origin on the inferior angle of the scapula. Its action on the spine will not be considered here.

Origin	Thoracolumbar fascia
	Supraspinal ligament
	Dorsal third of the iliac crest
	Ribs 9–12
	Inferior angle of the scapula (small part)
Insertion	Crest of the lesser tubercle
Innervation	Thoracodorsal nerve, C6–C8
Special features	The latissimus dorsi muscle forms the posterior axillary fold

Functions

 Synergists   Antagonists

Glenohumeral joint

Internal rotation

Subscapularis muscle
Pectoralis major muscle
Deltoideus muscle (clavicular part)
Teres major muscle

Infraspinatus muscle
Teres minor muscle
Deltoideus muscle (spinal part)
Biceps brachii muscle (long head)

Adduction

Pectoralis major muscle
Teres major muscle
Teres minor muscle
Coracobrachialis muscle
Biceps brachii muscle (short head)
Deltoideus muscle (spinal and clavicular parts, with arm adducted)
Infraspinatus muscle (caudal part)
Triceps brachii muscle (long head)

Deltoideus muscle (acromial part)
Deltoideus muscle (spinal and clavicular parts, with arm abducted)
Infraspinatus muscle (cranial part)
Biceps brachii muscle (long head)
Subscapularis muscle (cranial part)

Retroversion

Triceps brachii muscle (long head)
Teres major muscle
Deltoideus muscle (spinal part)
Subscapularis muscle (caudal part)

Pectoralis major muscle
Deltoideus muscle (clavicular part)
Biceps brachii muscle
Coracobrachialis muscle
Infraspinatus muscle (cranial part)

Acromioclavicular and sternoclavicular joints

Displacement of the scapula in the caudal direction

Trapezius muscle (ascending part)
Serratus anterior muscle (caudal part)
Pectoralis minor muscle
Indirectly via insertion at the humerus via adduction
Pectoralis major muscle

Trapezius muscle (descending part)
Levator scapulae muscle
Rhomboid muscles
Serratus anterior muscle (cranial part)

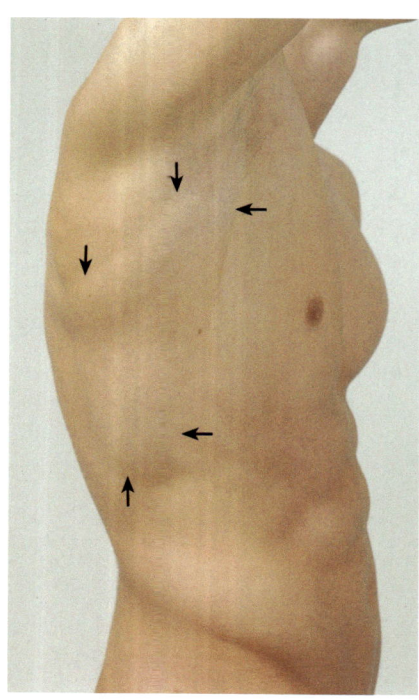

Muscle function testing

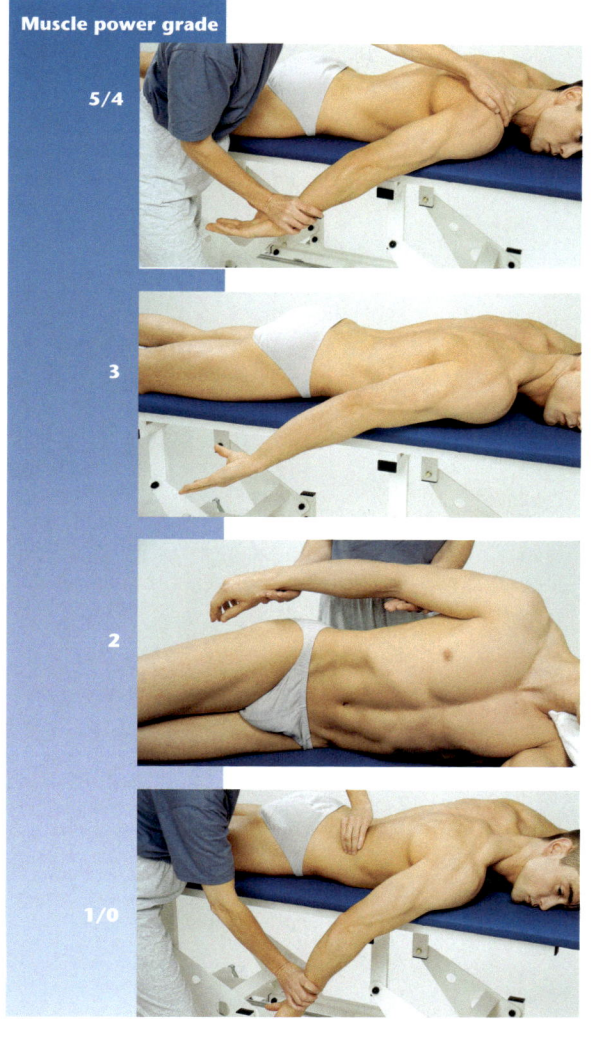

Muscle power grade

5/4

3

2

1/0

Starting position: The patient lies on his front, right at the edge of the examination couch. His arm hangs down.

Test procedure: The examiner holds the shoulder on the same side in place and applies pressure to the distal arm in the direction of anteversion, abduction and external rotation of the shoulder.

Instruction: "Raise your arm backwards and towards your body against my resistance and hold this position."

Starting position: The patient lies on his front, right at the edge of the examination couch. His arm hangs down.

Test procedure: The examiner looks at the movement of the arm.

Instruction: "Raise your arm backwards and towards your body."

Starting position: The patient lies on his side

Test procedure: The examiner supports the arm.

Instruction: "Move your arm towards your back."

Starting position: The patient lies on his front.

Test procedure: The examiner palpates the latissimus dorsi muscle.

Instruction: "Try to raise your arm backwards and towards your body."

 Clinical relevance

- The latissimus dorsi is an important muscle when using walking aids or for any activity in which the trunk is pulled towards the arms (climbing etc.).
- Shortening of this muscle impairs elevation of the arm during shoulder flexion and abduction. This is often seen in scoliosis, kyphosis or after the prolonged use of walking aids.
- The muscle is needed for forced expiration (coughing, sneezing) and for deep inspiration if the arms are fixed. The lateral margin of the latissimus dorsi ("coughing muscle") often becomes hypertrophic in chronic obstructive pulmonary disease

Problems/comments

- The latissimus dorsi is the driving force in swimming, rowing or chopping/hacking movements.

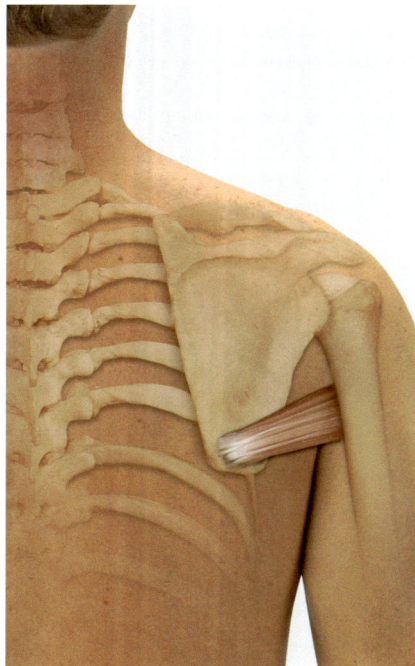

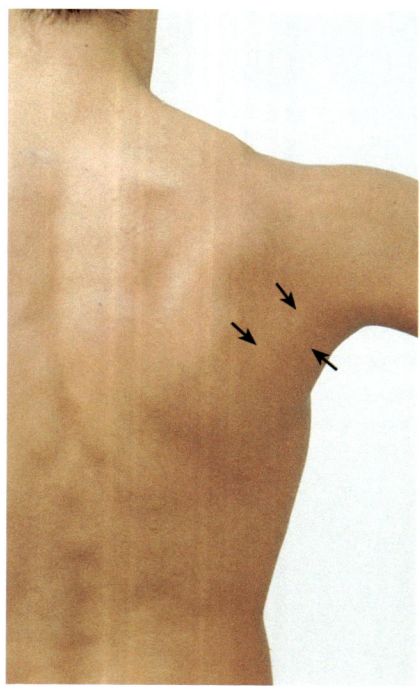

Teres major muscle

The teres major muscle adducts the upper arm and rotates it internally. It retroverts the humerus from anteversion back into its neutral position.

Origin	Dorsal surface of the inferior angle of the scapula and the adjoining section of the lateral margin of the scapula
Insertion	Crest of the lesser tubercle of the humerus
Innervation	Thoracodorsal nerve, C5–C7

Functions

 Synergists

Antagonists

Glenohumeral joint

Retroversion

Latissimus dorsi muscle	Pectoralis major muscle
Triceps brachii muscle (long head)	Deltoideus muscle (clavicular part)
Deltoideus muscle (spinal part)	Biceps brachii muscle
Subscapularis muscle (caudal part)	Coracobrachialis muscle
	Infraspinatus muscle (cranial part)

Adduction

Pectoralis major muscle	Deltoideus muscle (acromial part)
Latissimus dorsi muscle	Deltoideus muscle (spinal and clavicular parts,
Teres minor muscle	with arm abducted)
Coracobrachialis muscle	Infraspinatus muscle (cranial part)
Biceps brachii muscle (short head)	Biceps brachii muscle (long head)
Deltoideus muscle (spinal and clavicular parts, with arm adducted)	Subscapularis muscle (cranial part)
Infraspinatus muscle (caudal part)	
Triceps brachii muscle (long head)	

Internal rotation

Subscapularis muscle	Infraspinatus muscle
Pectoralis major muscle	Teres minor muscle
Latissimus dorsi muscle	Deltoideus muscle (spinal part)
Deltoideus muscle (clavicular part)	

Muscle function testing

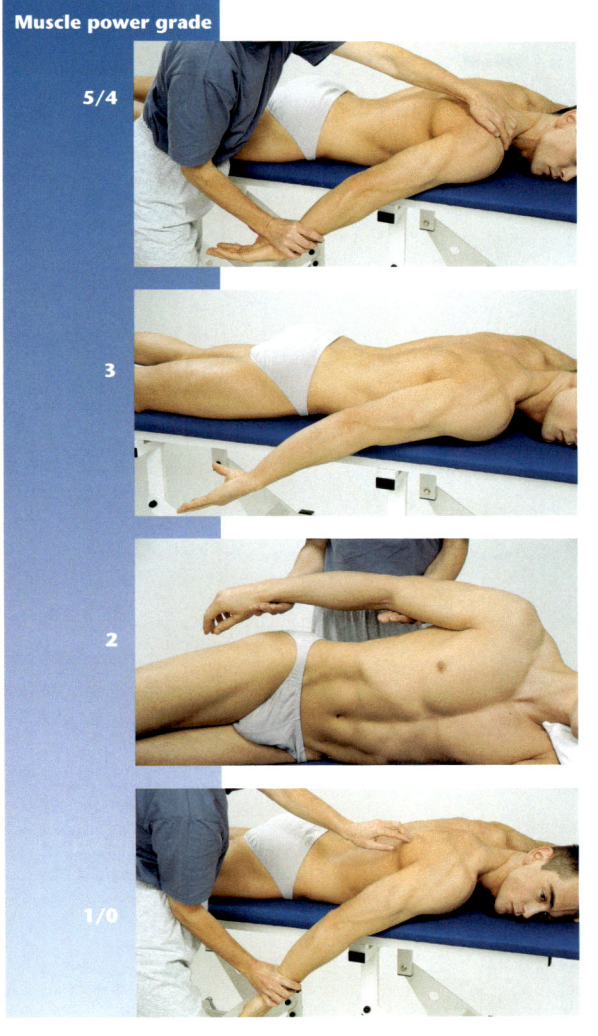

Muscle power grade

5/4

3

2

1/0

Starting position: The patient lies on his front, right at the edge of the examination couch. His arm hangs down.

Test procedure: The examiner holds the shoulder on the same side in place and applies pressure to the distal arm in the direction of anteversion, abduction and external rotation at the shoulder.

Instruction: "Raise your arm backwards and towards your body against my resistance and hold this position."

Starting position: The patient lies on his front, right at the edge of the examination couch. His arm hangs down.

Test procedure: The examiner looks at the movement of the arm.

Instruction: "Raise your arm backwards and towards your body."

Starting position: The patient lies on his side.

Test procedure: The examiner supports the arm.

Instruction: "Move your arm towards your back."

Starting position: The patient lies on his front.

Test procedure: The examiner palpates the latissimus dorsi muscle.

Instruction: "Try to raise your arm backwards and towards your body."

Clinical relevance

- Unequal movements of the scapulae (lateral rotation of the scapula angle), which occur during the anteversion and abduction of the arms at the shoulders, may be caused by shortening of the teres major and subscapularis muscles.
- The teres major muscle is sometimes referred to as the "small latissimus," because it produces the same movements.

Problems/comments

- The teres major muscle cannot always be palpated because of the presence of the overlying latissimus dorsi.
- The terms "extension" and "hyperextension" are often used instead of the term "retroversion" when describing the dorsal movement of the arm at the shoulder, analogously to similar movements in other joints.

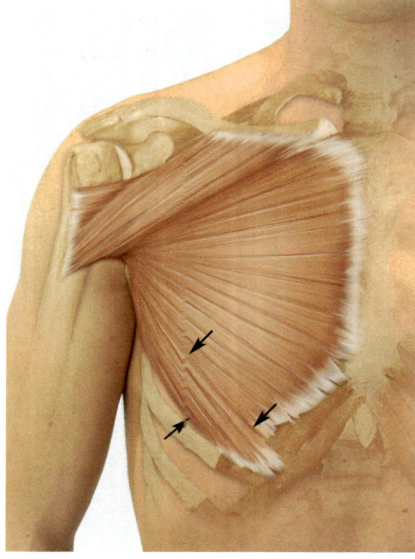

Pectoralis major muscle, abdominal part

The abdominal part of the pectoralis major (greater pectoral) muscle adducts the humerus at the shoulder and can rotate it internally. This part of the pectoralis major muscle can also retrovert the humerus from anteversion back into its neutral position and, acting indirectly via its insertion at the humerus, depress the shoulder blade.

Origin	Anterior lamina of the rectus sheath
Insertion	Crest of the lesser tubercle of the humerus
Innervation	Medial pectoral nerve, C8-T1
Special features	This muscle forms the anterior axillary fold

Functions

 Synergists   Antagonists

Glenohumeral joint

Adduction

Synergists	Antagonists
Latissimus dorsi muscle	Deltoideus muscle (acromial part)
Teres major muscle	Deltoideus muscle (spinal and clavicular parts,
Teres minor muscle	with arm abducted)
Coracobrachialis muscle	Infraspinatus muscle (cranial part)
Biceps brachii muscle (short head)	Biceps brachii muscle (long head)
Deltoideus muscle (spinal and clavicular parts,	Subscapularis muscle (cranial part)
with arm adducted)	
Infraspinatus muscle (caudal part)	
Triceps brachii muscle (long head)	

Internal rotation

Synergists	Antagonists
Subscapularis muscle	Infraspinatus muscle
Deltoideus muscle (clavicular part)	Teres minor muscle
Latissimus dorsi muscle	Deltoideus muscle (spinal part)
Teres major muscle	Biceps brachii muscle (long head)

Retroversion into neutral position

Synergists	Antagonists
Latissimus dorsi muscle	Deltoideus muscle (clavicular part)
Triceps brachii muscle (long head)	Biceps brachii muscle
Teres major muscle	Coracobrachialis muscle
Deltoideus muscle (spinal part)	Infraspinatus muscle (cranial part)
Subscapularis muscle (caudal part)	

Displacement of the scapula in the caudal direction (indirectly via the insertion at the humerus)

Synergists	Antagonists
Serratus anterior muscle (caudal part)	Trapezius muscle (descending part)
Pectoralis minor muscle	Levator scapulae muscle
Trapezius muscle (ascending part)	Rhomboid muscles
Latissimus dorsi muscle	Serratus anterior muscle (cranial part)

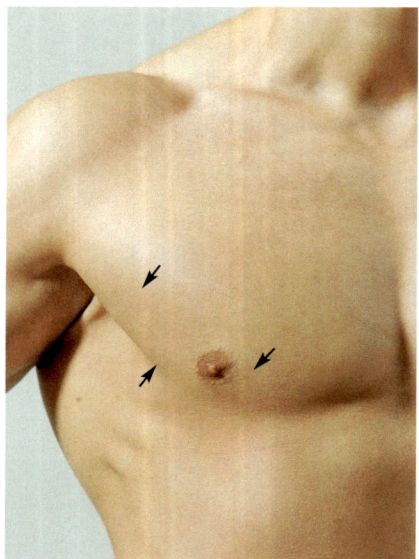

Muscle function testing

Muscle power grade	
5/4	
3	
2	
1/0	

Starting position: The patient lies on his back. The arm is abducted about 120° at the shoulder.

Test procedure: The examiner holds the shoulder in place with one hand and, with the other, applies downward pressure (towards the couch) to the distal upper arm.

Instruction: "Pull your arm towards your abdomen against my resistance and hold your final position."

Starting position: The patient lies on his back. The arm is abducted about 120° at the shoulder.

Test procedure: The examiner looks at the movement of the arm.

Instruction: "Raise your arm until it is vertical."

Starting position: The patient sits. The arm is abducted 90° at the shoulder and rests on the examination couch.

Test procedure: The examiner looks at the movement of the arm at the shoulder.

Instruction: "Push your arm forwards along the couch."

Starting position: The patient lies on his back. The arm is abducted about 120° at the shoulder.

Test procedure: The examiner palpates the origin of the abdominal part of the pectoralis major muscle.

Instruction: "Try to raise your arm."

⚕ Clinical relevance

- The abdominal part of the pectoralis major muscle can be used as an accessory respiratory muscle when the arms are fixed.

❗ Problems/comments

- All parts of the pectoralis major are included when testing for muscle power grade 2.

Pectoralis major muscle, sternocostal part

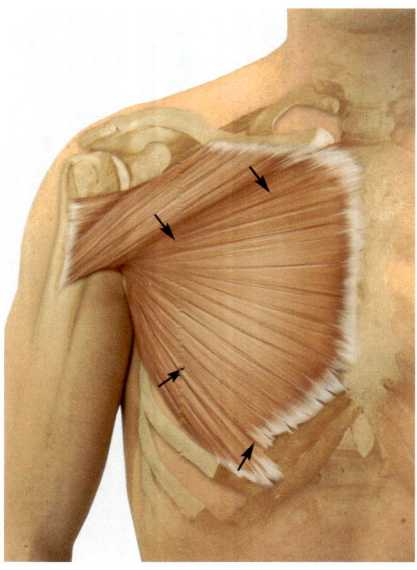

The sternocostal part of the pectoralis major (greater pectoral) muscle adducts the humerus at the shoulder and can rotate it internally. This part of the pectoralis major muscle can also retrovert the humerus from anteversion back into its neutral position.

Origin	Ventral surface of the sternum Cartilage of the upper ribs 6 and 7 Aponeurosis of the obliquus externus abdominis (anterior lamina of the rectus sheath)
Insertion	Crest of the greater tubercle of the humerus (the lower muscle strands are twisted and are attached further dorsally and cranially)
Innervation	Lateral pectoral nerve, C5–C7 Medial pectoral nerve, C8–T1
Special features	The sternocostal part of the pectoralis major forms the anterior axillary fold.

Functions

 Synergists   Antagonists

Glenohumeral joint

Adduction

Synergists	Antagonists
Latissimus dorsi muscle	Deltoideus muscle (acromial part)
Teres major muscle	Deltoideus muscle (spinal and clavicular parts,
Teres minor muscle	with arm abducted)
Coracobrachialis muscle	Infraspinatus muscle (cranial part)
Biceps brachii muscle (short head)	Biceps brachii muscle (long head)
Deltoideus muscle (spinal and clavicular parts,	Subscapularis muscle (cranial part)
with arm adducted)	
Infraspinatus muscle (caudal part)	
Triceps brachii muscle (long head)	

Internal rotation

Synergists	Antagonists
Subscapularis muscle	Infraspinatus muscle
Deltoideus muscle (clavicular part)	Teres minor muscle
Latissimus dorsi muscle	Deltoideus muscle (spinal part)
Teres major muscle	Biceps brachii muscle (long head)

Retroversion into neutral position

Synergists	Antagonists
Latissimus dorsi muscle	Deltoideus muscle (clavicular part)
Triceps brachii muscle (long head)	Biceps brachii muscle
Teres major muscle	Coracobrachialis muscle
Deltoideus muscle (spinal part)	Infraspinatus muscle (cranial part)
Subscapularis muscle (caudal part)	

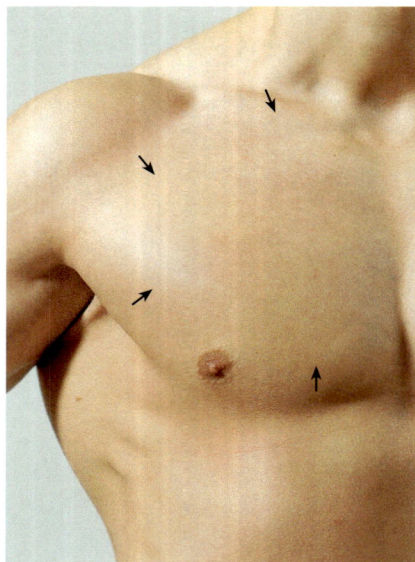

Muscle function testing

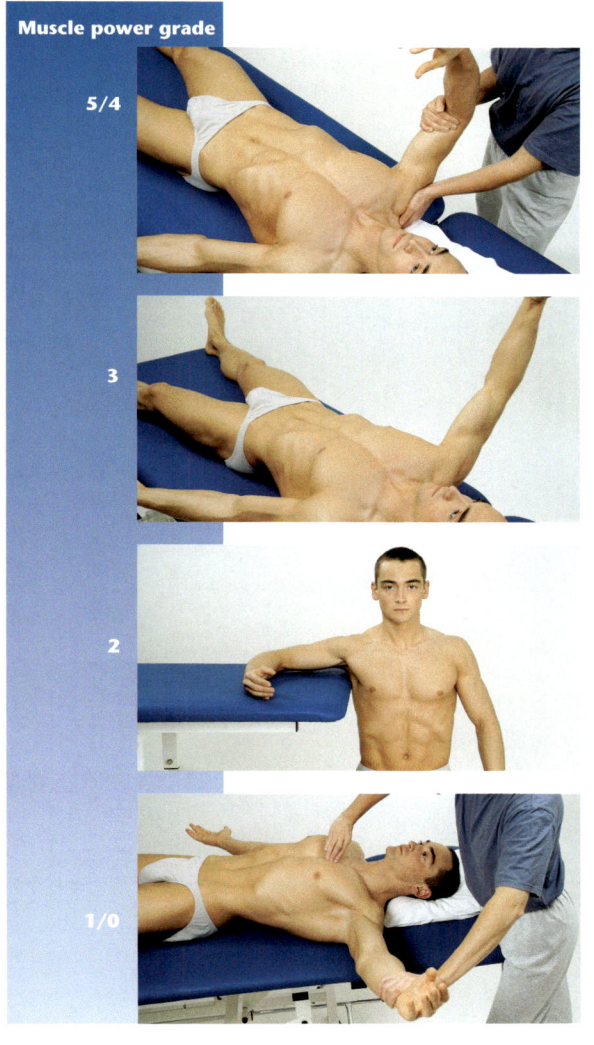

Muscle power grade

5/4

3

2

1/0

Starting position: The patient lies on his back. The arm is abducted 90° at the shoulder.

Test procedure: The examiner holds the shoulder in place with one hand and with the other applies downward pressure (towards the couch) to the distal upper arm.

Instruction: "Pull your arm towards the other side of your body against my resistance and hold your final position."

Starting position: The patient lies on his back. The arm is abducted 90° at the shoulder.

Test procedure: The examiner looks at the movement of the arm.

Instruction: "Raise your arm until it is vertical."

Starting position: The patient sits. The arm is abducted 90° at the shoulder and rests on the examination couch.

Test procedure: The examiner looks at the movement of the arm at the shoulder.

Instruction: "Push your arm forwards along the couch."

Starting position: The patient lies on his back. The arm is abducted 90° at the shoulder.

Test procedure: The examiner palpates the origin of the sternocostal part of the pectoralis major muscle.

Instruction: "Try to raise your arm."

⚕ Clinical relevance

- The sternocostal part of the pectoralis major muscle is important when using walking aids and exercising on parallel bars.

- Weakness of the sternocostal part of the pectoralis major muscle makes it difficult to carry out chopping/hacking and striking/beating movements.

- Weakness of the sternocostal part of the pectoralis major muscle makes it difficult to hold large or heavy objects in both hands at hip height.

- Absence of the sternocostal part of the pectoralis major muscle means that there is no anterior axillary fold, and the nipple is sited lower down.

⚠ Problems/comments

- All parts of the pectoralis major are included when testing for muscle power grade 2.

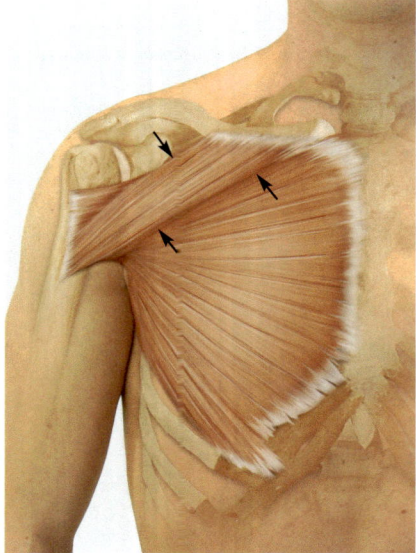

Pectoralis major muscle, clavicular part

Isolated contraction of the clavicular part of the pectoralis major (greater pectoral) muscle can lead to anteversion of the humerus in the shoulder joint, via the adductive and internal rotatory action of the sternocostal part of that muscle.

Origin	Anterior surface of the medial half of the clavicle
Insertion	Crest of the greater tubercle of the humerus
Innervation	Lateral pectoral nerve, C5–C8
Special features	The clavicular part of the pectoralis major demarcates the infraclavicular fossa

Functions

 Synergists   Antagonists

Glenohumeral joint

Adduction

Latissimus dorsi muscle	Deltoideus muscle (acromial part)
Teres major muscle	Deltoideus muscle (spinal and clavicular parts,
Teres minor muscle	with arm abducted)
Coracobrachialis muscle	Infraspinatus muscle (cranial part)
Biceps brachii muscle (short head)	Biceps brachii muscle (long head)
Deltoideus muscle (spinal and clavicular parts, with arm adducted)	Subscapularis muscle (cranial part)
Infraspinatus muscle (caudal part)	
Triceps brachii muscle (long head)	

Internal rotation

Subscapularis muscle	Infraspinatus muscle
Deltoideus muscle (clavicular part)	Teres minor muscle
Latissimus dorsi muscle	Deltoideus muscle (spinal part)
Teres major muscle	

Anteversion, clavicular part

Deltoideus muscle (clavicular part)	Latissimus dorsi muscle
Biceps brachii muscle	Triceps brachii muscle (long head)
Coracobrachialis muscle	Teres major muscle
Infraspinatus muscle (cranial part)	Deltoideus muscle (spinal part)
	Subscapularis muscle (caudal part)

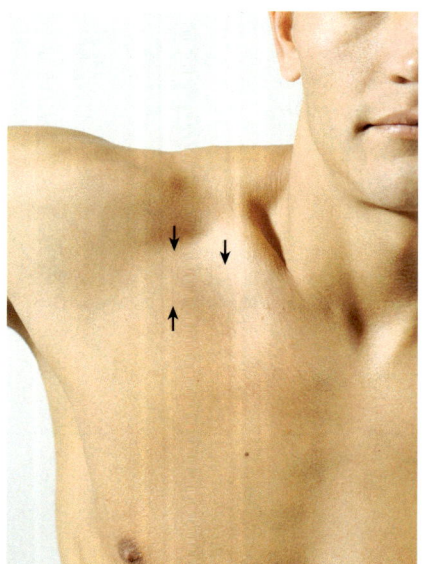

Muscle function testing

Muscle power grade

5/4

Starting position: The patient lies on his back. The arm is abducted 45° at the shoulder.

Test procedure: The examiner holds the shoulder in place with one hand and with the other applies downward pressure (towards the couch) to the distal upper arm.

Instruction: "Pull your arm towards your nose against my resistance and hold your final position."

3

Starting position: The patient lies on his back. The arm is abducted 45° at the shoulder.

Test procedure: The examiner looks at the movement of the arm.

Instruction: "Raise your arm until it is vertical."

2

Starting position: The patient sits. The arm is abducted 90° at the shoulder and rests on the examination couch.

Test procedure: The examiner looks at the movement of the arm in the shoulder.

Instruction: "Push your arm forwards along the couch."

1/0

Starting position: The patient lies on his back. The arm is abducted 45° at the shoulder.

Test procedure: The examiner palpates the origin and attachment of the clavicular part of the pectoralis major muscle.

Instruction: "Try to raise your arm."

Clinical relevance

- A patient with weakness of the clavicular part of the pectoralis major muscle may not be able to touch the opposite shoulder.

Problems/comments

- All parts of the pectoralis major are included when testing for muscle power grade 2.
- The activities of the clavicular part of the pectoralis major muscle and of the coracobrachialis muscle are not distinguishable.

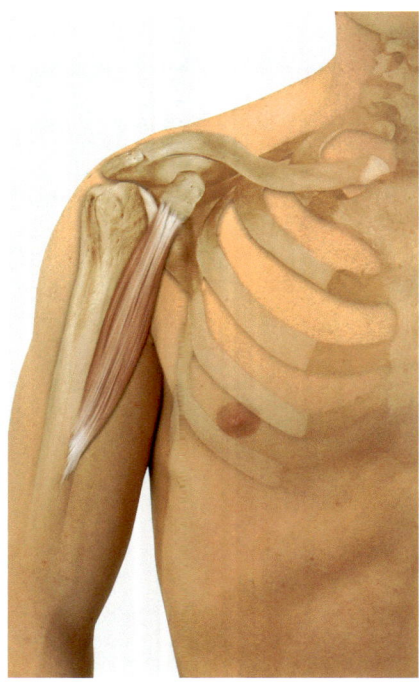

Coracobrachialis muscle

The coracobrachialis (corachobrachial) muscle adducts the humerus at the shoulder and contributes weakly to its anteversion. Its main task is to fix the head of the humerus in the socket of the joint, for example when loads are carried.

Origin	Tip of the coracoid process of the scapula
Insertion	Medial surface of the mid-humerus, at the level of the deltoid tuberosity
Innervation	Musculocutaneous nerve, C5–C7

Functions

Synergists

Antagonists

Glenohumeral joint

Adduction

Pectoralis major muscle
Latissimus dorsi muscle
Teres major muscle
Teres minor muscle
Biceps brachii muscle (short head)
Deltoideus muscle (spinal and clavicular parts, with arm adducted)
Infraspinatus muscle (caudal part)
Triceps brachii muscle (long head)

Deltoideus muscle (acromial part)
Deltoideus muscle (spinal and clavicular parts, with arm abducted)
Infraspinatus muscle (cranial part)
Biceps brachii muscle (long head)
Subscapularis muscle (cranial part)

Anteversion, clavicular part

Pectoralis major muscle
Deltoideus muscle (clavicular part)
Biceps brachii muscle
Infraspinatus muscle (cranial part)

Latissimus dorsi muscle
Triceps brachii muscle (long head)
Teres major muscle
Deltoideus muscle (spinal part)
Subscapularis muscle (caudal part)

Muscle function testing

Muscle power grade

5/4

Starting position: The patient lies on his back. The arm is abducted 45° at the shoulder.

Test procedure: The examiner holds the shoulder in place with one hand and with the other applies downward pressure (towards the couch) to the distal upper arm.

Instruction: "Pull your arm towards your nose against my resistance and hold the final position."

3

Starting position: The patient lies on his back. The arm is abducted 45° at the shoulder.

Test procedure: The examiner looks at the movement of the arm.

Instruction: "Raise your arm until it is vertical."

2

Starting position: The patient sits. The arm is abducted 90° at the shoulder and rests on the examination couch.

Test procedure: The examiner looks at the movement of the arm in the shoulder.

Instruction: "Push your arm forwards along the couch."

1/0

Starting position: The patient lies on his back. The arm is abducted 45° at the shoulder.

Test procedure: The examiner looks at the movement of the arm in the shoulder.

Instruction: "Try to raise your arm."

Clinical relevance

- The musculocutaneous nerve may be injured in the region of the coracobrachialis muscle. In these rare cases, the coracobrachialis muscle usually retains its own innervation, but the nerve supply to the biceps brachii and the brachialis muscles may be damaged.

Problems/comments

- The activities of the clavicular part of the pectoralis major muscle and of the coracobrachialis muscle are not distinguishable.

Stretch tests

Pectoralis major muscle

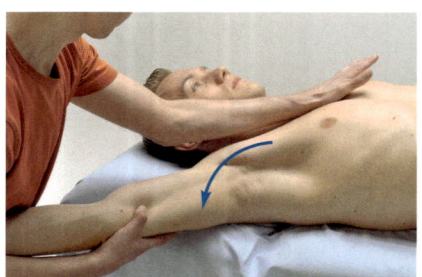

Method
The therapist moves the patient's arm to achieve maximum horizontal abduction and external rotation of the shoulder. The pectoral girdle should be in retraction and elevation.

Differentiating between the three parts of the muscle
Carrying out horizontal abduction and external rotation:
• clavicular part: less than 70° abduction
• sternocostal part: at c. 90° abduction
• abdominal part: more than 120° abduction

Finding
Muscle shortening is present if the movement cannot be performed to its full extent and if the endpoint feels soft and elastic.
The patient reports a stretching sensation along the course of the muscle.

Teres major muscle

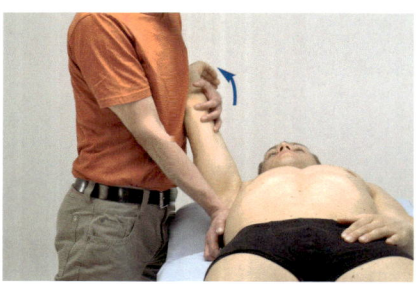

Method
The therapist holds the scapula in place and moves the patient's arm to achieve maximum flexion and external rotation of the shoulder.

Finding
Muscle shortening is present if the movement cannot be performed to its full extent and if the endpoint feels soft and elastic.
The patient reports a stretching sensation along the course of the muscle.

Latissimus dorsi muscle

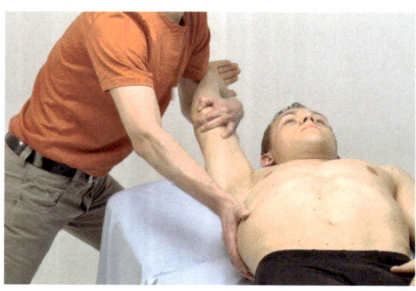

Method
The therapist moves the patient's arm to achieve maximum flexion, abduction and external rotation of the shoulder. The trunk should be in lateral flexion to the opposite side.

Finding
Muscle shortening is present if the movement cannot be performed to its full extent and if the endpoint feels soft and elastic.
The patient reports a stretching sensation along the course of the muscle.

Subscapularis muscle

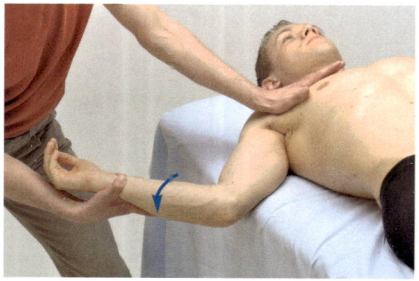

Method
The therapist moves the patient's arm to achieve maximum external rotation of the shoulder.

Finding
Muscle shortening is present if the movement cannot be performed to its full extent and if the endpoint feels soft and elastic.
The patient reports a stretching sensation along the course of the muscle.

2 Upper extremity

Muscles of the elbow

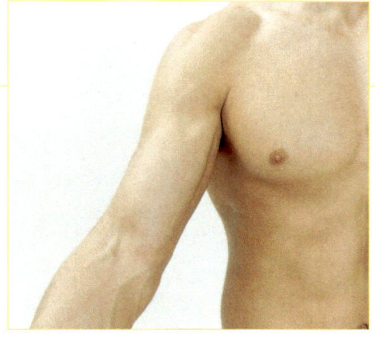

Biceps brachii muscle

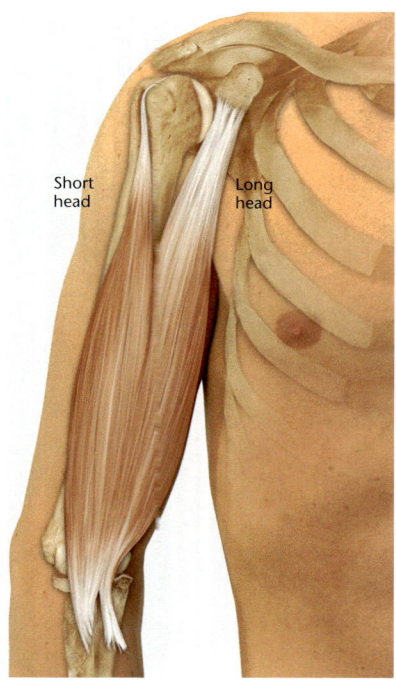

Short head

Long head

Short head

Long head

Firstly, the biceps brachii muscle (biceps muscle of the arm), with both its heads, can bring about anteversion of the shoulder; secondly, with its long head, it also causes abduction, particularly when the arm is already in external rotation. Furthermore, the tendon of the long head, secured at the intertubercular sulcus of the humerus via the coracohumeral ligament, secures the shoulder joint to the muscles of the rotator cuff. The actions of this muscle on the elbow are always performed by both its heads; these actions include flexion and supination. When the elbow is fully extended, the biceps brachii possesses no significant supinatory effect.

Origin	Long head: supraglenoid tubercle of the scapula Short head: coracoid process of the scapula
Insertion	Radial tuberosity and via the aponeurosis of the biceps brachii muscle at the fascia of the forearm
Innervation	Musculocutaneous nerve, C5–C6
Special features	The biceps brachii is an indicator muscle for segment C6

Functions

 Synergists

Antagonists

Glenohumeral joint

Anteversion, long head

Pectoralis major muscle	Latissimus dorsi muscle
Deltoideus muscle (clavicular part)	Triceps brachii muscle (long head)
Coracobrachialis muscle	Teres major muscle
Infraspinatus muscle (cranial part)	Deltoideus muscle (spinal part)
	Subscapularis muscle (caudal part)

Abduction (with the upper arm externally rotated)

Deltoideus muscle (acromial part)	Pectoralis major muscle
Deltoideus muscle (spinal and clavicular parts with arm abducted)	Latissimus dorsi muscle
	Teres major muscle
Infraspinatus muscle (cranial part)	Teres minor muscle
Subscapularis muscle	Coracobrachialis muscle
	Deltoideus muscle (spinal and clavicular parts with arm adducted)
	Triceps brachii muscle (long head)

Humeroradial and humeroulnar joints

Flexion

Brachialis muscle	Triceps brachii muscle
Brachioradialis muscle	Anconeus muscle
Pronator teres muscle	
Extensor carpi radialis longus muscle	
Flexor carpi ulnaris muscle (weak)	
Flexor carpi radialis muscle (weak)	

Proximal and distal radioulnar joints, humeroradial joint

Supination

Supinator muscle	Pronator quadratus muscle
Brachioradialis muscle (from pronation to the mid-point)	Pronator teres muscle
	Brachioradialis muscle (from supination to the mid-point)
	Flexor carpi radialis muscle (with the elbow extended)
	Extensor carpi radialis longus muscle

Muscle function testing

Muscle power grade	
5/4	

Starting position: The patient lies on his back. The arm is close to the body with the elbow extended and in supination.

Test procedure: The examiner holds the upper arm in place with one hand and with the other applies pressure to the forearm in the direction of elbow extension .

Instruction: "Flex your arm against my resistance and hold this position."

Starting position: The patient lies on his back or sits. The arm is close to the body with the elbow extended and in supination.

Test procedure: The examiner looks at the flexion of the elbow.

Instruction: "Flex your arm as far as possible."

Starting position: The patient sits next to the examination couch and rests his extended arm on it.

Test procedure: The examiner looks at the flexion of the elbow.

Instruction: "Flex your arm as far as possible."

Starting position: The patient lies on his back. The arm is close to the body with the elbow extended and in supination.

Test procedure: The examiner palpates the biceps brachii muscle.

Instruction: "Try to flex your arm."

⚕ Clinical relevance

- If there is a lesion of the musculocutaneous nerve, the forearm cannot be supinated by the action of the brachioradialis muscle when the elbow is flexed.
- A painless rupture of the tendon of the long head of the biceps is often associated with increasing age and strain and is visible as a marked protuberance.
- Pull-ups with the forearm in pronation are more difficult, as the biceps brachii produces less force for mechanical reasons.

⚠ Problems/comments

- The wrist flexors should remain relaxed, as they might otherwise aid flexion of the arm at the elbow.

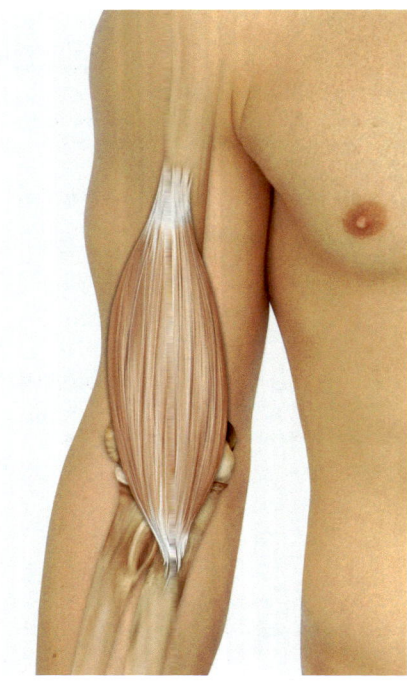

Brachialis muscle

The brachialis (brachial) muscle is the most important flexor of the elbow. It has no effect on the radioulnar joints via its insertion at the ulna.

Origin	Distal two thirds of the ventral surface of the humerus Intermuscular septum between the brachialis and triceps brachii muscles
Insertion	Ulnar tuberosity Coronoid process of the ulna
Innervation	Musculocutaneous nerve, C5–C7 Radial nerve, C5–C6
Special features	As the brachialis muscle is a flexor, a small part of its innervation still comes from the radial nerve

Functions

 Synergists   Antagonists

Humeroradial and humeroulnar joints

Flexion

Biceps brachii muscle	Triceps brachii muscle
Brachioradialis muscle	Anconeus muscle
Pronator teres muscle	
Extensor carpi radialis longus muscle	
Flexor carpi ulnaris muscle (weak)	
Flexor carpi radialis muscle (weak)	

Muscle function testing

Muscle power grade

5/4

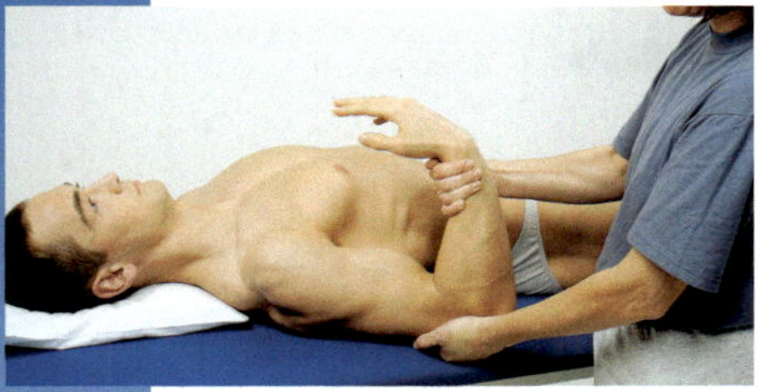

Starting position: The patient lies on his back or sits. The arm is close to the body with the elbow extended and in supination.

Test procedure: The examiner holds the upper arm in place with one hand and, with the other, applies pressure to the forearm in the direction of elbow extension.

Instruction: "Flex your arm against my resistance and hold this position."

3

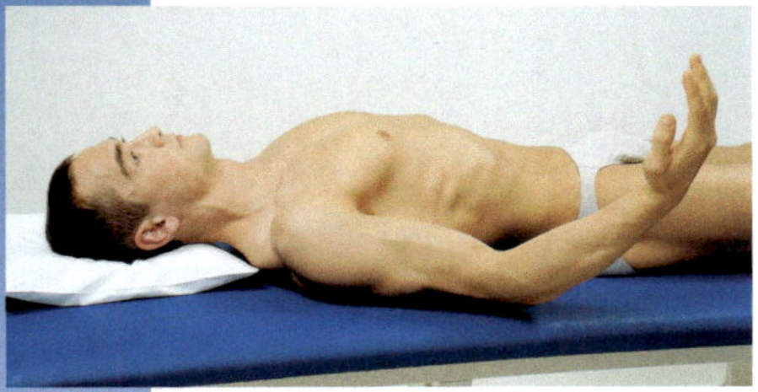

Starting position: The patient lies on his back or sits. The arm is close to the body with the elbow extended and in supination.

Test procedure: The examiner looks at the flexion of the elbow.

Instruction: "Flex your arm as far as possible."

2

Starting position: The patient sits next to the examination couch and rests his extended arm on it.

Test procedure: The examiner looks at the flexion of the elbow.

Instruction: "Flex your arm as far as possible."

1/0

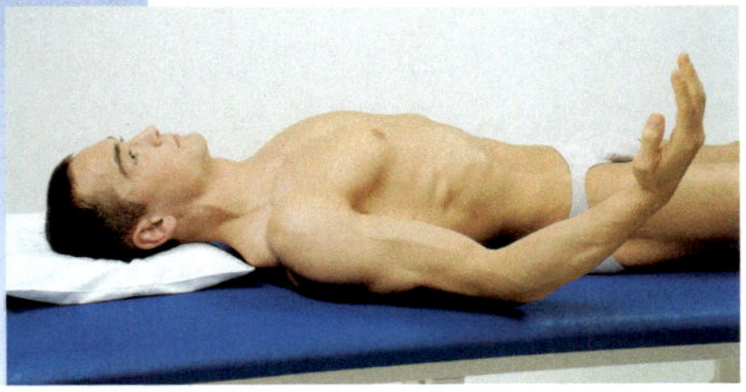

Starting position: The patient lies on his back. The arm is close to the body with the elbow extended and in supination.

Test procedure: The examiner looks at the flexion of the elbow.

Instruction: "Try to flex your elbow."

Clinical relevance

- The distal, lateral part of the brachialis muscle is innervated by a branch of the radial nerve.

Problems/comments

- The wrist flexors should remain relaxed, as they might otherwise aid flexion of the arm at the elbow.

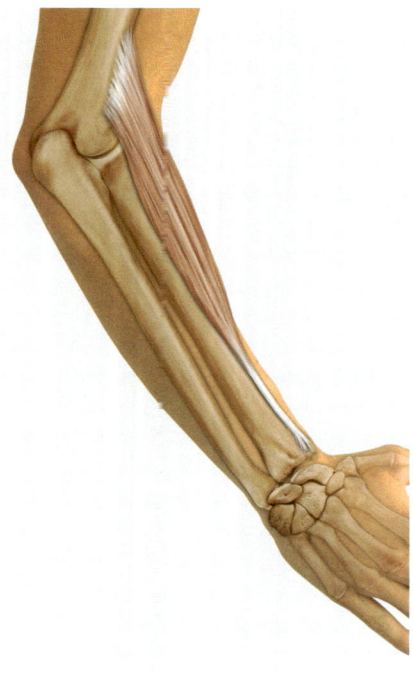

Brachioradialis muscle

The brachioradialis (brachioradial) muscle flexes the elbow and, in addition, takes the forearm back to its mid-point from extreme pronation or supination. When heavy loads are carried, its contraction reduces bending strain on the radius.

Origin	Lateral supracondylar crest of the humerus Lateral intermuscular septum of the arm
Insertion	Lateral surface of the radius proximally of the base of the styloid process
Innervation	Radial nerve, C5–C6

Functions

 Synergists   Antagonists

Humeroradial and humeroulnar joints

Flexion	
Brachialis muscle	Triceps brachii muscle
Biceps brachii muscle	Anconeus muscle
Pronator teres muscle	
Extensor carpi radialis longus muscle	
Flexor carpi ulnaris muscle (weak)	
Flexor carpi radialis muscle (weak)	

Proximal and distal radioulnar joints, humeroradial joint

Supination (from pronation to the mid-point)	
Biceps brachii muscle	Pronator quadratus muscle
Supinator muscle	Pronator teres muscle
	Brachioradialis muscle (from supination to the mid-point)
	Flexor carpi radialis muscle (with the elbow extended)
	Extensor carpi radialis longus muscle
Pronation (from supination to the mid-point)	
Pronator quadratus muscle	Supinator muscle
Pronator teres muscle	Biceps brachii muscle
Flexor carpi radialis muscle (with the elbow extended)	Brachioradialis muscle (from pronation to the mid-point)
Extensor carpi radialis longus muscle	

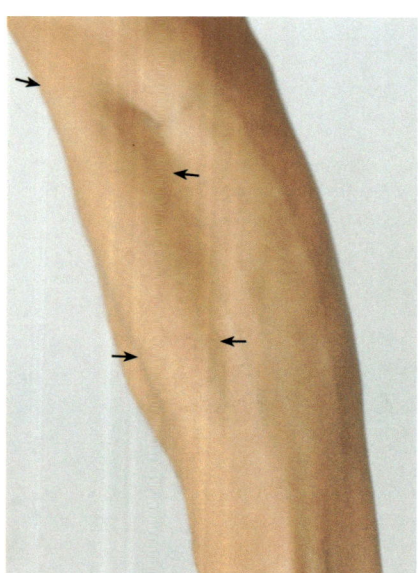

Muscle function testing

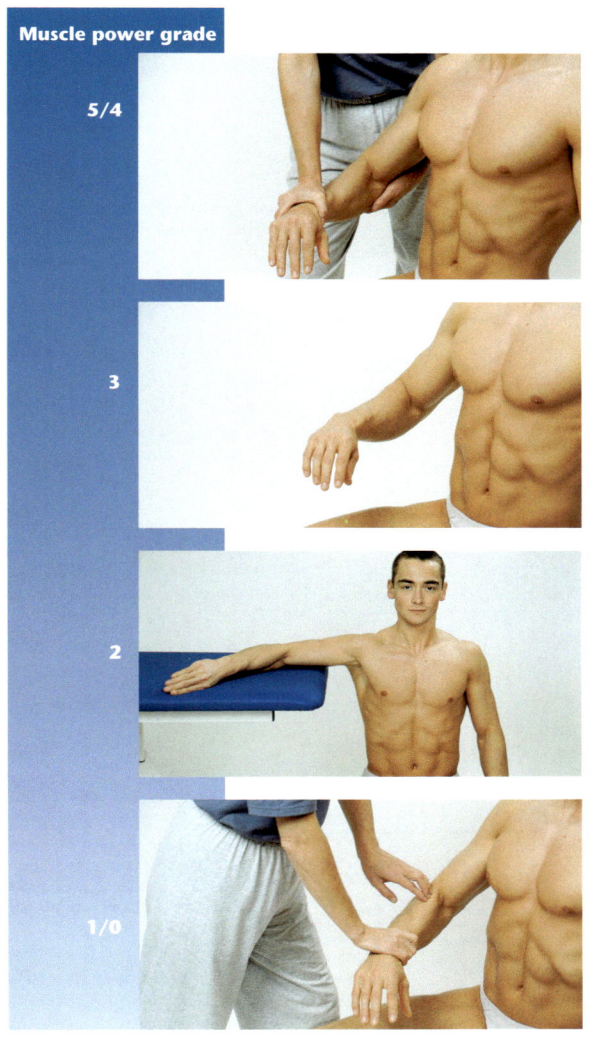

Muscle power grade

5/4

3

2

1/0

Starting position: The patient sits. The arm hangs down alongside the body, the forearm is in pronation.

Test procedure: The examiner holds the elbow in place with one hand and, with the other, applies pressure to the distal forearm in the direction of elbow extension.

Instruction: "Flex your arm against my resistance and hold this position."

Starting position: The patient sits. The arm hangs down alongside the body, the forearm is in pronation.

Test procedure: The examiner looks at the movement of the arm.

Instruction: "Flex your arm."

Starting position: The patient sits. The extended arm rests on the examination couch or other flat surface with the forearm in pronation.

Test procedure: The examiner looks at the movement of the arm.

Instruction: "Flex your arm by pulling it across the couch."

Starting position: The patient sits. The arm hangs down alongside the body, the forearm is in pronation.

Test procedure: The examiner palpates the brachioradialis muscle.

Instruction: "Try to flex your arm."

[!] **Problems/comments**

- The biceps brachii is more active during supination, while the brachioradialis is more active during pronation.

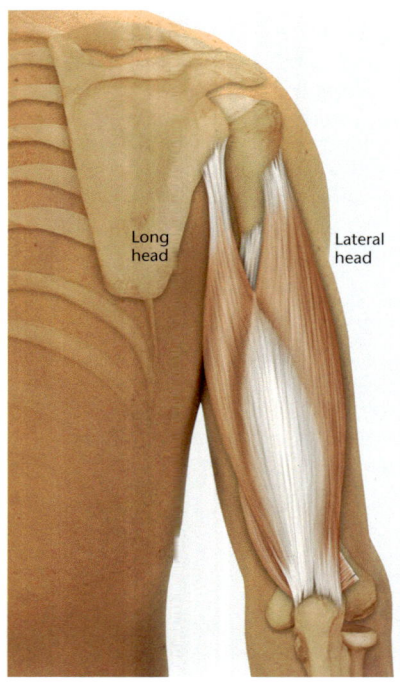

Triceps brachii muscle

The triceps brachii muscle (triceps muscle of the arm) powerfully extends the elbow (e.g. during push-ups.) The long head also has an adductive action on the shoulder joint.

Origin

Long head: infraglenoid tubercle of the scapula
Lateral head: dorsolateral surface of the humerus, laterally of the sulcus of the radial nerve
Medial head: dorsomedial side of the distal two thirds of the humerus, medially of the sulcus of the radial nerve and dorsally of the medial intermuscular septum

Insertion

Olecranon
Posterior wall of the joint capsule

Innervation

Radial nerve, C6–C8

Special features

The long head of the triceps brachii forms the medial boundary and the lateral head the lateral boundary of the lateral axilla
The triceps brachii is an indicator muscle for spinal cord segment C7

Functions

 Synergists   Antagonists

Glenohumeral joint

Retroversion (long head only)

Synergists	Antagonists
Latissimus dorsi muscle	Pectoralis major muscle
Teres major muscle	Deltoideus muscle (clavicular part)
Deltoideus muscle (spinal part)	Biceps brachii muscle
Subscapularis muscle (caudal part)	Coracobrachialis muscle
	Infraspinatus muscle (cranial part)

Adduction (with the arm abducted)

Synergists	Antagonists
Pectoralis major muscle	Deltoideus muscle (acromial part)
Latissimus dorsi muscle	Deltoideus muscle (spinal and clavicular parts
Teres major muscle	with arm abducted)
Teres minor muscle	Infraspinatus muscle (cranial part)
Coracobrachialis muscle	Biceps brachii muscle (long head)
Biceps brachii (short head)	Subscapularis muscle (cranial part)
Deltoideus muscle (spinal and clavicular parts with arm adducted)	
Infraspinatus muscle (caudal part)	

Humeroradial and humeroulnar joints

Extension

Synergists	Antagonists
Anconeus muscle	Brachialis muscle
	Biceps brachii muscle
	Brachioradialis muscle
	Pronator teres muscle
	Extensor carpi radialis longus muscle
	Flexor carpi ulnaris muscle (weak)
	Flexor carpi radialis muscle (weak)

Muscle function testing

Muscle power grade

5/4

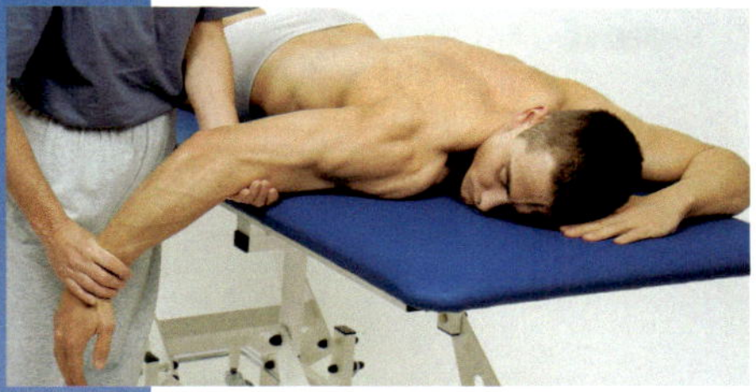

Starting position: The patient lies on his front, with the arm abducted 90° at the shoulder.

Test procedure: The examiner applies pressure to the forearm in the direction of elbow flexion.

Instruction: "Extend your elbow against my resistance and hold this position."

3

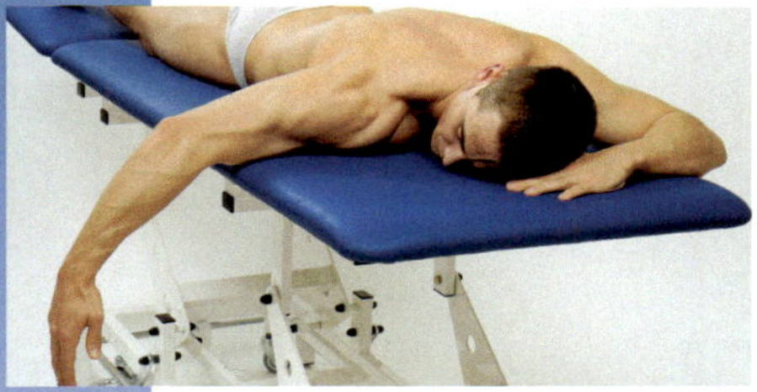

Starting position: The patient lies on his front, with the arm abducted 90° at the shoulder.

Test procedure: The examiner looks at the extension movement of the elbow.

Instruction: "Extend your elbow fully."

2

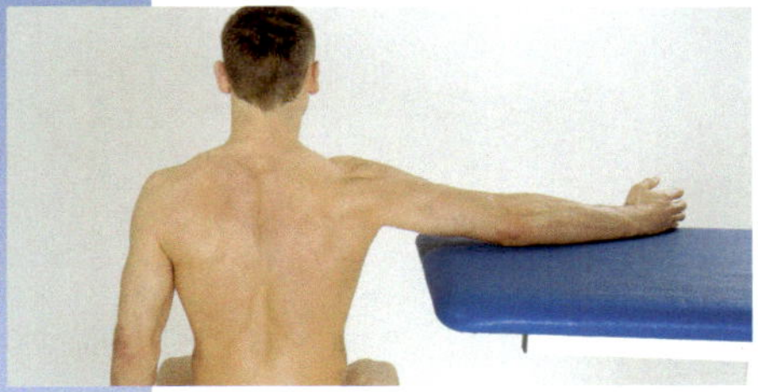

Starting position: The patient sits, his arm resting on a flat surface, with the shoulder abducted and the elbow flexed.

Test procedure: The examiner looks at the extension of the elbow.

Instruction: "Extend your elbow."

1/0

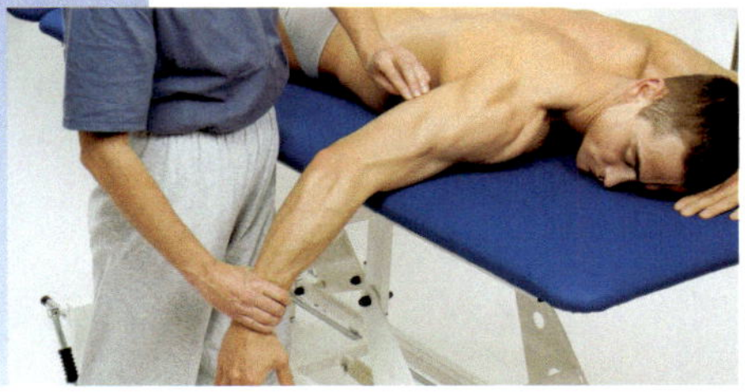

Starting position: The patient lies on his front.

Test procedure: The examiner palpates the triceps brachii muscle.

Instruction: "Try to extend your elbow."

⚕ Clinical relevance

• A fracture of the humerus frequently results in damage to the radial nerve. However, the branch which supplies the triceps brachii often remains intact, because it branches off above the sulcus of the radial nerve. Thus, the function of the triceps brachii is preserved, whereas all the other muscles supplied by the radial nerve fail.

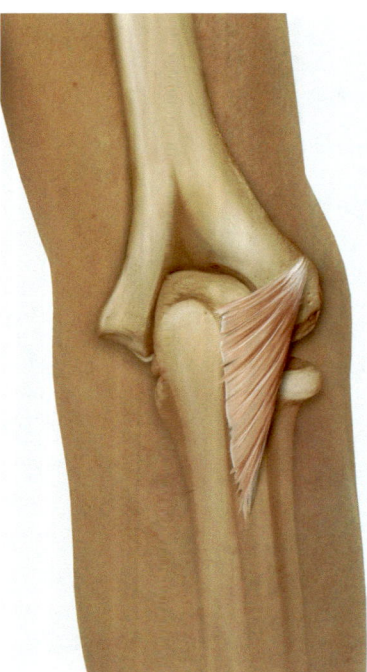

Anconeus muscle

The anconeus muscle extends the arm at the elbow. Its contraction may cause abduction of the ulna in the radial (lateral) direction; the result of this, particularly during pronation, is that the wrist cannot perform an arcuate movement around the capitulum of the ulna. In this case, the axis of movement for pronation and supination runs through the middle of the proximal wrist.

Origin   Dorsal surface of the lateral epicondyle of the humerus
                      Dorsal part of the elbow joint capsule

Insertion   Lateral side of the olecranon
                      Proximal, dorsal quarter of the ulna

Innervation   Radial nerve, C5–C6

Functions

 Synergists

 Antagonists

Humeroradial and humeroulnar joints

Extension
Triceps brachii muscle

Brachialis muscle
Biceps brachii muscle
Brachioradialis muscle
Pronator teres muscle
Extensor carpi radialis longus muscle
Flexor carpi ulnaris muscle (weak)
Flexor carpi radialis muscle (weak)

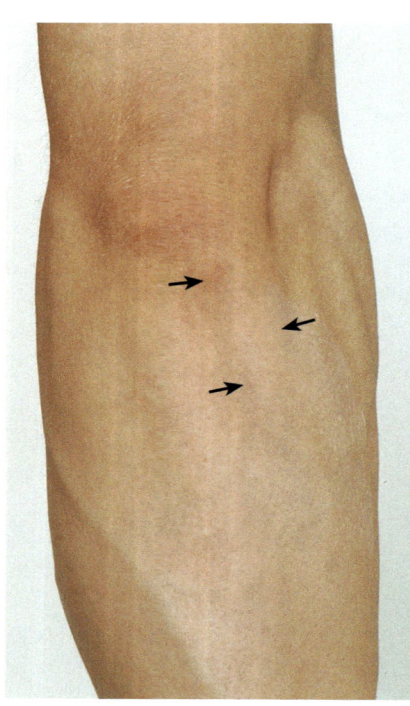

Muscle function testing

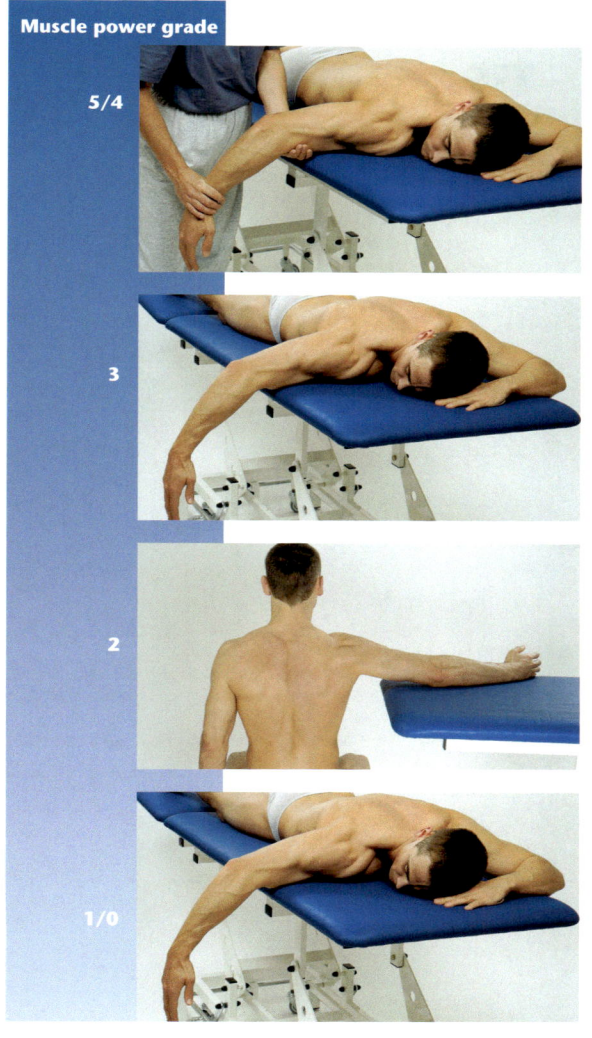

Muscle power grade

5/4

3

2

1/0

Starting position: The patient lies on his front, with the arm abducted 90° at the shoulder.

Test procedure: The examiner applies pressure to the forearm in the direction of elbow flexion.

Instruction: "Extend your elbow against my resistance and hold this position."

Starting position: The patient lies on his front, with the arm abducted 90° at the shoulder.

Test procedure: The examiner looks at the extension movement of the elbow.

Instruction: "Extend your elbow fully."

Starting position: The patient sits, his arm resting on a flat surface, with the shoulder abducted and the elbow flexed.

Test procedure: The examiner looks at the extension of the elbow.

Instruction: "Extend your elbow."

Starting position: The patient lies on his front.

Test procedure: The examiner looks at the extension of the elbow.

Instruction: "Try to extend your elbow."

 **Clinical relevance**

• A fracture of the humerus frequently results in damage to the radial nerve. However, the branch which supplies the triceps brachii often remains intact because it branches off above the sulcus of the radial nerve. Thus, the function of the triceps brachii is preserved. All the other muscles supplied by the radial nerve fail.

 Problems/comments

• The anconeus muscle is tested together with the triceps muscle

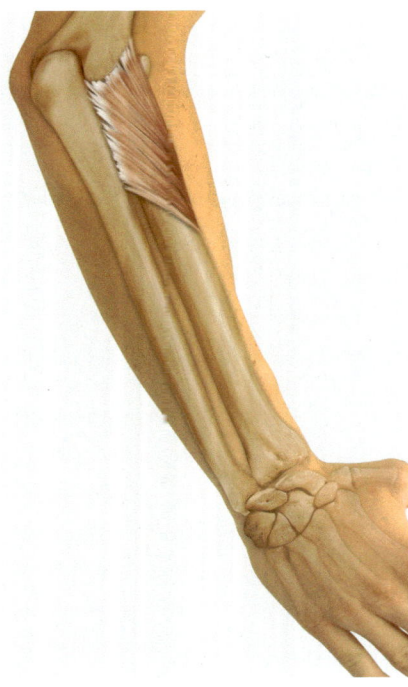

Supinator muscle

The supinator muscle carries out the supination movement of the elbow joint from any of its positions, i.e. it positions the radius so that it is parallel to the ulna. It is the most important muscle during supination when the elbow is fully extended or if little force is required; otherwise, it is aided by the stronger biceps brachii muscle.

Origin	Lateral epicondyle of the humerus
	Supinator crest of the ulna
	Annular ligament of the radius
	Radial collateral ligament
Insertion	Proximal third of the radius (across a broad area)
Innervation	Radial nerve, deep branch, C5–C6

Functions

 Synergists

Proximal and distal radioulnar joints, humeroradial joint

Supination
Biceps brachii muscle
Brachioradialis muscle (from pronation to the mid-point)

Antagonists

Pronator teres muscle
Pronator quadratus muscle
Extensor carpi radialis longus muscle
Brachioradialis muscle (from supination to the mid-point)
Flexor carpi radialis muscle (with the elbow extended)

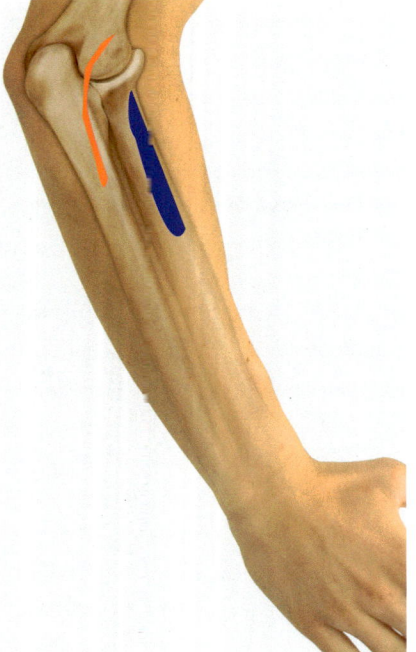

Muscle function testing

Muscle power grade

5/4

Starting position: The patient sits. The arm hangs down alongside the body and is flexed 90° at the elbow. The muscles of the fingers and wrist are relaxed.

Test procedure: The examiner grasps the patient's wrist and forearm with both hands and applies pressure in the direction of pronation.

Instruction: "Twist your forearm outwards against my resistance without extending your upper arm and hold your final position."

3

Starting position: The patient sits. The arm hangs down alongside the body and is flexed 90° at the elbow. The muscles of the fingers and wrist are relaxed.

Test procedure: The examiner looks at the movement of the forearm.

Instruction: "Twist your forearm outwards without extending your upper arm."

2

Starting position: The patient sits with his upper arm extended and resting on a flat surface. The arm is flexed at the elbow and the forearm stands vertically.

Test procedure: The examiner looks at the movement of the forearm.

Instruction: "Twist your forearm with the palm of your hand facing inwards."

1/0

Starting position: The patient sits. The arm hangs down alongside the body and is flexed 90° at the elbow. The muscles of the fingers and wrist are relaxed.

Test procedure: The examiner looks at the movement of the forearm.

Instruction: "Try to twist your forearm outwards."

Clinical relevance

- In contrast to the biceps brachii muscle, the supinator muscle can also supinate the arm when it is extended at the elbow

Problems/comments

- It is almost impossible to distinguish the function of the supinator muscle from that of the biceps brachii muscle.

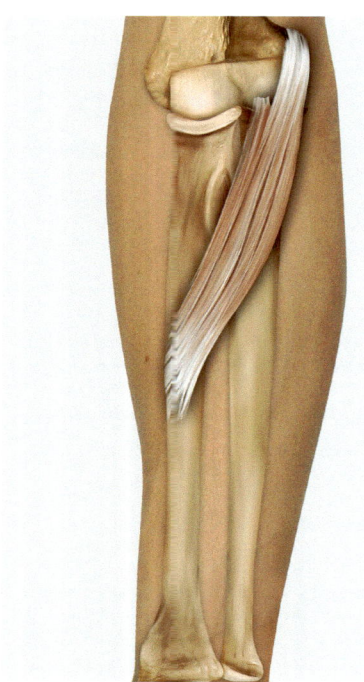

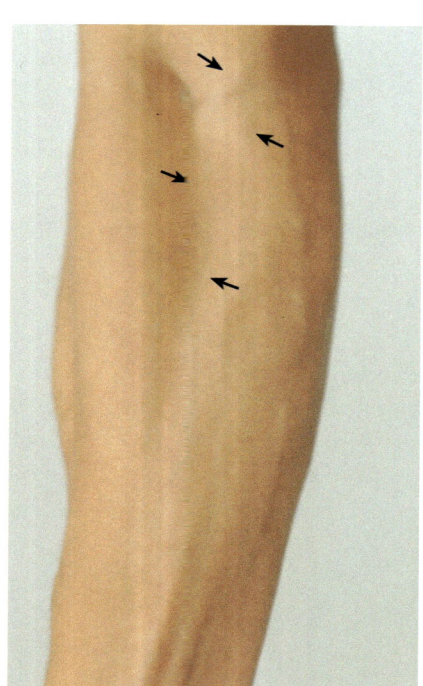

Pronator teres muscle

The pronator teres (round pronator) muscle brings the bones of the forearm into pronation and is a weak extensor of the elbow.

Origin

Head of the humerus: medial epicondyle of the humerus
Head of the ulna: coronoid process of the ulna

Insertion

Lateral surface of the mid-radius, pronator tuberosity

Innervation

Median nerve, C6–C7
Sometimes also the musculocutaneous nerve

Functions

 Synergists

 Antagonists

Humeroradial and humeroulnar joints

Flexion
Brachialis muscle
Biceps brachii muscle
Brachioradialis muscle
Extensor carpi radialis longus muscle
Flexor carpi ulnaris muscle (weak)
Flexor carpi radialis muscle (weak)

Triceps brachii muscle
Anconeus muscle

Proximal and distal radioulnar joints, humeroradial joint

Pronation
Pronator quadratus muscle
Brachioradialis muscle (from supination to the mid-point)
Flexor carpi radialis muscle (with the elbow extended)
Extensor carpi radialis longus muscle

Supinator muscle
Biceps brachii muscle
Brachioradialis muscle (from pronation to the mid-point)

Muscle function testing

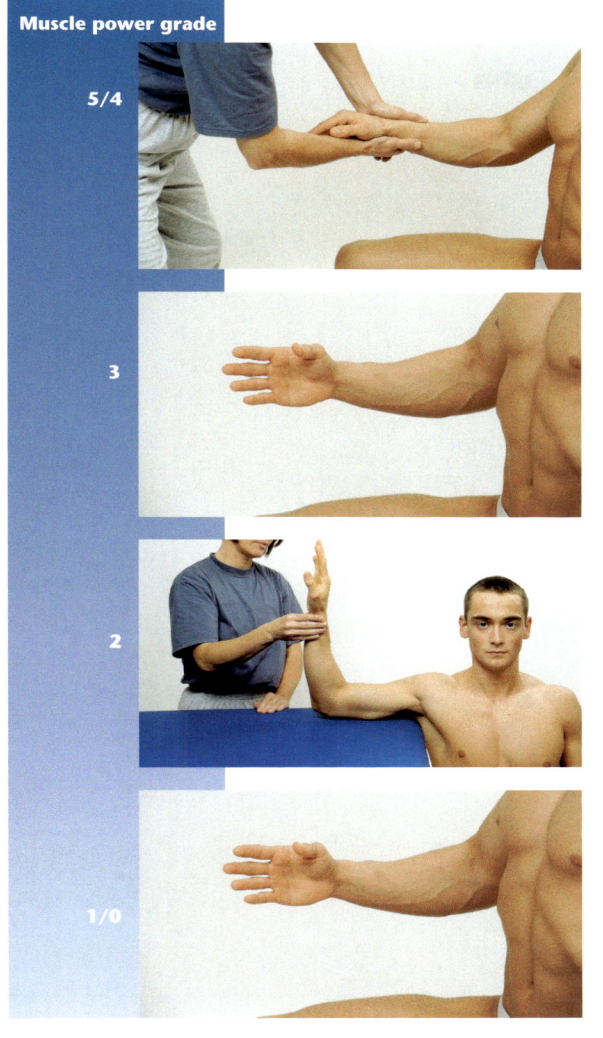

Muscle power grade

5/4

3

2

1/0

Starting position: The patient sits. The arm hangs down alongside the body and is flexed 90° at the elbow. The muscles of the fingers and wrist are relaxed.

Test procedure: The examiner grasps the patient's wrist and forearm with both hands and applies pressure in the direction of supination.

Instruction: "Twist your forearm inwards against my resistance without extending your upper arm and hold your final position."

Starting position: The patient sits. The arm hangs down alongside the body and is flexed 90° at the elbow. The muscles of the fingers and wrist are relaxed.

Test procedure: The examiner looks at the movement of the forearm.

Instruction: "Twist your forearm inwards without extending your upper arm."

Starting position: The patient sits with his upper arm extended and resting on a flat surface. The arm is flexed at the elbow and the forearm stands vertically.

Test procedure: The examiner looks at the movement of the forearm.

Instruction: "Twist your forearm inwards."

Starting position: The patient sits. The arm hangs down alongside the body and is flexed 90° at the elbow. The muscles of the fingers and wrist are relaxed.

Test procedure: The examiner looks at the movement of the forearm.

Instruction: "Try to twist your forearm inwards."

 Problems/comments

- The pronator quadratus and the pronator teres muscles cannot be distinguished functionally.

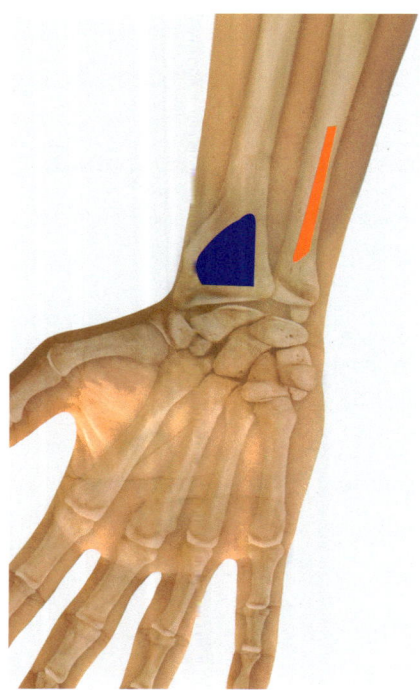

Pronator quadratus muscle

The pronator quadratus (quadrate pronator) muscle initiates the pronation movement and holds the two bones of the forearm together at the distal end, stabilizing the radioulnar joint in similar fashion to the interosseous membrane.

Origin	Distal quarter of the ventral surface of the ulna
Insertion	Distal quarter of the ventral surface of the radius
Innervation	Anterior interosseous nerve from the median nerve, C6–T1

Functions

 Synergists   Antagonists

Proximal and distal radioulnar joints, humeroradial joint

Pronation
Pronator teres muscle
Brachioradialis muscle (from supination to the mid-point)
Flexor carpi radialis muscle (with the elbow extended)
Extensor carpi radialis longus muscle

Supinator muscle
Biceps brachii muscle
Brachioradialis muscle (from pronation to the mid-point)

Muscle function testing

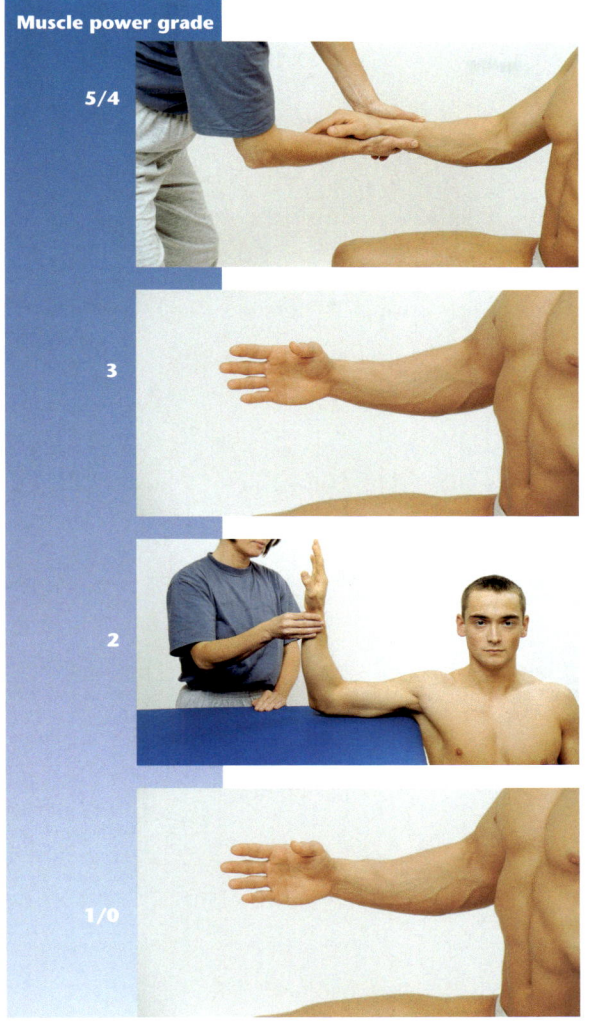

Muscle power grade

5/4

3

2

1/0

Starting position: The patient sits. The arm hangs down alongside the body and is flexed 90° at the elbow. The muscles of the fingers and wrist are relaxed.

Test procedure: The examiner grasps the patient's wrist and forearm with both hands and applies pressure in the direction of supination.

Instruction: "Twist your forearm inwards against my resistance without extending your upper arm and hold your final position."

Starting position: The patient sits. The arm hangs down alongside the body and is flexed 90° at the elbow. The muscles of the fingers and wrist are relaxed.

Test procedure: The examiner looks at the movement of the forearm.

Instruction: "Twist your forearm inwards without extending your upper arm."

Starting position: The patient sits with his upper arm extended and resting on a flat surface. The arm is flexed at the elbow and the forearm stands vertically.

Test procedure: The examiner looks at the movement of the forearm.

Instruction: "Twist your forearm inwards."

Starting position: The patient sits. The arm hangs down alongside the body and is flexed 90° at the elbow. The muscles of the fingers and wrist are relaxed.

Test procedure: The examiner looks at the movement of the forearm.

Instruction: "Try to twist your forearm inwards."

 Problems/comments

- The pronator quadratus and the pronator teres muscles cannot be distinguished functionally. The pronator quadratus muscle cannot be palpated.

Stretch tests

Triceps brachii muscle

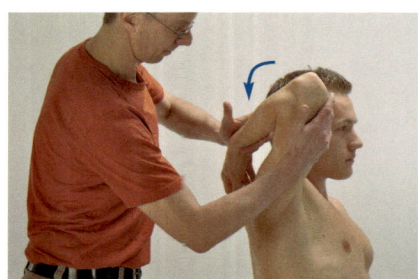

Method
The therapist moves the patient's forearm into maximum elbow flexion. The arm is flexed at the shoulder as far as possible.

Finding
Muscle shortening is present if the movement cannot be performed to its full extent and if the endpoint feels soft and elastic.
The patient reports a stretching sensation along the course of the muscle.

Biceps brachii muscle

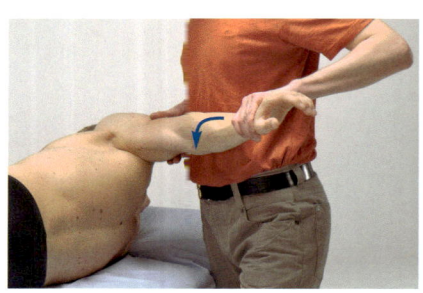

Method
The therapist moves the patient's forearm into maximum elbow extension and pronation. The arm is extended at the shoulder.

Finding
Muscle shortening is present if the movement cannot be performed to its full extent and if the endpoint feels soft and elastic.
The patient reports a stretching sensation along the course of the muscle.

Supinator and brachioradialis muscles

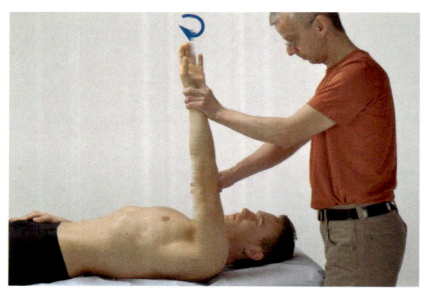

Method
The therapist moves the patient's forearm into maximum pronation. The elbow is also extended to its maximum.

Finding
Muscle shortening is present if the movement cannot be performed to its full extent and if the endpoint feels soft and elastic.
The patient reports a stretching sensation along the course of the muscle.

Pronator teres and pronator quadratus muscles

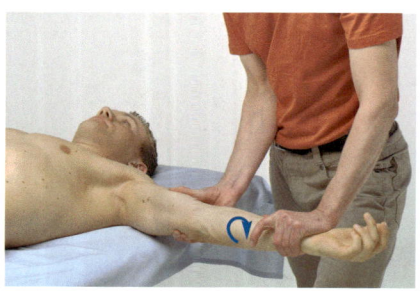

Method
The therapist moves the patient's forearm into maximum supination. The elbow is also extended to its maximum.

Finding
Muscle shortening is present if the movement cannot be performed to its full extent and if the endpoint feels soft and elastic.
The patient reports a stretching sensation along the course of the muscle.

Note
For a strech test of the head of the ulna and the pronator quadratus muscle in isolation, perform the test with the elbow flexed 90°.

2 Upper extremity

Muscles of the wrist

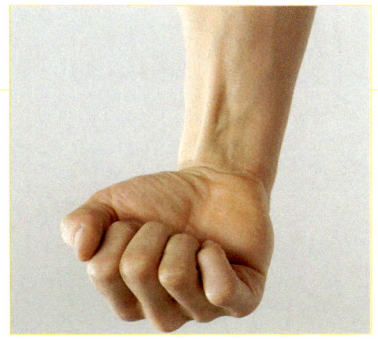

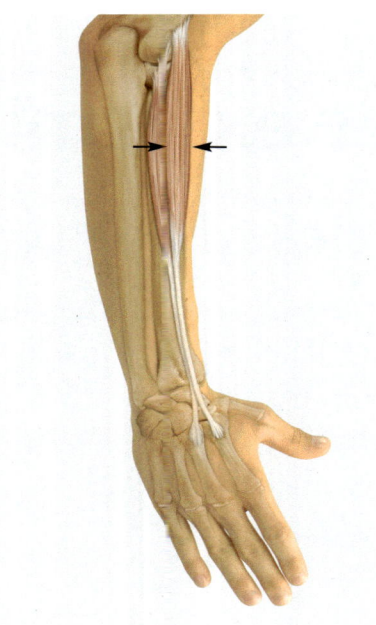

Extensor carpi radialis longus muscle

The extensor carpi radialis longus muscle (long radial extensor muscle of the wrist) extends the joints of the wrist. One important function of this muscle is to prevent flexion of the wrist when the long flexor muscles of the forearm cause powerful flexion of the fingers. When contracting in conjunction with the flexor carpi radialis muscle, it causes radial abduction of the hand. This muscle also has a weak pronating effect when the wrist is in supination.

Origin	Lateral supracondylar crest of the humerus
Insertion	Dorsal surface of the base of the second metacarpal bone
Innervation	Radial nerve, C6–C7

Functions

Synergists

Antagonists

Proximal and distal radioulnar joints, humeroradial joint

Pronation
Pronator quadratus muscle
Pronator teres muscle
Brachioradialis muscle (from supination to the mid-point)
Flexor carpi radialis muscle (with the elbow extended)

Supinator muscle
Biceps brachii muscle (with the elbow flexed)
Brachioradialis muscle (from pronation to the mid-point)

Radiocarpal joint and mediocarpal joint

Extension
Extensor digitorum muscle
Extensor carpi radialis brevis muscle
Extensor carpi ulnaris muscle
Extensor indicis muscle
Extensor digiti minimi muscle
Extensor pollicis longus muscle

Flexor digitorum superficialis muscle
Flexor digitorum profundus muscle
Flexor carpi ulnaris muscle
Flexor carpi radialis muscle
Flexor pollicis longus muscle
Abductor pollicis longus muscle
Palmaris longus muscle

Radial abduction
Flexor carpi radialis muscle
Extensor carpi radialis brevis muscle
Flexor pollicis longus muscle
Abductor pollicis longus muscle
Extensor pollicis brevis muscle
Extensor pollicis longus muscle

Flexor carpi ulnaris muscle
Extensor carpi ulnaris muscle
Flexor digitorum profundus muscle
Extensor digitorum muscle
Extensor digiti minimi muscle

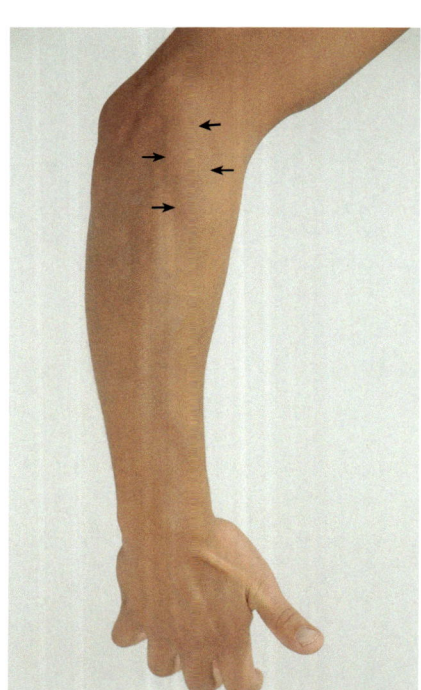

Muscle function testing

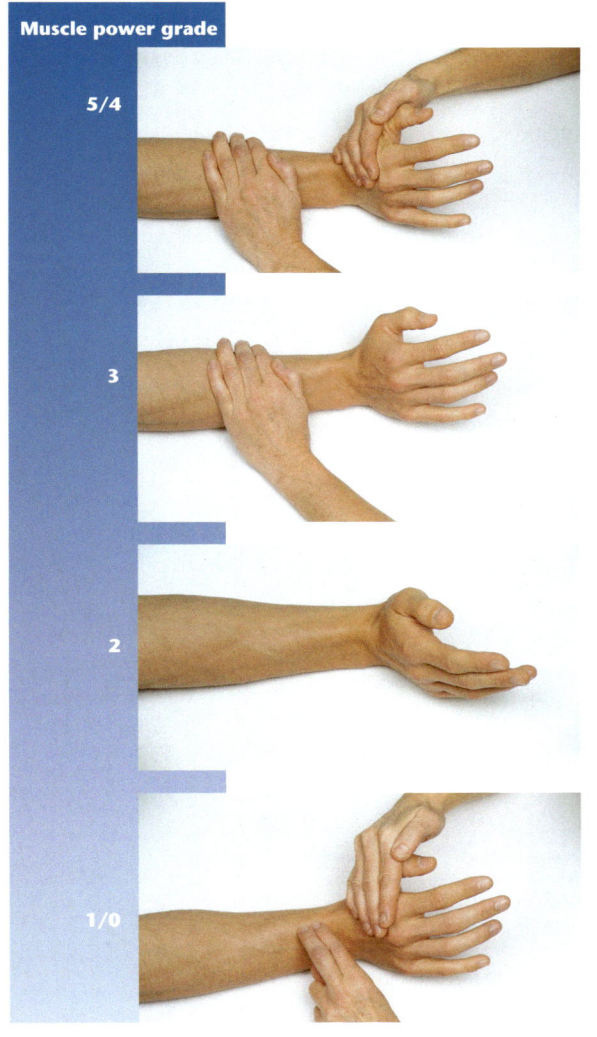

Muscle power grade

5/4

3

2

1/0

Starting position: The patient sits. The forearm and hand rest on a table or other flat surface, with the palm down.

Test procedure: The examiner holds the patient's forearm in place with one hand and with the other applies downward pressure to the medial side of the back of the hand.

Instruction: "Raise your hand off the table against my resistance. Keep your fingers loose while doing so and hold your final position.

Starting position: The patient sits. The forearm and hand rest on a table or other flat surface, with the palm down.

Test procedure: The examiner holds the patient's forearm in place with one hand.

Instruction: "Raise your hand off the table, thumb side first. Keep your fingers loose while doing so."

Starting position: The patient sits. The forearm and hand rest on a table or other flat surface, with the hand upright on its outer edge (the little finger side).

Test procedure: The examiner looks at the movement of the hand.

Instruction: "Pull your hand back along the table, so that the back of your hand moves towards your forearm."

Starting position: The patient sits. The forearm and hand rest on a table or other flat surface, with the palm down.

Test procedure: The examiner palpates the tendon of the extensor carpi radialis longus muscle.

Instruction: "Try to raise your hand off the table."

 Clinical relevance

- The extensor carpi radialis longus muscle aids fist closure, since slight dorsal extension is needed to achieve the full power of the finger flexors.

 Problems/comments

- The extensor carpi radialis longus and the extensor carpi radialis brevis muscles are tested together.
- The hand extensors that are palpable above the wrist, going from the radial to the ulnar side, are as follows:
  – extensor carpi radialis longus muscle
  – extensor carpi radialis brevis muscle
  – extensor digitorum muscle
  – extensor digiti minimi muscle
  – extensor carpi ulnaris muscle.

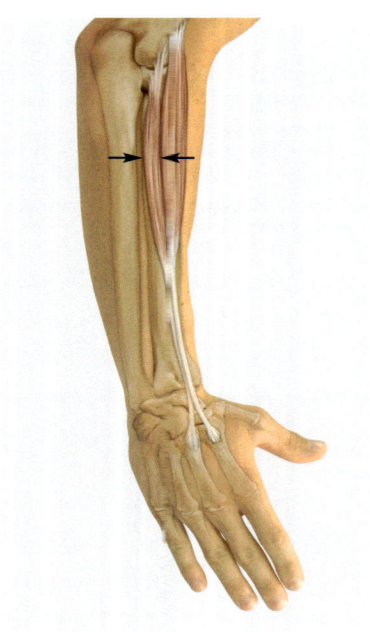

Extensor carpi radialis brevis muscle

The extensor carpi radialis brevis muscle (short radial extensor muscle of the wrist) extends the joints of the wrist and prevents flexion in these joints when the long flexor muscles are active. When contracting in conjunction with the flexor carpi radialis muscle, it also causes radial abduction of the hand.

Origin	Lateral epicondyle of the humerus
Insertion	Dorsal surface of the base of the third metacarpal bone
Innervation	Radial nerve, deep branch, C6–C7

Functions

Synergists   Antagonists

Proximal and distal radioulnar joints, humeroradial joint

Pronation

Pronator quadratus muscle	Supinator muscle
Pronator teres muscle	Biceps brachii muscle
Brachioradialis muscle (from supination to the mid-point)	Brachioradialis muscle (from pronation to the mid-point)
Flexor carpi radialis muscle (with the elbow extended)	
Extensor carpi radialis longus muscle	

Radiocarpal joint and mediocarpal joint

Extension

Extensor digitorum muscle	Flexor digitorum superficialis muscle
Extensor carpi radialis longus muscle	Flexor digitorum profundus muscle
Extensor carpi ulnaris muscle	Flexor carpi ulnaris muscle
Extensor indicis muscle	Flexor carpi radialis muscle
Extensor digiti minimi muscle	Flexor pollicis longus muscle
Extensor pollicis longus muscle	Abductor pollicis longus muscle
	Palmaris longus muscle

Radial abduction

Flexor carpi radialis muscle	Flexor carpi ulnaris muscle
Extensor carpi radialis longus muscle	Extensor carpi ulnaris muscle
Flexor pollicis longus muscle	Flexor digitorum profundus muscle
Abductor pollicis longus muscle	Extensor digitorum muscle
Extensor pollicis brevis muscle	Extensor digiti minimi muscle
Extensor pollicis longus muscle	

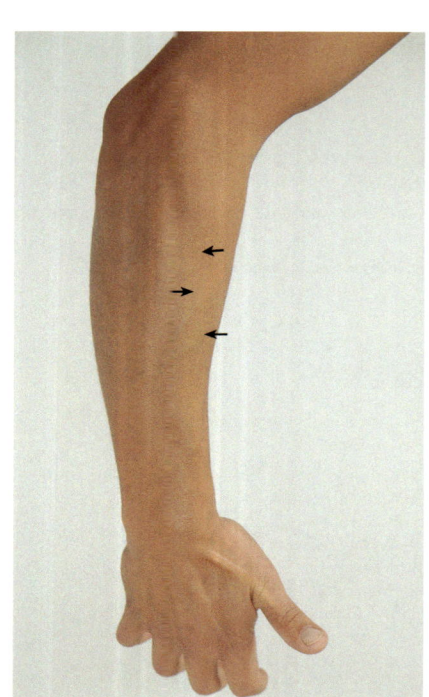

Muscle function testing

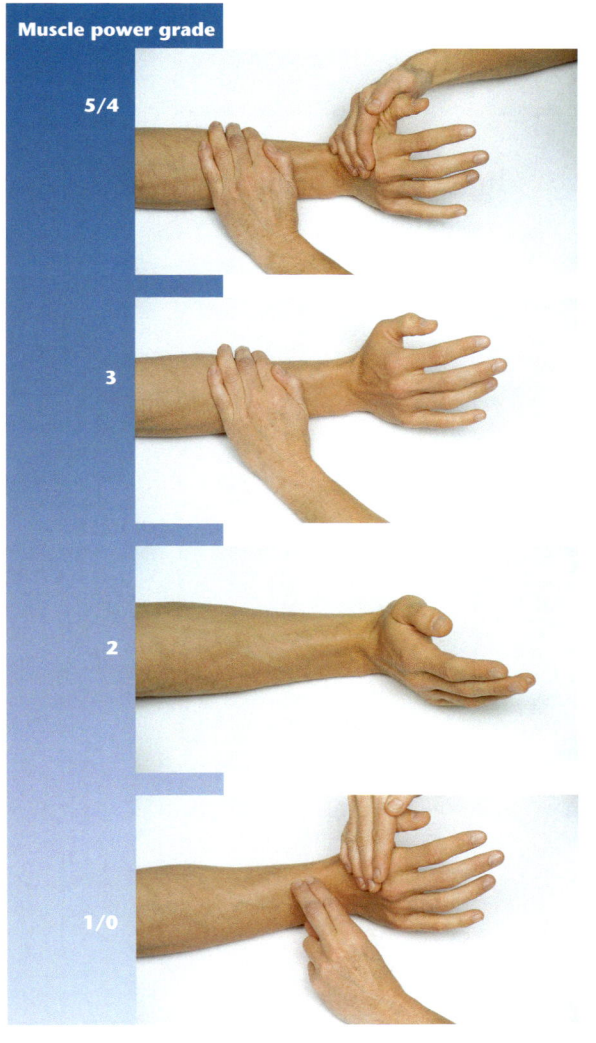

Muscle power grade

5/4

3

2

1/0

Starting position: The patient sits. The forearm and hand rest on a table or other flat surface, with the palm down.

Test procedure: The examiner holds the patient's forearm in place with one hand and with the other applies downward pressure to the medial side of the back of the hand.

Instruction: "Raise your hand off the table against my resistance. Keep your fingers loose while doing so and hold your final position.

Starting position: The patient sits. The forearm and hand rest on a table or other flat surface, with the palm down.

Test procedure: The examiner holds the patient's forearm in place with one hand.

Instruction: "Raise your hand off the table, thumb side first. Keep your fingers loose while doing so."

Starting position: The patient sits. The forearm and hand rest on a table or other flat surface, with the hand upright on its outer edge (the little finger side).

Test procedure: The examiner looks at the movement of the hand.

Instruction: "Pull your hand back along the table, so that the back of your hand moves towards your forearm."

Starting position: The patient sits. The forearm and hand rest on a table or other flat surface, with the palm down.

Test procedure: The examiner palpates the tendon of the extensor carpi radialis brevis muscle.

Instruction: "Try to raise your hand off the table."

! Problems/comments

- The extensor carpi radialis brevis and the extensor carpi radialis longus muscles are tested together.
- The hand extensors that are palpable above the wrist, going from the radial to the ulnar side, are as follows:
  extensor carpi radialis longus muscle
  extensor carpi radialis brevis muscle
  extensor digitorum muscle
  extensor digiti minimi muscle
  extensor carpi ulnaris muscle.

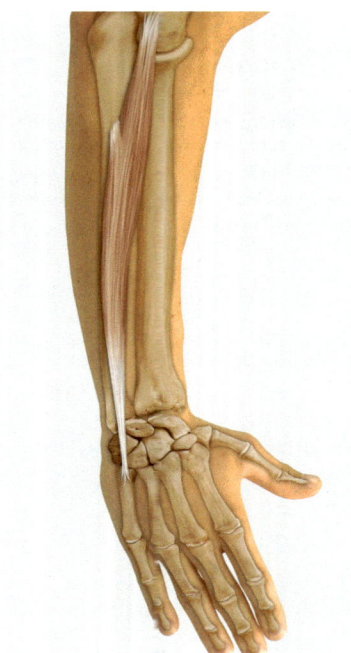

Extensor carpi ulnaris muscle

The extensor carpi ulnaris muscle (ulnar extensor muscle of the wrist) extends the joints of the wrist or causes ulnar abduction, if acting in conjunction with the flexor ulnaris muscle. Furthermore, it fixes the joints of the wrist, thus conducting the force of the long finger flexors to the joints of the fingers.

Origin	Humeral head: lateral epicondyle of the humerus, antebrachial fascia
	Ulnar head: dorsal part of the ulna
Insertion	Dorsal surface of the base of the fourth metacarpal bone
Innervation	Radial nerve, deep branch, C6–C8

Functions

 Synergists

 Antagonists

Radiocarpal joint and mediocarpal joint

Extension

Synergists	Antagonists
Extensor digitorum muscle	Flexor digitorum superficialis muscle
Extensor carpi radialis longus muscle	Flexor digitorum profundus muscle
Extensor carpi radialis brevis muscle	Flexor carpi ulnaris muscle
Extensor indicis muscle	Flexor carpi radialis muscle
Extensor digiti minimi muscle	Flexor pollicis longus muscle
Extensor pollicis longus muscle	Abductor pollicis longus muscle
	Palmaris longus muscle

Ulnar abduction

Synergists	Antagonists
Flexor carpi ulnaris muscle	Flexor carpi radialis muscle
Flexor digitorum profundus muscle	Extensor carpi radialis longus muscle
Extensor digitorum muscle	Extensor carpi radialis brevis muscle
Extensor digiti minimi muscle	Flexor pollicis longus muscle
	Abductor pollicis longus muscle
	Extensor pollicis brevis muscle
	Extensor pollicis longus muscle

Muscle function testing

Muscle power grade

5/4

Starting position: The patient sits. The forearm and hand rest on a table or other flat surface with the radial side down.

Test procedure: The examiner holds the patient's forearm in place with one hand and with the other applies pressure to the outer (little finger) side of the hand towards the inner (thumb) side.

Instruction: "Move your hand towards the little finger side against my resistance and hold the final position.

3

Starting position: The patient sits. The forearm and hand rest on a table or other flat surface with the radial side down.

Test procedure: The examiner holds the patient's forearm in place with one hand.

Instruction: "Raise your hand. When you do so, the little finger side should move towards your forearm."

2

Starting position: The patient sits. The forearm and hand rest on a table or other flat surface with the palm down.

Test procedure: The examiner looks at the movement of the hand.

Instruction: "Move your hand along the table. When you do so, the little finger side should move towards your forearm."

1/0

Starting position: The patient sits. The forearm and hand rest on a table or other flat surface, with the palm down.

Test procedure: The examiner palpates the tendon of the extensor carpi ulnaris muscle.

Instruction: "Try to move the little finger side of your hand towards your forearm."

 Problems/comments

- Because of the course of its tendon at the wrist, the extensor carpi ulnaris muscle is responsible more for the ulnar extension of the hand than for its dorsal extension.
- The hand extensors that are palpable above the wrist, going from the radial to the ulnar side, are as follows:
  – extensor carpi radialis longus muscle
  – extensor carpi radialis brevis muscle
  – extensor digitorum muscle
  – extensor digiti minimi muscle
  – extensor carpi ulnaris muscle.

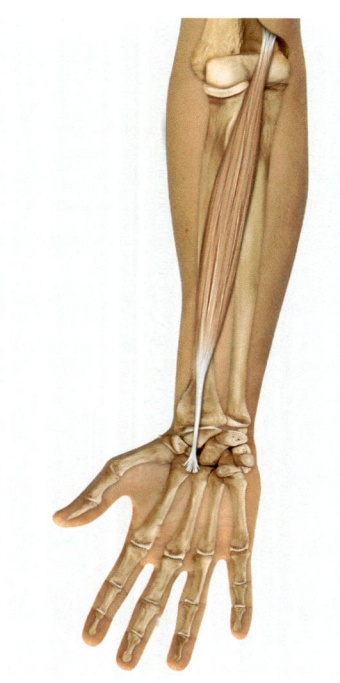

Flexor carpi radialis muscle

The flexor carpi radialis muscle (radial flexor muscle of the wrist) flexes the joints of the wrist or causes radial abduction. Above all, like the flexor carpi ulnaris, it can deploy the full force of the long finger extensors to the joints of the fingers by fixing the joints of the wrists.

Origin	Medial epicondyle of the humerus Antebrachial fascia
Insertion	Ventral surface of the base of the second metacarpal bone, with a small part also at the base of the third metacarpal bone
Innervation	Median nerve, C6–C8

Functions

 Synergists   Antagonists

Proximal and distal radioulnar joints, humeroradial joint

Pronation
Pronator quadratus muscle
Pronator teres muscle
Brachioradialis muscle (from supination to the mid-point)
Extensor carpi ulnaris muscle
Extensor carpi radialis longus muscle (from supination)

Supinator muscle
Biceps brachii muscle
Brachioradialis muscle (from pronation to the mid-point)

Humeroradial joint and humeroulnar joint

Flexion (weak)
Brachialis muscle
Biceps brachii muscle
Brachioradialis muscle
Pronator teres muscle
Extensor carpi radialis longus muscle
Flexor carpi ulnaris muscle

Triceps brachii muscle
Anconeus muscle

Radiocarpal joint and mediocarpal joint

Flexion
Flexor digitorum superficialis muscle
Flexor digitorum profundus muscle
Flexor carpi ulnaris muscle
Flexor pollicis longus muscle
Abductor pollicis longus muscle
Palmaris longus muscle

Extensor digitorum muscle
Extensor carpi radialis longus muscle
Extensor carpi radialis brevis muscle
Extensor carpi ulnaris muscle
Extensor indicis muscle
Extensor digiti minimi muscle
Extensor pollicis longus muscle

Radial abduction
Extensor carpi radialis longus muscle
Extensor carpi radialis brevis muscle
Flexor pollicis longus muscle
Abductor pollicis longus muscle
Extensor pollicis brevis muscle
Extensor pollicis longus muscle

Flexor carpi ulnaris muscle
Extensor carpi ulnaris muscle
Flexor digitorum profundus muscle
Extensor digitorum muscle
Extensor digiti minimi muscle

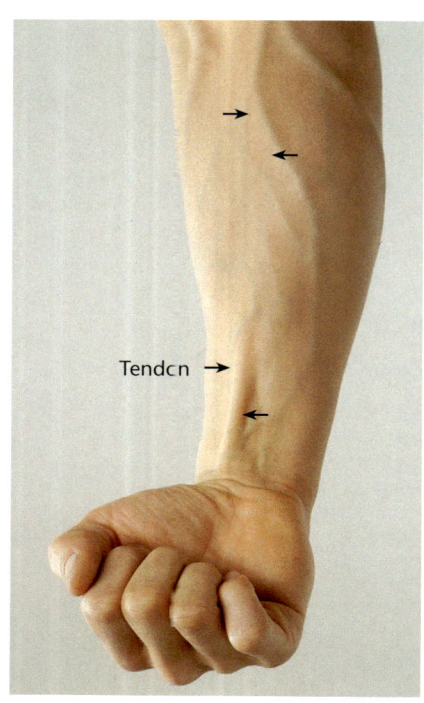

Tendon →

Muscle function testing

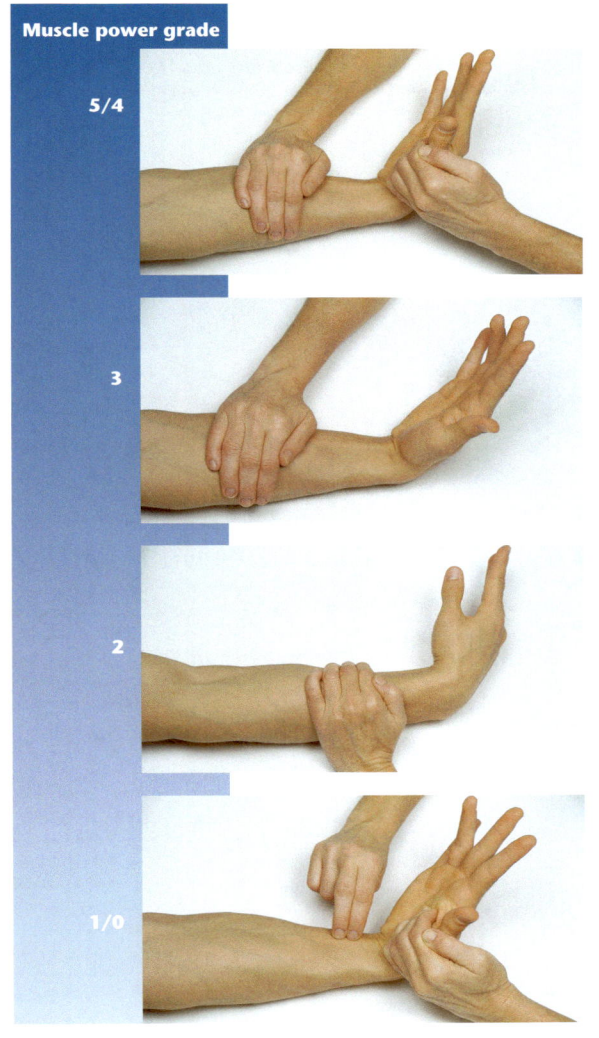

Muscle power grade

5/4

3

2

1/0

Starting position: The patient sits. The forearm and hand rest on a table or other flat surface in supination.

Test procedure: The examiner holds the patient's forearm in place with one hand and with the other applies downward pressure to the radial side of the palm.

Instruction: "Raise your hand off the table against my resistance and hold your final position.

Starting position: The patient sits. The forearm and hand rest on a table or other flat surface in supination.

Test procedure: The examiner holds the patient's forearm in place with one hand.

Instruction: "Raise your hand off the table. When you do so, the thumb side of your hand should move closer to your forearm."

Starting position: The patient sits. The forearm and hand rest on a table or other flat surface with the hand upright on its outer edge (the little finger side).

Test procedure: The examiner looks at the movement of the hand.

Instruction: "Move your hand along the table. When you do so, your palm should move closer to your forearm."

Starting position: The patient sits. The forearm and hand rest on a table or other flat surface in supination.

Test procedure: The examiner palpates the tendon of the flexor carpi radialis muscle.

Instruction: "Try to lift your hand off the table."

 Problems/comments

- The hand flexors that are palpable above the wrist, going from the radial to the ulnar side, are as follows:
  – flexor pollicis longus muscle
  – flexor carpi radialis muscle
  – palmaris longus muscle
  – flexor digitorum superficialis muscle
  – flexor carpi ulnaris muscle.

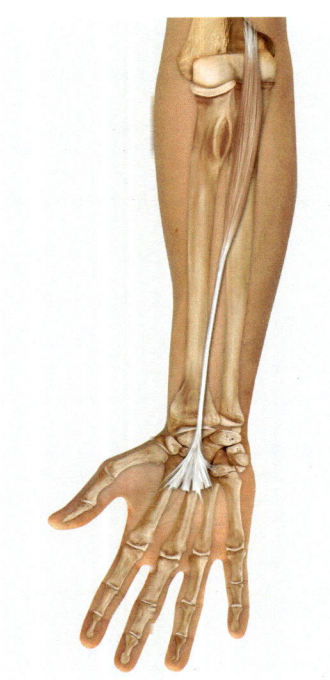

Palmaris longus muscle

The palmaris longus (long palmar) muscle is a negligibly weak flexor of the elbow and a weak flexor of the joints of the wrist. It tightens the palmar aponeurosis and tenses when the tips of the thumb and index finger are drawn together powerfully.

Origin	Medial epicondyle of the humerus
	Antebrachial fascia
Insertion	Palmar aponeurosis
Innervation	Median nerve, C7–T1

Functions

 Synergists

 Antagonists

Radiocarpal joint and mediocarpal joint

Flexion

Flexor digitorum superficialis muscle	Extensor digitorum muscle
Flexor digitorum profundus muscle	Extensor carpi radialis longus muscle
Flexor carpi ulnaris muscle	Extensor carpi radialis brevis muscle
Flexor carpi radialis muscle	Extensor carpi ulnaris muscle
Flexor pollicis longus muscle	Extensor indicis muscle
Abductor pollicis longus muscle	Extensor digiti minimi muscle
	Extensor pollicis longus muscle

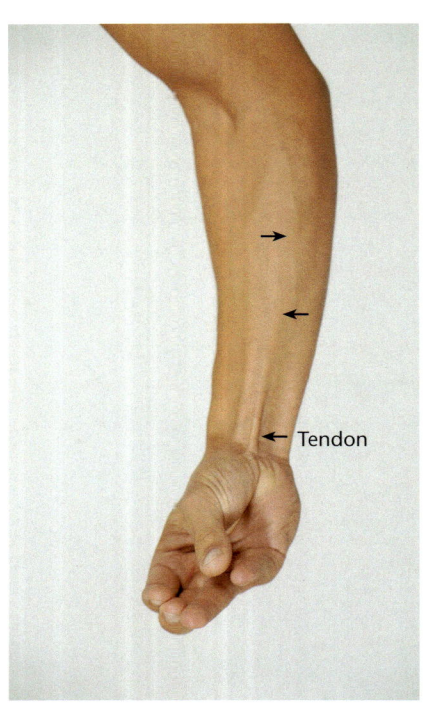

← Tendon

Muscle function testing

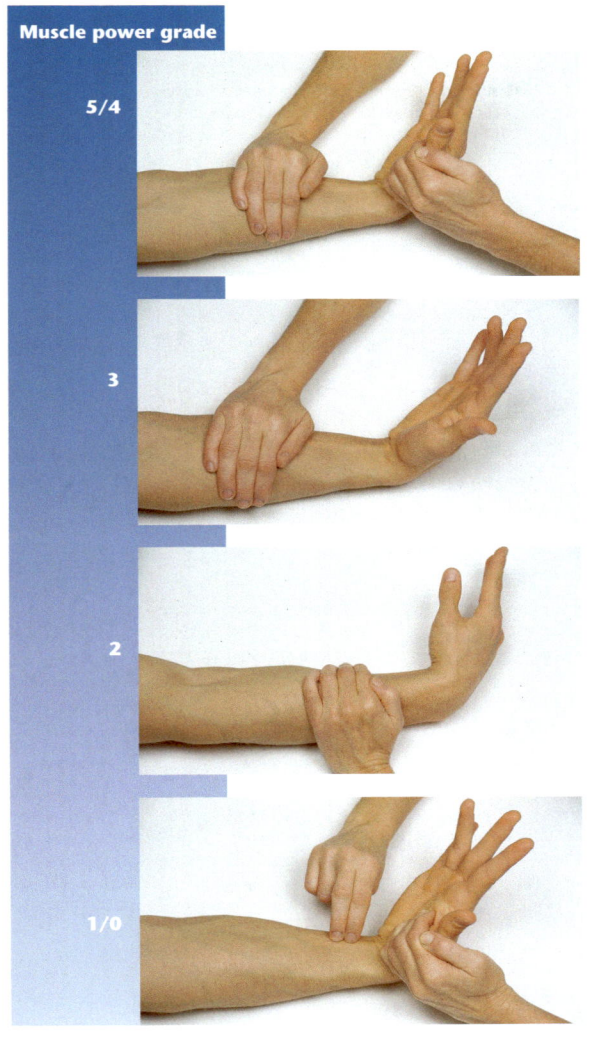

Muscle power grade

5/4

3

2

1/0

Starting position: The patient sits. The forearm and hand rest on a table or other flat surface in supination.

Test procedure: The examiner holds the patient's forearm in place with one hand and, with the other, applies downward pressure to the radial side of the palm.

Instruction: "Raise your hand off the table against my resistance and hold your final position.

Starting position: The patient sits. The forearm and hand rest on a table or other flat surface in supination.

Test procedure: The examiner holds the patient's forearm in place with one hand.

Instruction: "Raise your hand off the table. When you do so, the thumb side of your hand should move closer to your forearm."

Starting position: The patient sits. The forearm and hand rest on a table or other flat surface with the hand upright on its outer edge (the little finger side).

Test procedure: The examiner looks at the movement of the hand.

Instruction: "Move your hand along the table. When you do so, your palm should move closer to your forearm."

Starting position: The patient sits. The forearm and hand rest on a table or other flat surface in supination.

Test procedure: The examiner palpates the tendon of the palmaris longus muscle.

Instruction: "Try to lift your hand off the table."

 Problems/comments

- The palmaris longus muscle is an accessory muscle in volar flexion, which is effected principally by the flexor carpi ulnaris and flexor carpi radialis muscles.
- The hand flexors that are palpable above the wrist, going from the radial to the ulnar side, are as follows:
  – flexor pollicis longus muscle
  – flexor carpi radialis muscle
  – palmaris longus muscle
  – flexor digitorum superficialis muscle
  – flexor carpi ulnaris muscle.

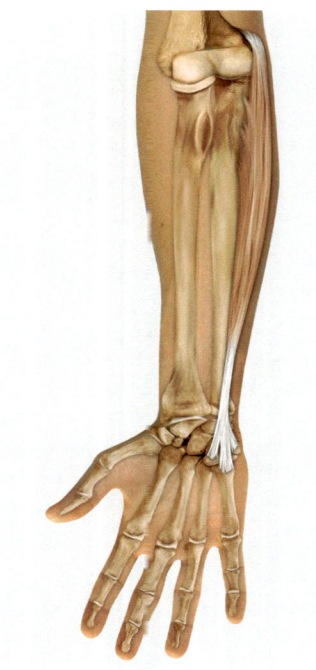

Flexor carpi ulnaris muscle

The flexor carpi ulnaris muscle (ulnar flexor muscle of the wrist) either flexes the joints of the wrist or causes radial abduction, depending on its synergists. Above all, like the flexor carpi radialis, it can conduct the force of the long finger extensors to the finger joints by fixing the joints of the wrist. It is therefore very difficult to define its synergists and antagonists at this point.

Origin	Humeral head: medial epicondyle of the humerus Ulnar head: olecranon, proximal two thirds of the ulna, antebrachial fascia
Insertion	Hamate (4th carpal) bone, pisiform bone, 5th metacarpal bone
Innervation	Ulnar nerve, C7–T1

Functions

Synergists

Antagonists

Radiocarpal joint and mediocarpal joint

Flexion

Flexor digitorum superficialis muscle	Extensor digitorum muscle
Flexor digitorum profundus muscle	Extensor carpi radialis longus muscle
Flexor carpi radialis muscle	Extensor carpi radialis brevis muscle
Flexor pollicis longus muscle	Extensor carpi ulnaris muscle
Abductor pollicis longus muscle	Extensor indicis muscle
Palmaris longus muscle	Extensor digiti minimi muscle
	Extensor pollicis longus muscle

Ulnar abduction

Extensor carpi ulnaris muscle	Flexor carpi radialis muscle
Flexor digitorum profundus muscle	Extensor carpi radialis longus muscle
Extensor digitorum muscle	Extensor carpi radialis brevis muscle
Extensor digiti minimi muscle	Flexor pollicis longus muscle
	Abductor pollicis longus muscle
	Extensor pollicis brevis muscle
	Extensor pollicis longus muscle

Humeroradial joint and humeroulnar joint

Flexion

Brachialis muscle	Triceps brachii muscle
Biceps brachii muscle	Anconeus muscle
Brachioradialis muscle	
Pronator teres muscle	
Extensor carpi radialis longus muscle	
Flexor carpi radialis muscle	

← Tendon

Muscle function testing

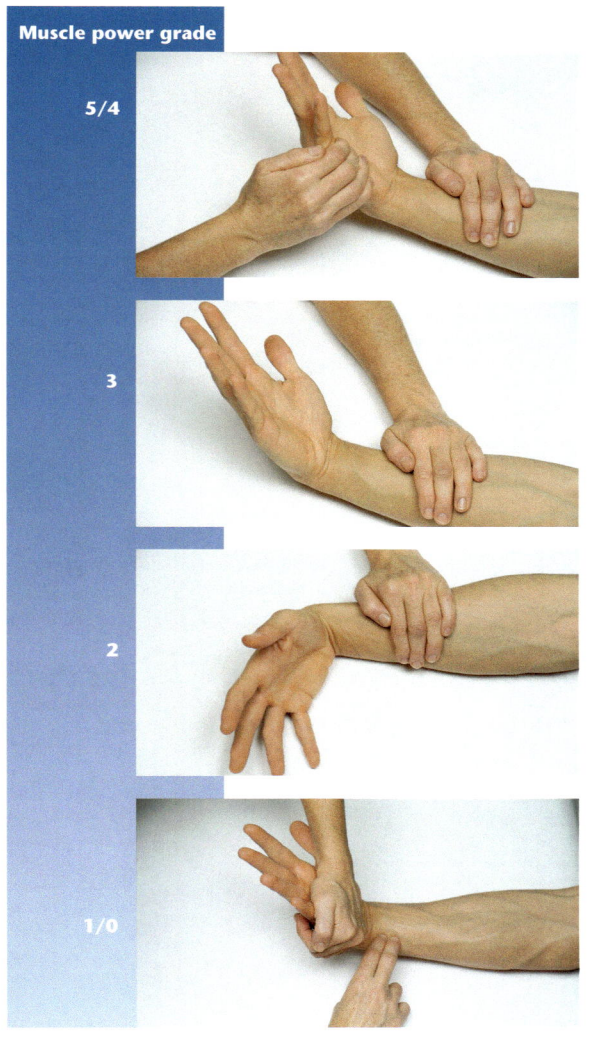

Muscle power grade

5/4

3

2

1/0

Starting position: The patient sits. The forearm and hand rest on a table or other flat surface in supination.

Test procedure: The examiner holds the patient's forearm in place with one hand and with the other applies downward pressure to the ulnar side of the palm.

Instruction: "Raise your hand off the table against my resistance and hold your final position.

Starting position: The patient sits. The forearm and hand rest on a table or other flat surface in supination.

Test procedure: The examiner holds the patient's forearm in place with one hand.

Instruction: "Raise your hand off the table. When you do so, the little finger side of your hand should move closer to your forearm."

Starting position: The patient sits. The forearm and hand rest on a table or other flat surface with the hand upright on its outer edge (the little finger side).

Test procedure: The examiner looks at the movement of the hand.

Instruction: "Move your hand along the table. When you do so, your palm should move closer to your forearm."

Starting position: The patient sits. The forearm and hand rest on a table or other flat surface in supination.

Test procedure: The examiner palpates the tendon of the flexor carpi ulnaris muscle.

Instruction: "Try to lift your hand off the table."

 Problems/comments

- The hand flexors that are palpable above the wrist, going from the radial to the ulnar side, are as follows:
  – flexor pollicis longus muscle
  – flexor carpi radialis muscle
  – palmaris longus muscle
  – flexor digitorum superficialis muscle
  – flexor carpi ulnaris muscle.

Stretch tests

Extensor carpi radialis and longus muscles, extensor carpi ulnaris muscle

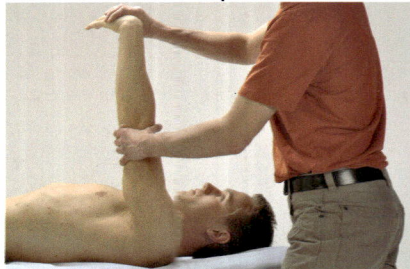

Method

The therapist moves the patient's hand to achieve maximum volar flexion of the wrist. The elbow should be in extension. For a stretch test of the extensor carpi radialis muscle, the wrist should also be in ulnar abduction; for a stretch test of the extensor carpi ulnaris, it should be in radial abduction.

Finding

Shortening of the relevant muscles is present if the movement cannot be performed to its full extent and if the endpoint feels soft and elastic.
The patient reports a stretching sensation along the course of the muscle.

Flexor carpi ulnaris muscle, flexor carpi radialis muscle and palmaris longus muscle

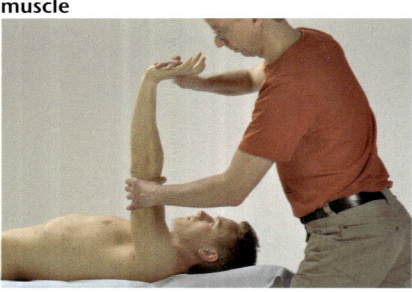

Method

The therapist moves the patient's hand to achieve maximum dorsal extension of the wrist. The elbow should be in extension. For a stretch test of the flexor carpi radialis muscle, the wrist should also be in ulnar abduction; for the stretch test of the flexor carpi ulnaris, it should be in radial abduction.

Finding

Shortening of the relevant muscles is present if the movement cannot be performed to its full extent and if the endpoint feels soft and elastic.
The patient reports a stretching sensation along the course of the muscle.

2 Upper extremity

Muscles of the finger joints

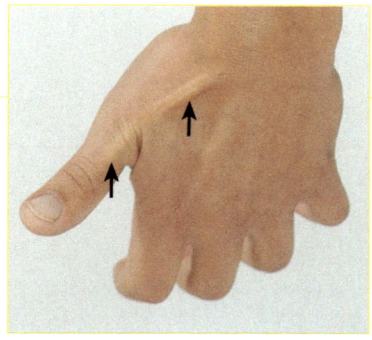

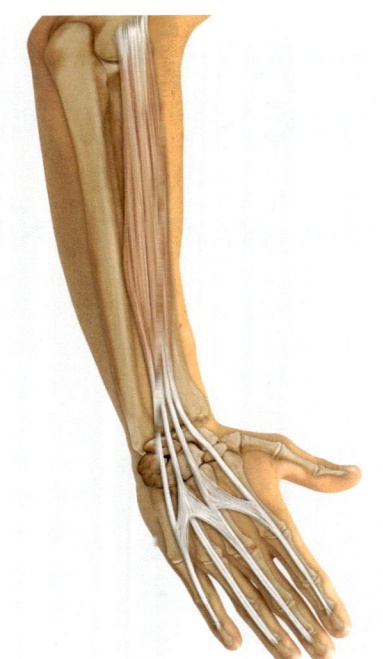

Extensor digitorum muscle

The extensor digitorum muscle (extensor muscle of the fingers) brings about extension of metacarpophalangeal joints II to V, of the interphalangeal joints of the same digits and also of the joints of the wrist. If it is to achieve its full effect on the interphalangeal joints, the wrists must first be stabilized by the carpal flexor muscles, and the metacarpophalangeal joints must be stabilized by the lumbricales and interossei muscles of the hand. The extensor digitorum aids ulnar abduction to a negligible extent.

Origin	Lateral epicondyle of the humerus, collateral ulnar ligament
	Antebrachial fascia
Insertion	The four tendons divide on the back of the hand and are inserted as follows at the dorsal surfaces of fingers II–V:
	– a middle piece of each tendon is inserted at the base of the middle phalanx
	– two lateral parts reconnect over the middle phalanx and then insert at the base of the distal phalanx
Innervation	Radial nerve, deep branch, C6–C8

Functions

 Synergists

Antagonists

Radiocarpal joint and mediocarpal joint

Extension

Extensor carpi radialis longus muscle	Flexor digitorum superficialis muscle
Extensor carpi radialis brevis muscle	Flexor digitorum profundus muscle
Extensor carpi ulnaris muscle	Flexor carpi ulnaris muscle
Extensor indicis muscle	Flexor carpi radialis muscle
Extensor digiti minimi muscle	Flexor pollicis longus muscle
Extensor pollicis longus muscle	Abductor pollicis longus muscle
	Palmaris longus muscle

Metacarpophalangeal joints II to V

Extension

Extensor indicis muscle (II)	Flexor digitorum superficialis muscle
Extensor digiti minimi muscle (V)	Flexor digitorum profundus muscle
	Palmar interossei muscles 1–3 (II, IV, V)
	Dorsal interossei muscles of the hand 1–4 (II, III, IV)
	Lumbricales muscles of the hand (II–V)
	Abductor digiti minimi muscle (V)
	Flexor digiti minimi brevis muscle (V)

Abduction (away from the middle finger)

Dorsal interossei muscles of the hand 1–4 (II, III, IV)	Palmar interossei muscles 1–3 (II, IV, V)
Abductor digiti minimi muscle (V)	Flexor digiti minimi brevis muscle (V)
	Flexor digitorum superficialis muscle
	Flexor digitorum profundus muscle

Proximal and distal interphalangeal joints II to V

Extension

Extensor digiti minimi muscle (V)	Flexor digitorum profundus muscle
Extensor indicis muscle (II)	Flexor digitorum superficialis muscle (only at
Lumbricales muscles of the hand (II-V)	the proximal interphalangeal joints)
Dorsal interossei muscles of the hand 1–4 (II, III, IV)	
Palmar interossei muscles 1–3 (II, IV, V)	

Muscle function testing

Muscle power grade	
5/4	

Starting position: The patient sits. The forearm and metacarpus rest on a flat surface in pronation, while the flexed fingers hang over the edge.

Test procedure: The examiner holds the patient's hand at its mid-point with one hand and, with the other, applies downward pressure to fingers II–V.

Instruction: "Extend your fingers against my resistance and hold your final position."

3

Starting position: The patient sits. The forearm and metacarpus rest on a flat surface in pronation, while the flexed fingers hang over the edge.

Test procedure: The examiner holds the patient's hand at its mid-point with one hand.

Instruction: "Extend your fingers."

2

Starting position: The patient sits. The forearm and hand rest on a table or other flat surface with the hand upright on its outer edge (the little finger side).

Test procedure: The examiner looks at the movement of the fingers.

Instruction: "Extend your fingers."

1/0

Starting position: The patient sits. The forearm and metacarpus rest on a flat surface in pronation, while the flexed fingers hang over the edge.

Test procedure: The examiner palpates the extensor digitorum muscle.

Instruction: "Try to extend your fingers."

Clinical relevance

- Finger extension is associated with abduction.
- During the test, it is essential for the wrist to remain at the mid-point between dorsal extension and palmar flexion, to avoid active insufficiency of the extensor digitorum muscle during dorsal flexion of the wrist.

Problems/comments

- The extensor digitorum muscle is tested together with the extensor digiti minimi and the extensor indicis muscles; optimum muscle power can only be deployed if these muscles act in conjunction.
- The hand extensors that are palpable above the wrist, going from the radial to the ulnar side, are as follows:
  – extensor carpi radialis longus muscle
  – extensor carpi radialis brevis muscle
  – extensor digitorum muscle
  – extensor digiti minimi muscle
  – extensor carpi ulnaris muscle.

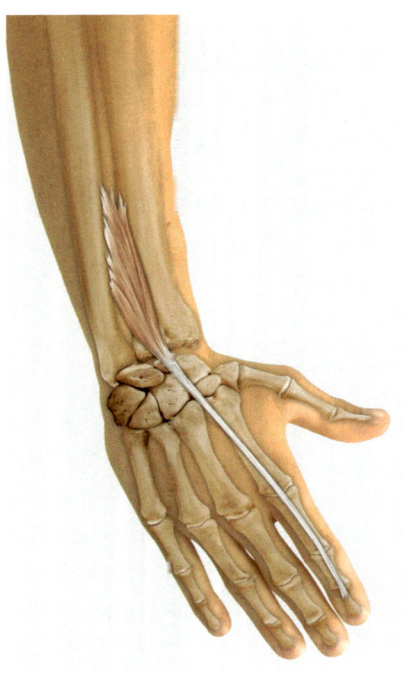

Extensor indicis muscle

The extensor indicis muscle (extensor muscle of the index finger) extends the index finger, its effect being additive to that of the extensor digitorum muscle. It can also extend the index finger in isolation. In addition, it extends the joints of the wrist.

Origin	Distal half of the dorsal surface of the ulna
	Interosseous membrane
Insertion	Ulnar side of the dorsal aponeurosis of the index finger
Innervation	Radial nerve, deep branch, C6–C8

Functions

 Synergists   Antagonists

Radiocarpal joint and mediocarpal joint

Extension

Extensor digitorum muscle	Flexor digitorum superficialis muscle
Extensor carpi radialis longus muscle	Flexor digitorum profundus muscle
Extensor carpi radialis brevis muscle	Flexor carpi ulnaris muscle
Extensor carpi ulnaris muscle	Flexor carpi radialis muscle
Extensor digiti minimi muscle	Flexor pollicis longus muscle
Extensor pollicis longus muscle	Abductor pollicis longus muscle
	Palmaris longus muscle

Metacarpophalangeal joint II

Extension

Extensor digitorum muscle	Flexor digitorum superficialis muscle
	Flexor digitorum profundus muscle
	Palmar interosseus muscle [1]
	Dorsal interosseus muscle of the hand [1]
	Lumbricalis muscle of the hand [1]

Abduction (away from the middle finger)

Dorsal interosseus muscle of the hand 1	Palmar interosseus muscle [1]
	Flexor digitorum superficialis muscle
	Flexor digitorum profundus muscle

Proximal and distal interphalangeal joint II

Extension

Extensor digitorum muscle	Flexor digitorum profundus muscle
Lumbricalis muscle of the hand [1]	Flexor digitorum superficialis muscle (proximal
Dorsal interosseus muscle of the hand [1]	interphalangeal joint II only)
Palmar interosseus muscle [1]	

Tendon

Muscle function testing

Muscle power grade	
5/4	
3	
2	
1/0	

Starting position: The patient sits. The forearm and metacarpus rest on a flat surface in pronation, while the flexed fingers hang over the edge.

Test procedure: The examiner holds the patient's hand at its mid-point with one hand and with the other applies downward pressure to fingers II–V.

Instruction: "Extend your fingers against my resistance and hold your final position."

Starting position: The patient sits. The forearm and metacarpus rest on a flat surface in pronation, while the flexed fingers hang over the edge.

Test procedure: The examiner holds the patient's hand at its mid-point with one hand.

Instruction: "Extend your fingers."

Starting position: The patient sits. The forearm and hand rest on a table or other flat surface with the hand upright on its outer edge (the little finger side).

Test procedure: The examiner looks at the movement of the index finger.

Instruction: "Extend your fingers."

Starting position: The patient sits. The forearm and metacarpus rest on a flat surface in pronation.

Test procedure: The examiner looks at the movement of the index finger.

Instruction: "Try to extend your index finger."

 Clinical relevance

- Finger extension is associated with abduction.
- During the test, it is essential for the wrist to remain at the mid-point between dorsal extension and palmar flexion, to avoid active insufficiency of the extensor digitorum muscle during dorsal flexion of the wrist.

 Problems/comments

- The extensor indicis muscle is tested together with the extensor digitorum and the extensor digiti minimi muscles; optimum muscle power can only be deployed if these muscles act in conjunction.
- The extensor indicis muscle cannot be palpated.

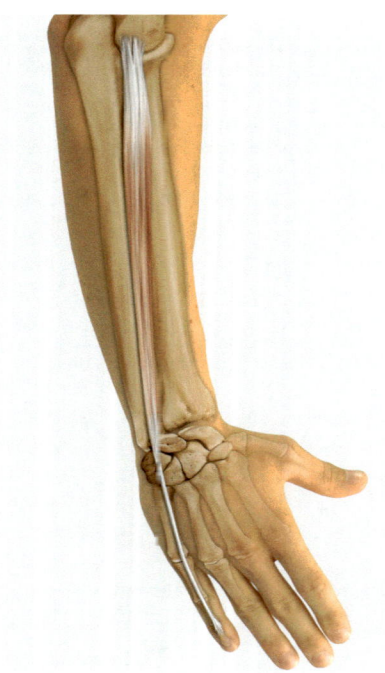

Extensor digiti minimi muscle

The extensor digiti minimi muscle (extensor muscle of the little finger) extends the little finger at the metacarpophalangeal and interphalangeal joints. Its extensor effect on the wrist joints is minimal.

Origin	Lateral epicondyle of the humerus Antebrachial fascia
Insertion	Dorsal aponeurosis of the little finger
Innervation	Radial nerve, deep branch, C6–C8

Functions

 Synergists   Antagonists

Radiocarpal and mediocarpal joints

Extension

Extensor digitorum muscle	Flexor digitorum superficialis muscle
Extensor carpi radialis longus muscle	Flexor digitorum profundus muscle
Extensor carpi radialis brevis muscle	Flexor carpi ulnaris muscle
Extensor carpi ulnaris muscle	Flexor carpi radialis muscle
Extensor indicis muscle	Flexor pollicis longus muscle
Extensor pollicis longus muscle	Abductor pollicis longus muscle
	Palmaris longus muscle

Metacarpophalangeal joint V

Extension

Extensor digitorum muscle	Flexor digitorum superficialis muscle
	Flexor digitorum profundus muscle
	Palmar interosseus muscle 3
	Lumbricalis muscle of the hand 4
	Abductor digiti minimi muscle
	Flexor digiti minimi brevis muscle

Abduction (away from the middle finger)

Abductor digiti minimi muscle	Palmar interosseus muscle 3
	Flexor digiti minimi brevis muscle
	Flexor digitorum superficialis muscle
	Flexor digitorum profundus muscle

Proximal and distal interphalangeal joint V

Extension

Extensor digitorum muscle	Flexor digitorum superficialis muscle (proximal interphalangeal joint V only)
Lumbricalis muscle of the hand 4	
Palmar interosseus muscle 3	
Flexor digitorum profundus muscle	

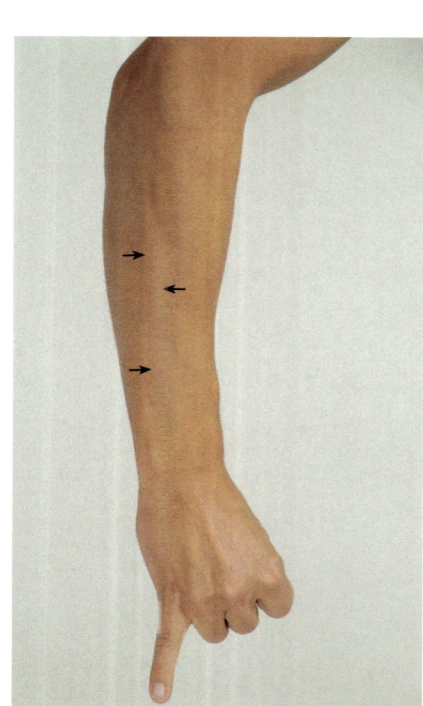

Muscle function testing

Muscle power grade	
5/4	

Starting position: The patient sits. The forearm and metacarpus rest on a flat surface in pronation, while the flexed fingers hang over the edge.

Test procedure: The examiner holds the patient's hand at its midpoint with one hand and with the other applies downward pressure to fingers II-V.

Instruction: "Extend your fingers against my resistance and hold your final position."

3	

Starting position: The patient sits. The forearm and metacarpus rest on a flat surface in pronation, while the flexed fingers hang over the edge.

Test procedure: The examiner holds the patient's hand at its midpoint with one hand.

Instruction: "Extend your fingers."

2	

Starting position: The patient sits. The forearm and hand rest on a table or other flat surface with the hand upright on its outer edge (the little finger side).

Test procedure: The examiner looks at the movement of the little finger.

Instruction: "Extend your fingers."

1/0	

Starting position: The patient sits. The forearm and metacarpus rest on a flat surface in pronation, while the flexed fingers hang over the edge.

Test procedure: The examiner palpates the extensor digiti minimi muscle.

Instruction: "Try to extend your little finger."

 Clinical relevance

- Finger extension is associated with abduction.
- During the test, it is essential for the wrist to remain at the mid-point between dorsal extension and palmar flexion, to avoid active insufficiency of the extensor digitorum muscle during dorsal flexion of the wrist.

 Problems/comments

- The extensor digiti minimi muscle is tested together with the extensor digitorum and the extensor indicis muscles; optimum muscle power can only be deployed if these muscles act in conjunction.
- The hand extensors that are palpable above the wrist, going from the radial to the ulnar side, are as follows:
  – extensor carpi radialis longus muscle
  – extensor carpi radialis brevis muscle
  – extensor digitorum muscle
  – extensor digiti minimi muscle
  – extensor carpi ulnaris muscle.

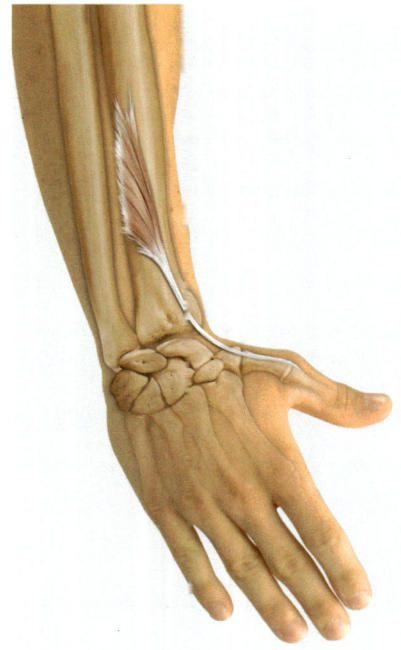

Extensor pollicis brevis muscle

The action of the extensor pollicis brevis muscle corresponds largely to that of the extensor pollicis longus muscle, although the former does not act as a supinator. It extends the joints of the wrist and the thumb, but not interphalangeal joint I.

Origin	Distal third of the dorsal surface of the radius Interosseous membrane
Insertion	Dorsal surface of the base of the proximal phalanx of the thumb
Innervation	Radial nerve, deep branch, C6–C8

Functions

 Synergists   Antagonists

Radiocarpal joint and mediocarpal joint

Radial abduction

Synergists	Antagonists
Flexor carpi radialis muscle	Flexor carpi ulnaris muscle
Extensor carpi radialis longus muscle	Extensor carpi ulnaris muscle
Extensor carpi radialis brevis muscle	Flexor digitorum profundus muscle
Flexor pollicis longus muscle	Extensor digitorum muscle
Abductor pollicis longus muscle	Extensor digiti minimi muscle
Extensor pollicis longus muscle	

Carpometacarpal joint I

Extension

Synergists	Antagonists
Extensor pollicis longus muscle	Flexor pollicis longus muscle
Abductor pollicis longus muscle	Flexor pollicis brevis muscle
	Adductor pollicis muscle
	Opponens pollicis muscle

Abduction

Synergists	Antagonists
Abductor pollicis brevis muscle	Adductor pollicis muscle
Abductor pollicis longus muscle	Dorsal interosseus muscle of the hand 1
Flexor pollicis brevis muscle (superficial head)	Opponens pollicis muscle
	Extensor pollicis longus muscle
	Flexor pollicis brevis muscle (deep head)

Metacarpophalangeal joint I

Extension

Synergists	Antagonists
Extensor pollicis longus muscle	Flexor pollicis longus muscle
Abductor pollicis brevis muscle	Flexor pollicis brevis muscle
	Adductor pollicis muscle

Muscle function testing

Muscle power grade

5/4

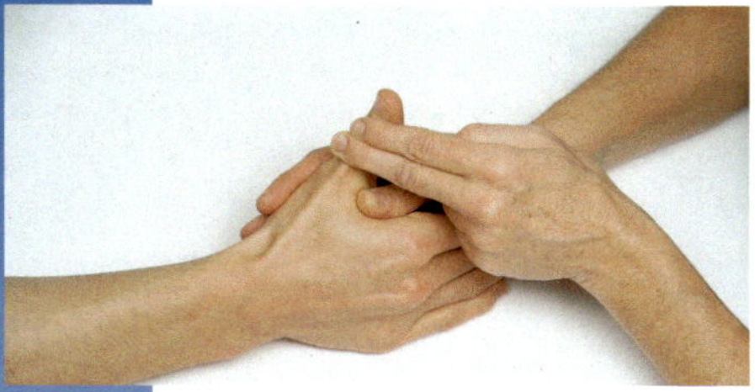

Starting position: The patient sits. The forearm and hand rest on a flat surface with the hand upright on its outer edge (the little finger side).

Test procedure: The examiner holds the patient's first metacarpal bone in place with one hand and, with one finger of other hand, applies pressure to the dorsal side of the proximal phalanx of the thumb in the direction of flexion.

Instruction: "Extend your thumb against my resistance and hold it in its final position."

3/2

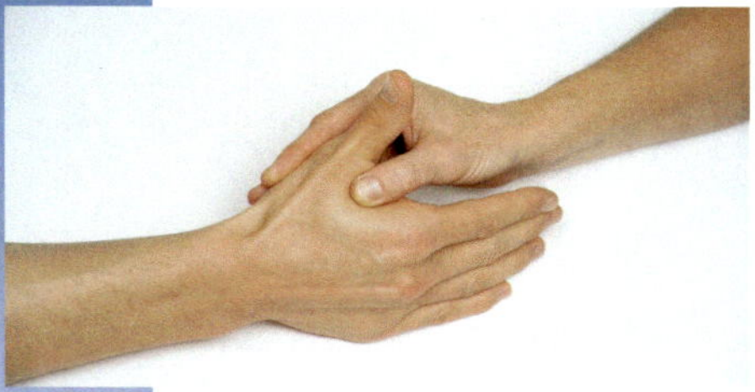

Starting position: The patient sits. The forearm and hand rest on a flat surface with the hand upright on its outer edge (the little finger side).

Test procedure: The examiner holds the patient's first metacarpal bone in place with one hand.

Instruction: "Extend your thumb."

1/0

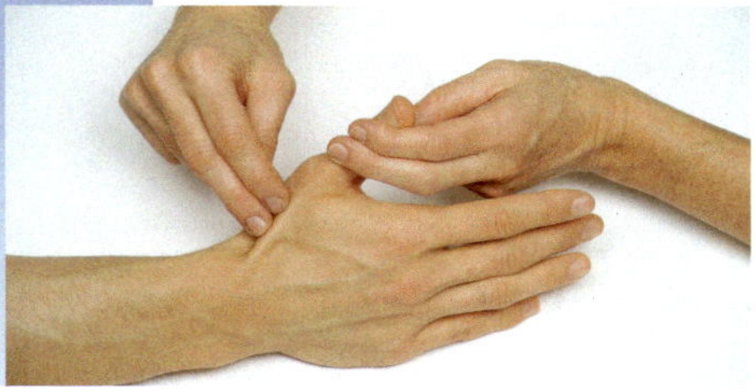

Starting position: The patient sits. The forearm and hand rest on a flat surface with the hand upright on its outer edge (the little finger side).

Test procedure: The examiner palpates the extensor pollicis brevis muscle.

Instruction: "Try to extend your thumb."

 Problems/comments

- The extensor pollicis brevis and the extensor pollicis longus are both involved in extension of the metacarpophalangeal joint of the thumb.

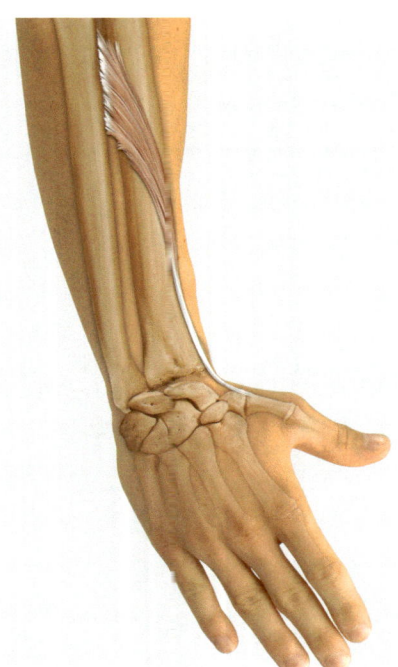

Extensor pollicis longus muscle

The extensor pollicis longus muscle extends all the thumb joints and can also abduct the thumb. Furthermore, it can aid extension and radial abduction at the joints of the wrist and has an added effect as a supinator, as a result of its course from the ulna across the forearm and to the thumb.

Origin	Middle dorsal surface of the ulna, interosseous membrane
Insertion	Dorsal surface of the base of the distal phalanx of the thumb
Innervation	Radial nerve, deep branch, C6–C8

Functions

 Synergists   Antagonists

Proximal and distal radioulnar joints, humeroradial joint

Supination

Supinator muscle	Pronator quadratus muscle
Biceps brachii muscle (with the elbow flexed)	Pronator teres muscle
Brachioradialis muscle (from pronation to the neutral position)	Brachioradialis muscle (from supination to the neutral position)
	Flexor carpi radialis muscle
	Extensor carpi radialis longus muscle

Radiocarpal joint and mediocarpal joint

Extension

Extensor digitorum muscle	Flexor carpi ulnaris muscle
Extensor carpi radialis longus muscle	Extensor carpi ulnaris muscle
Extensor carpi radialis brevis muscle	Flexor digitorum profundus muscle
Extensor carpi ulnaris muscle	Extensor digitorum muscle
Extensor indicis muscle	Extensor digiti minimi muscle
Extensor digiti minimi muscle	

Radial abduction

Flexor carpi radialis muscle	Flexor carpi ulnaris muscle
Extensor carpi radialis longus muscle	Extensor carpi ulnaris muscle
Extensor carpi radialis brevis muscle	Flexor digitorum profundus muscle
Flexor pollicis longus muscle	Extensor digitorum muscle
Abductor pollicis longus muscle	Extensor digiti minimi muscle
Extensor pollicis brevis muscle	

Carpometacarpal joint I

Extension

Extensor pollicis brevis muscle	Flexor pollicis longus muscle
Abductor pollicis longus muscle	Flexor pollicis brevis muscle
	Abductor pollicis brevis muscle
	Adductor pollicis muscle
	Opponens pollicis muscle

Adduction

Adductor pollicis muscle	Abductor pollicis longus muscle
Dorsal interosseus muscle of the hand 1	Abductor pollicis brevis muscle
Opponens pollicis muscle	Extensor pollicis brevis muscle
Flexor pollicis brevis muscle (deep head)	Flexor pollicis brevis muscle (superficial head)

Metacarpophalangeal joint I

Extension

Extensor pollicis brevis muscle	Flexor pollicis longus muscle
Abductor pollicis brevis muscle	Flexor pollicis brevis muscle
	Adductor pollicis muscle

Interphalangeal joint I

Extension

None	Flexor pollicis longus muscle

Muscle function testing

Muscle power grade	
5/4	
3/2	
1/0	

Starting position: The patient sits. The forearm and hand rest on a flat surface with the hand upright on its outer edge (the little finger side).

Test procedure: The examiner holds the proximal phalanx of the patient's thumb in place with one hand and with one finger of the other hand applies pressure to the dorsal side of the distal phalanx of the thumb in the direction of flexion.

Instruction: "Extend your thumb against my resistance and hold it in its final position."

Starting position: The patient sits. The forearm and hand rest on a flat surface with the hand upright on its outer edge (the little finger side).

Test procedure: The examiner holds the proximal phalanx of the patient's thumb in place with one hand.

Instruction: "Extend your thumb."

Starting position: The patient sits. The forearm and hand rest on a table or other flat surface with the hand upright on its outer edge (the little finger side).

Test procedure: The examiner palpates the extensor pollicis longus muscle.

Instruction: "Try to extend your thumb."

 Problems/comments

- The extensor pollicis brevis and the extensor pollicis longus are both involved in extension of the metacarpophalangeal joint of the thumb.
- The extensor pollicis longus alone is responsible for extension of the interphalangeal joint of the thumb.

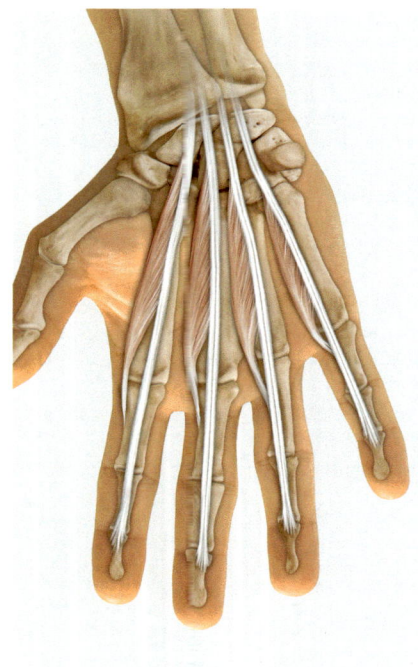

Lumbricales muscles of the hand

The lumbricales (lumbrical) muscles flex metacarpophalangeal joints II to V and extend the same fingers at the proximal interphalangeal joints. This movement is important in activities such as writing and holding cutlery.

Origin	Tendons of the deep finger flexors
Insertion	Radial side of each finger in the region of the extensor aponeuroses
Innervation	Lumbricales muscles 1 and 2: median nerve, C8–T1
	Lumbricales muscles 3 and 4: ulnar nerve, deep branch, C8–T1

Functions

 Synergists

 Antagonists

Metacarpophalangeal joints II to V

Flexion

Synergists	Antagonists
Flexor digitorum superficialis muscle	Extensor digitorum muscle
Flexor digitorum profundus muscle	Extensor digiti minimi muscle (V)
Dorsal interossei muscles of the hand 1–4 (II–IV)	Extensor indicis muscle (II)
Palmar interossei muscles 1–3	
Flexor digiti minimi brevis muscle (V)	
Abductor digiti minimi muscle (V)	

Proximal interphalangeal joints II to V

Extension

Synergists	Antagonists
Extensor digitorum muscle	Flexor digitorum superficialis muscle
Extensor digiti minimi muscle (V)	Flexor digitorum profundus muscle
Extensor indicis muscle (II)	Flexor digiti minimi brevis (V)
Dorsal interossei muscles of the hand 1–4 (II–IV)	
Palmar interossei muscles 1–3	

Muscle function testing

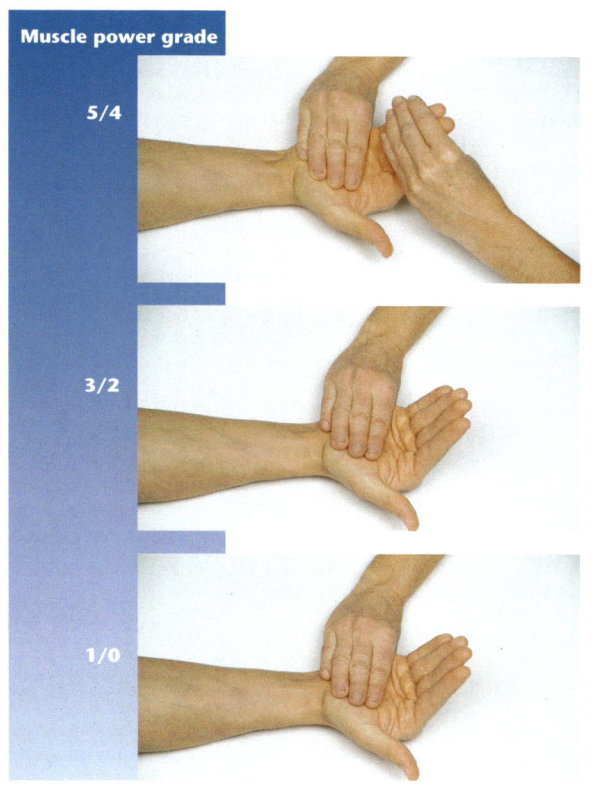

Muscle power grade	
5/4	
3/2	
1/0	

Starting position: The patient sits. The forearm and hand rest on the examination couch in supination. The fingers are extended.

Test procedure: The examiner holds the patient's metacarpus in place with one hand and with the other applies pressure to the proximal phalanges of fingers II to V in the direction of extension.

Instruction: "Flex your already extended fingers at the knuckles (metacarpophalangeal joints) against my resistance."

Starting position: The patient sits. The forearm and hand rest on the examination couch in supination. The fingers are extended.

Test procedure: The examiner holds the patient's metacarpus in place with one hand.

Instruction: "Flex your already extended fingers at the knuckles (metacarpophalangeal joints)."

Starting position: The patient sits. The forearm and hand rest on the examination couch in supination. The fingers are extended.

Test procedure: The examiner looks at the movements of the fingers.

Instruction: "Try to flex your already extended fingers at the knuckles (metacarpophalangeal joints)."

 Problems/comments

- The lumbricales muscles 1–3 are aided in their function by the dorsal and palmar interossei muscles of the hand.

Flexor digitorum superficialis muscle

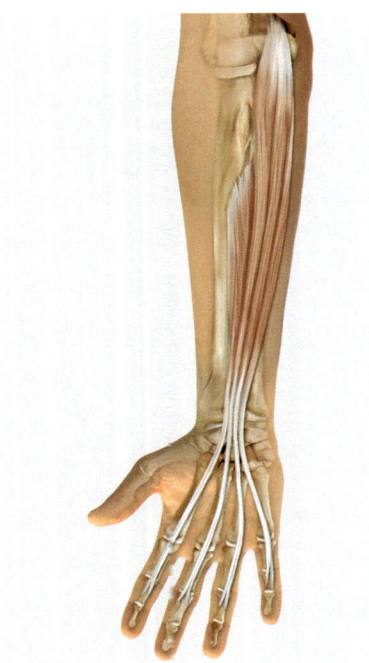

The main function of the flexor digitorum superficialis muscle is to flex metacarpophalangeal joints II to V and the proximal interphalangeal joints of the same fingers. In doing so, it is particularly efficient when the extensor carpi muscles fix the wrist, which this muscle would otherwise also flex. It also exerts a negligible flexor effect on the elbow.

Origin	Humeroulnar head: medial epicondyle of the humerus, coronoid process of the ulna
	Radial head: ventral surface of the radius
Insertion	Each of its four tendons divides into two parts, which insert at the sides of the bases of the middle phalanges of fingers II–V
	The tendons of the flexor digitorum profundus muscle run to their respective distal phalanx between each of these two parts
Innervation	Median nerve, C7–T1

Functions

Synergists   Antagonists

Radiocarpal and mediocarpal joints

Flexion
Flexor digitorum profundus muscle
Flexor carpi ulnaris muscle
Flexor carpi radialis muscle
Flexor pollicis longus muscle
Abductor pollicis longus muscle
Palmaris longus muscle

Extensor digitorum muscle
Extensor carpi radialis longus muscle
Extensor carpi radialis brevis muscle
Extensor carpi ulnaris muscle
Extensor indicis muscle
Extensor digiti minimi muscle
Extensor pollicis longus muscle

Metacarpophalangeal joints II to V

Flexion
Flexor digitorum profundus muscle
Palmar interossei muscles 1–3 (II, IV, V)
Dorsal interossei muscles of the hand 1–4 (II-IV)
Lumbricales muscles of the hand 1–4
Abductor digiti minimi muscle (V)
Flexor digiti minimi brevis muscle (V)

Extensor digitorum muscle
Extensor indicis muscle (II)
Extensor digiti minimi muscle (V)

Adduction (towards the middle finger)
Palmar interossei muscles 1–3 (II, IV, V)
Flexor digitorum profundus muscle
Flexor digiti minimi brevis muscle (V)

Dorsal interossei muscles
of the hand 1–4 (II–IV)
Extensor digitorum muscle
Abductor digiti minimi muscle (V)
Extensor digiti minimi muscle (V)

Proximal interphalangeal joints II to V

Extension
Flexor digitorum profundus muscle
(also flexes the distal interphalangeal joints)

Extensor digitorum muscle
Extensor digiti minimi muscle (V)
Extensor indicis muscle (II)
Lumbricales muscles of the hand 1–4 (II–V)
Dorsal interossei muscles
of the hand 1–4 (II–IV)
Palmar interossei muscles 1–3 (II, IV, V)

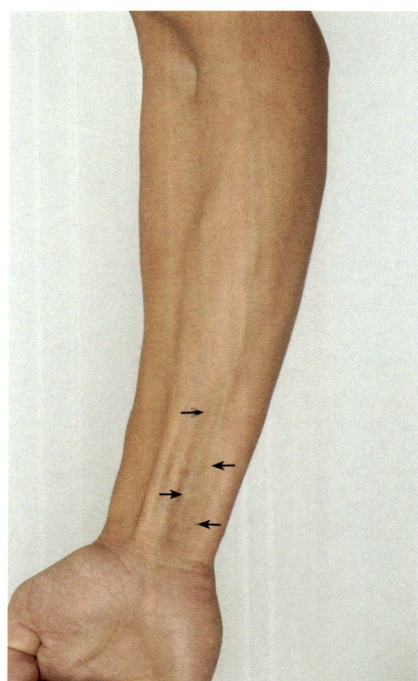

Muscle function testing

Muscle power grade	
5/4	

Starting position: The patient sits. The forearm and hand rest on a flat surface in supination.

Test procedure: The examiner holds the proximal phalanges of the patient's fingers II to V in place with one hand and with the other applies pressure to the volar side of the middle phalanges in the direction of extension.

Instruction: "Flex your fingers at the middle joints (proximal interphalangeal joints) against my resistance."

3/2

Starting position: The patient sits. The forearm and hand rest on a flat surface in supination.

Test procedure: The examiner holds the proximal phalanges of the patient's fingers II to V in place with one hand.

Instruction: "Flex your fingers at the middle joints (proximal interphalangeal joints)."

1/0

Starting position: The patient sits. The forearm and hand rest on a flat surface in supination.

Test procedure: The examiner palpates the flexor digitorum superficialis muscle.

Instruction: "Try to flex your fingers at the middle joints (proximal interphalangeal joints)."

⚠ Problems/comments

- The flexor digitorum superficialis and profundus work together to flex the proximal interphalangeal joints.
- Individual fingers may also be tested in isolation, if the others are held in place as appropriate; however, the flexor digitorum superficialis always moves fingers II to V simultaneously.
- The wrist should be at its mid-point, as the flexor digitorum superficialis becomes actively insufficient when the wrist is in volar flexion.

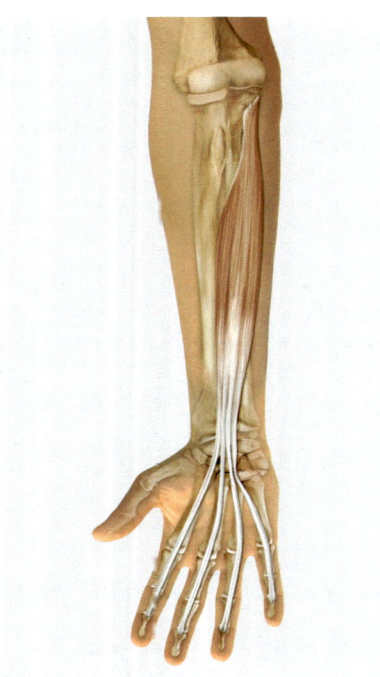

Flexor digitorum profundus muscle

The flexor digitorum profundus muscle acts primarily on metacarpophalangeal joints II to V and on the proximal and distal interphalangeal joints of the same digits. It is the only flexor of the distal interphalangeal joints. Its action is particularly efficient when the extensor carpi muscles fix the wrist. Otherwise, it would also cause flexion there.

Origin	Proximal anterior surface of the ulna Antebrachial fascia Interosseous membrane
Insertion	The four tendons run to fingers II–V, through the divided tendon of the flexor digitorum superficialis muscle and to the palmar surface of the base of each distal phalanx
Innervation	Anterior interosseous nerve from the median nerve (II, III), C5–T1 Ulnar nerve (IV, V), C8–T1

Functions

 Synergists

 Antagonists

Radiocarpal joint and mediocarpal joint

Flexion

Flexor digitorum superficialis muscle	Extensor digitorum muscle
Flexor carpi ulnaris muscle	Extensor carpi radialis longus muscle
Flexor carpi radialis muscle	Extensor carpi radialis brevis muscle
Flexor pollicis longus muscle	Extensor carpi ulnaris muscle
Abductor pollicis longus muscle	Extensor indicis muscle
Palmaris longus muscle	Extensor digiti minimi muscle
	Extensor pollicis longus muscle

Metacarpophalangeal joints II to V

Flexion

Flexor digitorum superficialis muscle	Extensor digitorum muscle
Palmar interossei muscles 1–3 (II, IV, V)	Extensor indicis muscle (II)
Dorsal interossei muscles of the hand 1-4	Extensor digiti minimi muscle (V)
Lumbricales muscles of the hand 1–3	
Abductor digiti minimi muscle (V)	
Flexor digiti minimi brevis muscle (V)	

Adduction (towards the middle finger)

Palmar interossei muscles 1–3 (II, IV, V)	Dorsal interossei muscles
Flexor digitorum superficialis muscle	of the hand 1–4 (II–IV)
Flexor digiti minimi brevis muscle (V)	Extensor digitorum muscle
	Abductor digiti minimi muscle (V)
	Extensor digiti minimi muscle (V)

Proximal and distal interphalangeal joints II to V

Extension

Flexor digitorum superficialis muscle (only at the proximal interphalangeal joints)	Extensor digitorum muscle
	Extensor digiti minimi muscle (V)
	Extensor indicis muscle (II)
	Lumbricales muscles of the hand 1–4
	Dorsal interossei muscles
	of the hand 1–4 (II–IV)
	Palmar interossei muscles 1–3 (II, IV, V)

Muscle function testing

Muscle power grade	
5/4	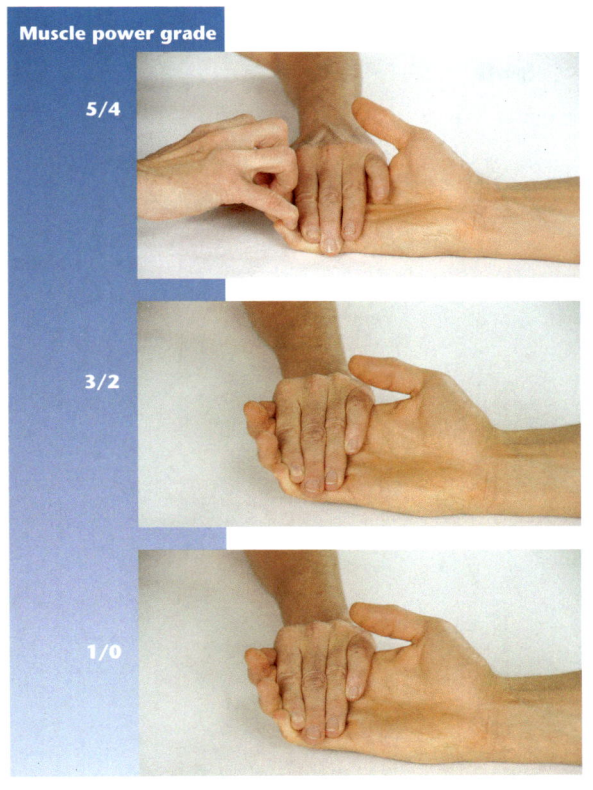
3/2	
1/0	

Starting position: The patient sits. The forearm and hand rest on a flat surface in supination.

Test procedure: The examiner holds the middle phalanges of the patient's fingers II to V in place with one hand and with the other applies pressure to the volar side of the distal phalanges in the direction of extension.

Instruction: "Flex your fingers at the end joints (distal interphalangeal joints) against my resistance."

Starting position: The patient sits. The forearm and hand rest on a flat surface in supination.

Test procedure: The examiner holds the middle phalanges of the patient's fingers II to V in place with one hand.

Instruction: "Flex your fingers at the end joints (distal interphalangeal joints)."

Starting position: The patient sits. The forearm and hand rest on a flat surface in supination.

Test procedure: The examiner holds the middle phalanges of the patient's fingers II to V in place with one hand.

Instruction: "Try to flex your fingers at the end joints (proximal interphalangeal joints)."

⚠ Problems/comments

- The flexor digitorum superficialis and profundus work together to flex the proximal interphalangeal joints.
- The flexor digitorum profundus muscle works alone to flex the distal interphalangeal joints
- Individual fingers may also be tested in isolation, if the others are held in place as appropriate; however, the flexor digitorum superficialis always moves fingers II to V simultaneously.
- The wrist should be at its mid-point, as the flexor digitorum profundus becomes actively insufficient when the wrist is in volar flexion.

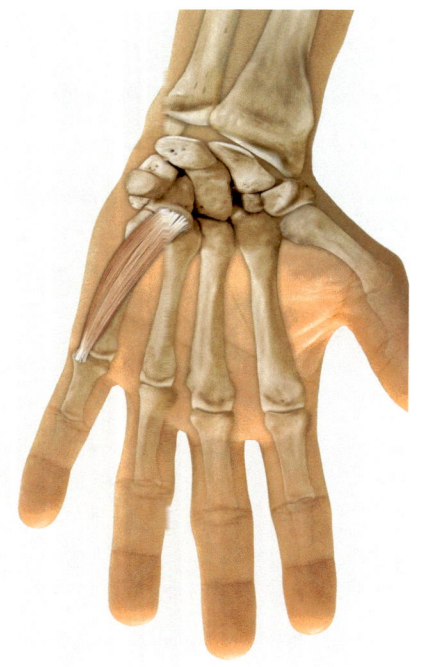

Flexor digiti minimi brevis muscle

The flexor digiti minimi brevis muscle (short flexor muscle of the little finger) flexes metacarpophalangeal joint V.

Origin   Hamate bone (fourth carpal bone)
                              Flexor retinaculum of the hand

Insertion   Ulnar edge of the base of proximal phalanx V

Innervation   Ulnar nerve, deep branch, C8–T1

Functions

🏃 Synergists   🏃 Antagonists

Metacarpophalangeal joint V

Flexion
Flexor digitorum superficialis muscle   Extensor digitorum muscle
Flexor digitorum profundus muscle   Extensor digiti minimi muscle (V)
Palmar interosseus muscle 3
Lumbricalis muscle of the hand 4

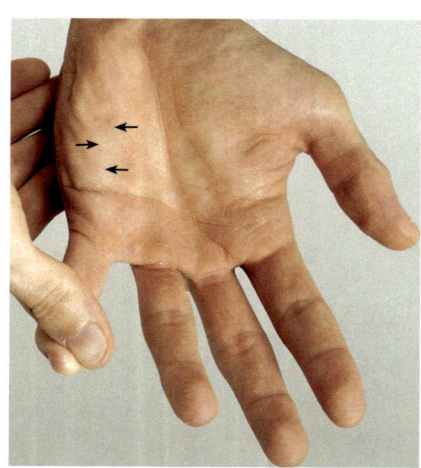

Muscle function testing

Muscle power grade	
5/4	

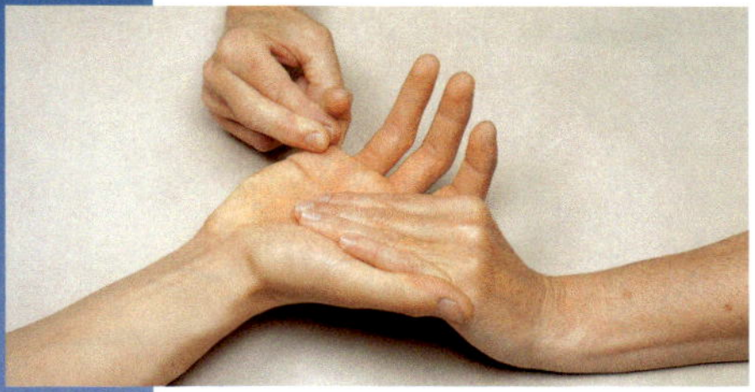

Starting position: The patient sits. The forearm and hand rest on a flat surface in supination.

Test procedure: The examiner holds the patient's metacarpus in place with one hand and with the other applies pressure to the volar side of the proximal phalanx of the little finger in the direction of extension.

Instruction: "Flex your little finger against my resistance."

3/2

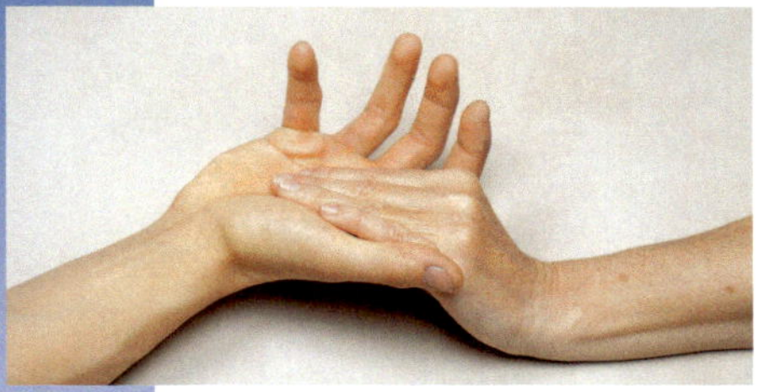

Starting position: The patient sits. The forearm and hand rest on a flat surface in supination.

Test procedure: The examiner looks at the movement of the little finger.

Instruction: "Flex your little finger."

1/0

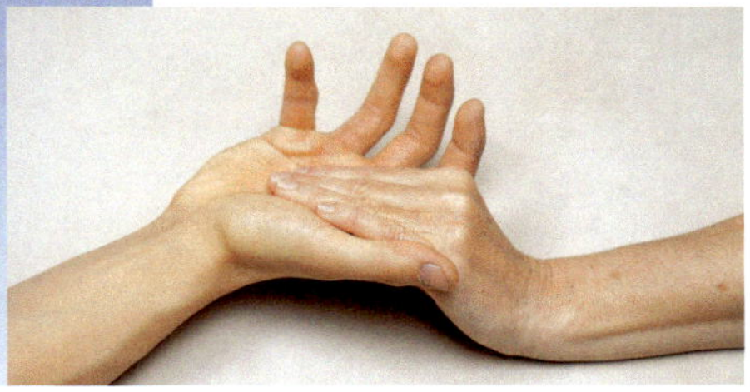

Starting position: The patient sits. The forearm and hand rest on a flat surface in supination.

Test procedure: The examiner looks at the movement of the little finger.

Instruction: "Try to flex your little finger."

 Problems/comments

- The function of the flexor digiti minimi brevis muscle cannot be distinguished from that of the flexor digitorum superficialis and flexor digitorum profundus muscles.
- It is often impossible to move the little finger in isolation in a coordinated way.

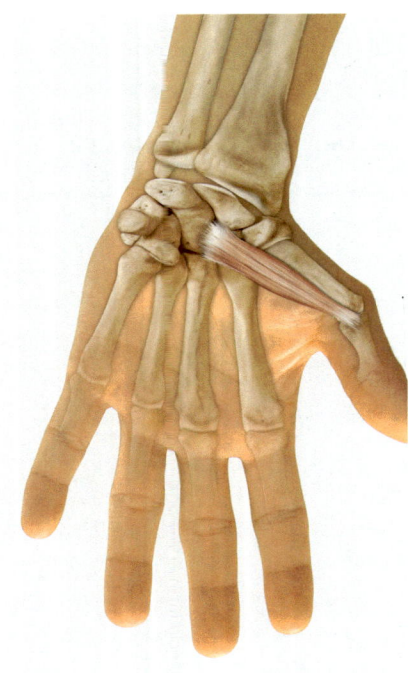

Flexor pollicis brevis muscle

The flexor pollicis brevis muscle (short flexor muscle of the thumb) brings about flexion and opposition at the carpometacarpal joint of the thumb and flexes the thumb's metacarpophalangeal joint.

Origin	Superficial head: trapezium (first carpal) bone, flexor retinaculum of the hand
	Deep head: trapezoid (second carpal) bone, capitate (third carpal) bone, palmar ligaments between the carpal bones
Insertion	Radial side of the base of the proximal phalanx of the thumb
Innervation	Superficial head: median nerve, C7–T1
	Deep head: ulnar nerve, C7–T1
Special features	The flexor pollicis brevis muscle has dual innervation

Functions

 Synergists   Antagonists

Carpometacarpal joint I

Flexion

Flexor pollicis longus muscle	Extensor pollicis longus muscle
Abductor pollicis brevis muscle	Extensor pollicis brevis muscle
Adductor pollicis muscle	Abductor pollicis longus muscle
Opponens pollicis muscle	

Abduction (superficial head)

Abductor pollicis longus muscle	Adductor pollicis muscle
Abductor pollicis brevis muscle	Dorsal interosseus muscle of the hand 1
Extensor pollicis brevis muscle	Opponens pollicis muscle
	Extensor pollicis longus muscle
	Flexor pollicis brevis muscle (deep head)

Adduction (deep head)

Adductor pollicis muscle	Abductor pollicis longus muscle
Dorsal interosseus muscle of the hand 1	Abductor pollicis brevis muscle
Opponens pollicis muscle	Extensor pollicis brevis muscle

Opposition

Opponens pollicis muscle	Extensor pollicis longus muscle
Flexor pollicis longus muscle	Extensor pollicis brevis muscle
Adductor pollicis muscle	

Metacarpophalangeal joint I

Flexion

Flexor pollicis longus muscle	Extensor pollicis longus muscle
Adductor pollicis muscle	Extensor pollicis brevis muscle
Opponens pollicis muscle	Abductor pollicis longus muscle
	Abductor pollicis brevis muscle

Muscle function testing

Muscle power grade	
5/4	
3/2	
1/0	

Starting position: The patient sits. The forearm and hand rest on a flat surface with the hand upright on its outer edge (the little finger side) and the wrist at mid-point. The thumb is extended and abducted.

Test procedure: The examiner holds the patient's first metacarpal bone in place with one hand and with the other applies pressure to the proximal phalanx of the thumb in the direction of extension.

Instruction: "Flex your thumb against my resistance."

Starting position: The patient sits. The forearm and hand rest on a flat surface with the hand upright on its outer edge (the little finger side) and the wrist at mid-point. The thumb is extended and abducted.

Test procedure: The examiner holds the patient's first metacarpal bone in place with one hand.

Instruction: "Flex your thumb."

Starting position: The patient sits. The forearm and hand rest on a flat surface with the hand upright on its outer edge (the little finger side) and the wrist at mid-point. The thumb is extended and abducted.

Test procedure: The examiner looks at the movement of the thumb.

Instruction: "Try to flex your thumb."

 Problems/comments

• The function of the flexor pollicis brevis is not distinguishable from that of the flexor pollicis longus.

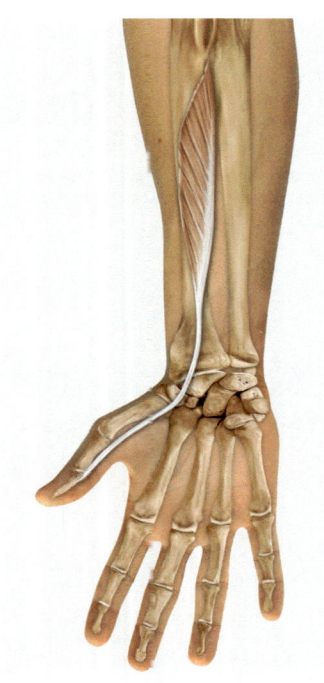

Flexor pollicis longus muscle

The flexor pollicis longus muscle (long flexor muscle of the thumb) flexes the car-pometacarpal, the metacarpophalangeal and the interphalangeal joints of the thumb. It is the only flexor of the last joint. It also adds its flexor effect to that of the other flexors of the wrist.

Origin	Ventral surface of the middle half of the radius
	Interosseous membrane
	Coronoid process of the ulna
	Medial epicondyle of the humerus in some cases
Insertion	Palmar surface of the base of the distal phalanx of the thumb
Innervation	Anterior interosseous nerve branch of the median nerve, C7–T1

Functions

 Synergists   Antagonists

Radiocarpal joint and mediocarpal joint

Flexion

Flexor digitorum superficialis muscle	Extensor digitorum muscle
Flexor digitorum profundus muscle	Extensor carpi radialis longus muscle
Flexor carpi ulnaris muscle	Extensor carpi radialis brevis muscle
Flexor carpi radialis muscle	Extensor carpi ulnaris muscle
Abductor pollicis longus muscle	Extensor indicis muscle
Palmaris longus muscle	Extensor digiti minimi muscle
	Extensor pollicis longus muscle

Carpometacarpal joint I

Flexion

Flexor pollicis brevis muscle	Extensor pollicis longus muscle
Adductor pollicis muscle	Extensor pollicis brevis muscle
Opponens pollicis muscle	Abductor pollicis longus muscle
	Abductor pollicis brevis muscle

Opposition

Opponens pollicis muscle	Extensor pollicis longus muscle
Flexor pollicis brevis muscle	Extensor pollicis brevis muscle
Adductor pollicis muscle	

Metacarpophalangeal joint I

Flexion

Flexor pollicis brevis muscle	Extensor pollicis longus muscle
Adductor pollicis muscle	Extensor pollicis brevis muscle
Opponens pollicis muscle	Abductor pollicis longus muscle
	Abductor pollicis brevis muscle

Interphalangeal joint I

Flexion

None	Extensor pollicis longus muscle

Muscle function testing

Muscle power grade	
5/4	
3/2	
1/0	

Starting position: The patient sits. The forearm and hand rest on a flat surface. The thumb is extended.

Test procedure: The examiner holds the metacarpophalangeal joint of the patient's thumb in extension with one hand and with the other applies pressure to the distal phalanx of the thumb in the direction of extension.

Instruction: "Flex your thumb against my resistance and hold the final position."

Starting position: The patient sits. The forearm and hand rest on a flat surface. The thumb is extended.

Test procedure: The examiner holds the metacarpophalangeal joint of the patient's thumb in extension with one hand.

Instruction: "Flex your thumb at the end joint (interphalangeal joint)."

Starting position: The patient sits. The forearm and hand rest on a flat surface. The thumb is extended.

Test procedure: The examiner palpates the flexor pollicis longus muscle.

Instruction: "Try to flex your thumb at the end joint (interphalangeal joint)."

Clinical relevance

- Weakness of the flexor pollicis longus makes it more difficult to write or to hold small objects between the thumb and index finger.

⚠ Problems/comments

- The flexor pollicis longus works alone when flexing the thumb at the interphalangeal joint; flexion of the thumb at the metacarpophalangeal joint is brought about by the flexor pollicis longus and brevis muscles together.
- The hand flexors that are palpable above the wrist, going from the radial to the ulnar side, are as follows:
  – flexor pollicis longus muscle
  – flexor carpi radialis muscle
  – palmaris longus muscle
  – flexor digitorum superficialis muscle
  – flexor carpi ulnaris muscle.

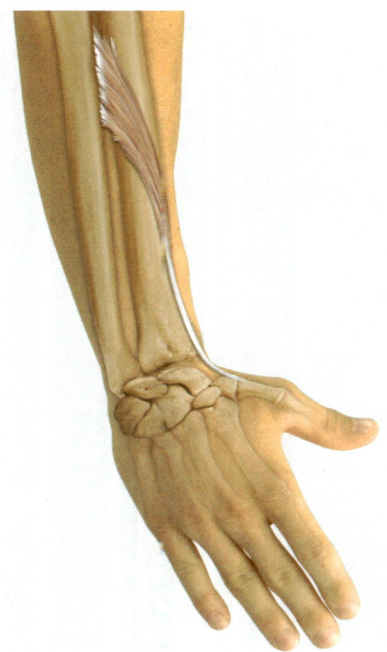

Abductor pollicis longus muscle

The abductor pollicis longus muscle (long abductor muscle of the thumb) extends and abducts the thumb at carpometacarpal joint I. It also brings about flexion and radial abduction of the wrist joints.

Origin	Middle third of the dorsal surface of the radius Interosseous membrane Distal two thirds of the dorsal surface of the ulna
Insertion	Radial side of the base of the first metacarpal bone
Innervation	Radial nerve, deep branch, C6–C8
Special features	The abductor pollicis longus and the extensor pollicis brevis muscles form the radial boundaries of the anatomical snuff box

Functions

 Synergists   Antagonists

Radiocarpal joint and mediocarpal joint

Flexion

Flexor digitorum superficialis muscle	Extensor digitorum muscle
Flexor digitorum profundus muscle	Extensor carpi radialis longus muscle
Flexor carpi ulnaris muscle	Extensor carpi radialis brevis muscle
Flexor carpi radialis muscle	Extensor carpi ulnaris muscle
Flexor pollicis longus muscle	Extensor indicis muscle
Palmaris longus muscle	Extensor digiti minimi muscle
	Extensor pollicis longus muscle

Radial abduction

Flexor carpi radialis muscle	Flexor carpi ulnaris muscle
Extensor carpi radialis longus muscle	Extensor carpi ulnaris muscle
Extensor carpi radialis brevis muscle	Flexor digitorum profundus muscle
Flexor pollicis longus muscle	Extensor digitorum muscle
Extensor pollicis brevis muscle	Extensor digiti minimi muscle
Extensor pollicis longus muscle	

Carpometacarpal joint I

Abduction

Abductor pollicis brevis muscle	Adductor pollicis muscle
Flexor pollicis brevis muscle (superficial head)	Dorsal interosseus muscle of the hand I
Extensor pollicis brevis muscle	Opponens pollicis muscle
	Extensor pollicis longus muscle
	Flexor pollicis brevis muscle (deep head)

Extension

Extensor pollicis longus muscle	Flexor pollicis longus muscle
Extensor pollicis brevis muscle	Flexor pollicis brevis muscle
Abductor pollicis brevis muscle	Adductor pollicis muscle
	Opponens pollicis muscle

← Tendon

Muscle function testing

Muscle power grade	
5/4	
3/2	
1/0	

Starting position: The patient sits. The forearm and hand rest on a flat surface in supination.

Test procedure: The examiner stabilizes the patient's wrist with one hand and with the other applies pressure from laterally in the direction of adduction of the carpometacarpal joint of the thumb.

Instruction: "Stick your thumb out against my resistance."

Starting position: The patient sits. The forearm and hand rest on a flat surface in supination.

Test procedure: The examiner stabilizes the patient's wrist with one hand.

Instruction: "Stick your thumb out."

Starting position: The patient sits. The forearm and hand rest on a flat surface in supination.

Test procedure: The examiner palpates the abductor pollicis longus muscle.

Instruction: "Try to stick your thumb out."

 Problems/comments

- Part of the tendon of the abductor pollicis longus muscle fuses with that of the extensor pollicis brevis and the abductor pollicis brevis muscles.

- It is difficult to distinguish between the functions of the abductor pollicis longus and brevis muscles. Both muscles work together to bring about abduction of the carpometacarpal joint, but abduction of the metacarpophalangeal joint is brought about by the abductor pollicis brevis muscle alone.

- Abduction and adduction movements take place at the carpometacarpal joint (basal joint) of the thumb. Its planes of movement are at an oblique angle to those of the fingers.

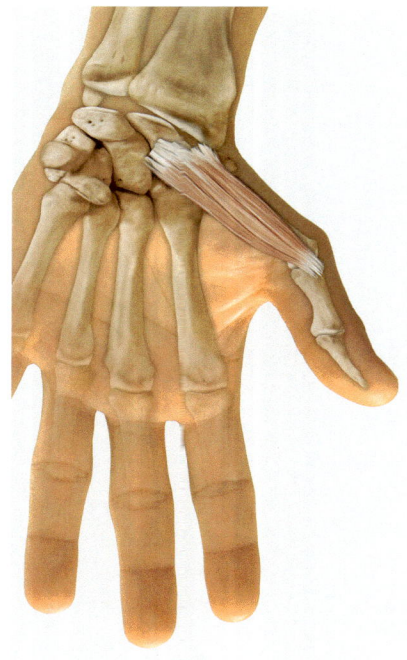

Abductor pollicis brevis muscle

The abductor pollicis brevis muscle (short abductor muscle of the thumb) abducts the thumb at the carpometacarpal joint and extends it at the metacarpophalangeal joint. It also runs into the aponeurosis of the extensors and has an extensor effect on the thumb's interphalangeal joint.

Origin	Scaphoid tubercle, flexor retinaculum of the hand, trapezium bone (first carpal bone)
Insertion	Palmar surface of the base of the proximal phalanx of the thumb
Innervation	Median nerve, C7–T1

Functions

 Synergists

 Antagonists

Carpometacarpal joint I

Abduction
Abductor pollicis longus muscle
Flexor pollicis brevis muscle (superficial head)
Extensor pollicis brevis muscle

Adductor pollicis muscle
Dorsal interosseus muscle of the hand I
Opponens pollicis muscle
Extensor pollicis longus muscle
Flexor pollicis brevis muscle (deep head)

Metacarpophalangeal joint I

Extension
Extensor pollicis longus muscle
Extensor pollicis brevis muscle

Flexor pollicis longus muscle
Flexor pollicis brevis muscle
Adductor pollicis muscle
Opponens pollicis muscle

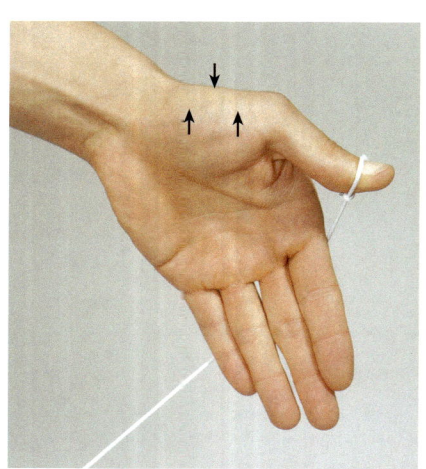

Muscle function testing

Muscle power grade	
5/4	
3/2	
1/0	

Starting position: The patient sits. The forearm and hand rest on the examination couch in supination. The fingers lie close together.

Test procedure: The examiner stabilizes the patient's metacarpus with one hand and with the other applies pressure to the outer side of the proximal phalanx of the thumb in the direction of adduction.

Instruction: "Stick your thumb out against my resistance."

Starting position: The patient sits. The forearm and hand rest on the examination couch in supination. The fingers lie close together.

Test procedure: The examiner stabilizes the patient's metacarpus with one hand.

Instruction: "Stick your thumb out."

Starting position: The patient sits. The forearm and hand rest on the examination couch in supination. The fingers lie close together.

Test procedure: The examiner palpates the abductor pollicis brevis muscle.

Instruction: "Try to stick your thumb out."

 Problems/comments

- It is difficult to distinguish between the functions of the abductor pollicis longus and brevis muscles. Both muscles work together to bring about abduction of the carpometacarpal joint, but abduction of the metacarpophalangeal joint is brought about by the abductor pollicis brevis muscle alone.
- Abduction and adduction movements take place at the carpometacarpal joint (basal joint) of the thumb. Its planes of movement are at an oblique angle to those of the fingers.

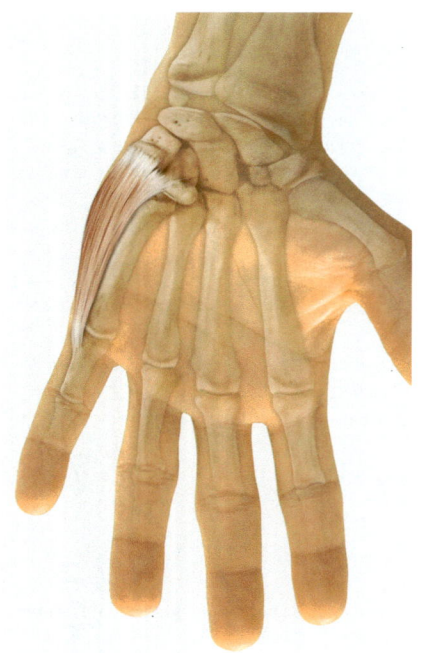

Abductor digiti minimi muscle

The abductor digiti minimi muscle (abductor muscle of the little finger) abducts and flexes the little finger at metacarpophalangeal joint V, for instance when grasping with the fingers spread out (e.g. playing the violin).

Origin	Pisiform bone, flexor retinaculum of the hand Tendon of the flexor carpi ulnaris muscle
Insertion	Ulnar edge of the base of proximal phalanx V, running into the flexor aponeurosis of the little finger
Innervation	Ulnar nerve, deep branch, C8–T1
Special features	The abductor digiti minimi muscle is one of the indicator muscles of spinal cord segment C8

Functions

 Synergists

 Antagonists

Metacarpophalangeal joint V

Abduction
Extensor digiti minimi muscle (V)

Palmar interosseus muscle 3

Flexion
Flexor digitorum superficialis muscle
Flexor digitorum profundus muscle
Palmar interosseus muscle 3
Lumbricalis muscle of the hand 4
Flexor digiti minimi brevis muscle
Opponens digiti minimi muscle

Extensor digitorum muscle
Extensor digiti minimi muscle

Muscle function testing

Muscle power grade	
5/4	

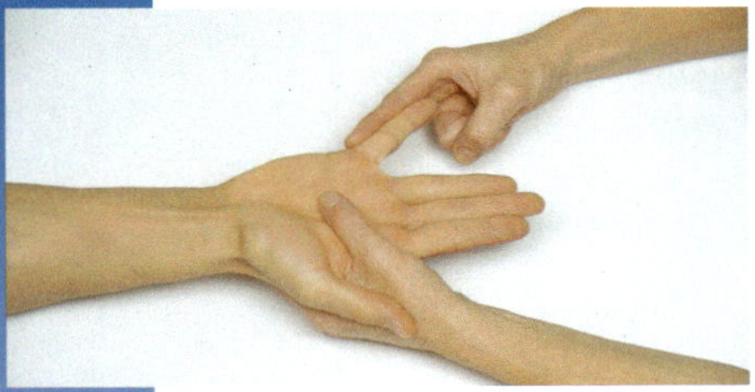

Starting position: The patient sits. The forearm and hand rest on the examination couch in supination. The fingers lie close together.

Test procedure: The examiner holds the patient's metacarpus in place with one hand and with the other applies pressure to the proximal phalanx of the little finger in the direction of adduction.

Instruction: "Stick your little finger out sideways against my resistance."

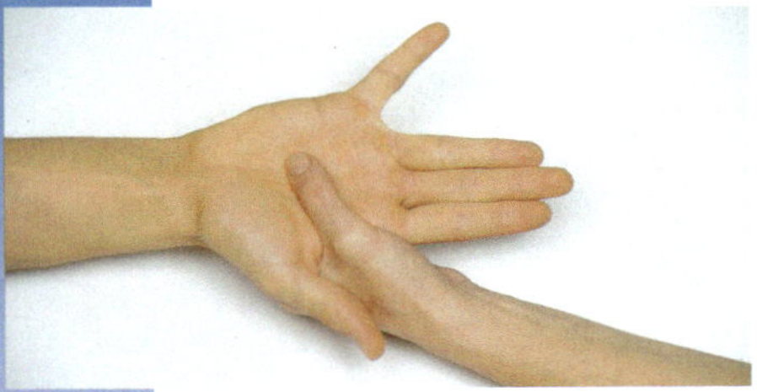

Starting position: The patient sits. The forearm and hand rest on the examination couch in supination. The fingers lie close together.

Test procedure: The examiner holds the patient's metacarpus in place with one hand.

Instruction: "Stick your little finger out sideways."

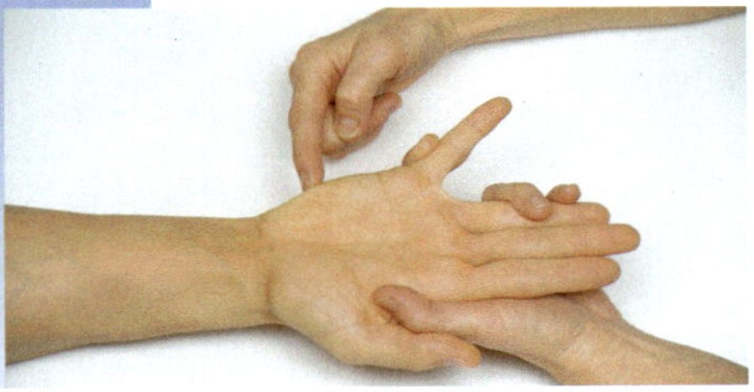

Starting position: The patient sits. The forearm and hand rest on the examination couch in supination. The fingers lie close together.

Test procedure: The examiner palpates the abductor digiti minimi muscle.

Instruction: "Try to stick your little finger out sideways."

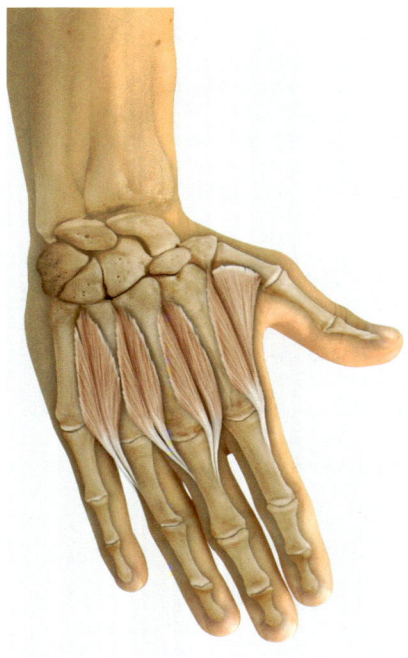

Dorsal interossei muscles of the hand

The dorsal interossei muscles of the hand spread out (abduct) fingers II to IV. They bring about flexion at the carpometacarpal joints and extension at the interphalangeal joints of these fingers.

Origin	Each muscle originates at the facing sides of two adjoining metacarpal bones
Insertion	Radial side of the base of the proximal phalanx of the index finger Radial and ulnar side of the middle finger (III) Ulnar side of the ring finger (V) Extensor aponeuroses of the fingers (II–IV)
Innervation	Ulnar nerve, deep branch, C8–T1

Functions

 Synergists   Antagonists

Metacarpophalangeal joints II to IV

Abduction

Extensor digitorum muscle

Palmar interossei muscles 1 and 2 (II, IV)
Flexor digitorum superficialis muscle
Flexor digitorum profundus muscle

Flexion

Flexor digitorum superficialis muscle
Flexor digitorum profundus muscle
Palmar interossei muscles 1 and 2 (II, IV)
Lumbricales muscles of the hand 1–3 (II–IV)

Extensor digitorum muscle
Extensor indicis muscle (II)

Proximal and distal interphalangeal joints II to IV

Extension

Extensor digitorum muscle
Extensor indicis muscle (II)
Lumbricales muscles of the hand 1-3 (II–IV)
Palmar interossei muscles 1 and 2 (II, IV)

Flexor digitorum superficialis muscle
Flexor digitorum profundus muscle

Muscle function testing

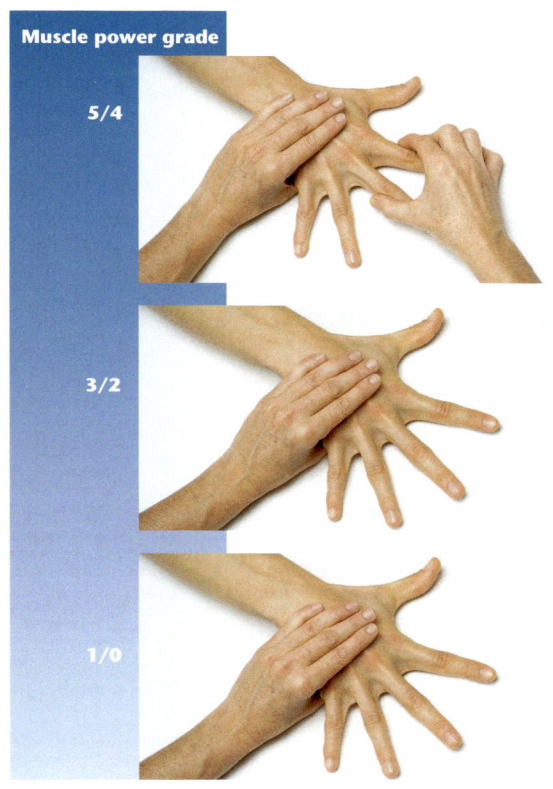

Muscle power grade

5/4

3/2

1/0

Starting position: The patient sits. The forearm and hand rest on the examination couch in pronation. The fingers lie close together.

Test procedure: The examiner holds the patient's metacarpus in place with one hand and, with the other, applies resistance at the radial side of the index finger (II) and the ulnar side of the middle finger (III).

Instruction: "Spread out your fingers against my resistance."

Starting position: The patient sits. The forearm and hand rest on the examination couch in pronation. The fingers lie close together.

Test procedure: The examiner looks at the movements of the fingers.

Instruction: "Spread out your fingers."

Starting position: The patient sits. The forearm and hand rest on the examination couch in pronation. The fingers lie close together.

Test procedure: The examiner looks at the movements of the fingers.

Instruction: "Try to spread out your fingers."

⚠ Problems/comments

- To test the other dorsal interossei muscles, resistance should be applied between the middle and ring fingers (III and IV), and between the ring finger and little finger (IV and V), respectively.

- The abductor digiti minimi contributes towards abduction of the little finger.

- The dorsal interossei muscles have an abductor effect relative to an axis running through the middle finger. They also bring about flexion of the metacarpophalangeal joint and extension of the proximal and distal interphalangeal joints. They share this function with the palmar interossei muscles and the lumbricales muscles of the hand.

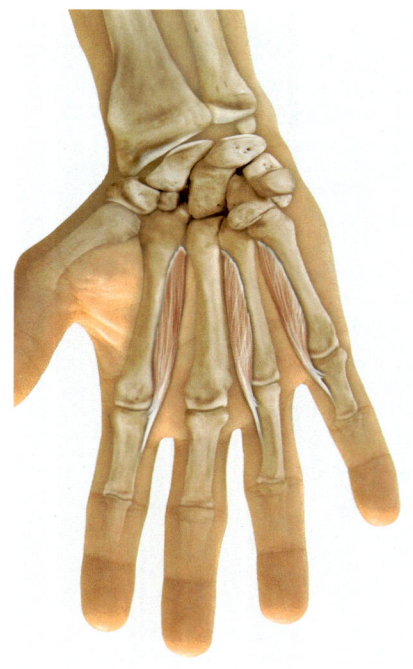

Palmar interossei muscles

The palmar interossei muscles adduct fingers II, IV and V towards the middle finger, flex the metacarpophalangeal joints and extend the interphalangeal joints of the same fingers.

Origin	Palmar interosseus muscle 1: ulnar side of the 2nd metacarpal bone Palmar interosseus muscle 2: radial side of the 4th metacarpal bone Palmar interosseus muscle 3: radial side of the 5th metacarpal bone
Insertion	Running into the extensor neuroses of the proximal phalanges of fingers II, IV and V
Innervation	Ulnar nerve, deep branch, C8–T1

Functions

 Synergists

Antagonists

Metacarpophalangeal joints II, IV, V

Adduction

Extensor indicis muscle (II)
Flexor digitorum superficialis muscle
Flexor digitorum profundus muscle

Dorsal interossei muscles of the hand 1 and 4 (II, IV)
Abductor digiti minimi muscle (V)
Extensor digitorum muscle (II, IV, V)
Extensor digiti minimi muscle (V)

Flexion

Flexor digitorum superficialis muscle
Flexor digitorum profundus muscle
Dorsal interossei muscles of the hand 1 and 4 (II, IV)
Lumbricales muscles of the hand 1, 3 and 4 (II, IV, V)
Flexor digiti minimi brevis muscle (V)
Opponens digiti minimi muscle (V)
Abductor digiti minimi muscle (V)

Extensor digitorum muscle
Extensor indicis muscle (II)
Extensor digiti minimi muscle (V)

Proximal interphalangeal joints II, IV, V

Extension

Extensor digitorum muscle
Extensor indicis muscle (II)
Extensor digiti minimi muscle (V)
Abductor digiti minimi muscle (V)
Lumbricales muscles of the hand 1, 3 and 4 (II, IV, V)
Dorsal interossei muscles of the hand 1 and 4 (II, IV)

Flexor digitorum superficialis muscle
Flexor digitorum profundus muscle

Muscle function testing

Muscle power grade

5/4

Starting position: The patient sits. The forearm and hand rest on the examination couch in pronation. The fingers are spread out.

Test procedure: The examiner holds the patient's metacarpus in place with one hand and, with the other, applies resistance between the index and middle finger (fingers II and III).

Instruction: "Squeeze your fingers together against my resistance."

3/2

Starting position: The patient sits. The forearm and hand rest on the examination couch in pronation. The fingers are spread out.

Test procedure: The examiner holds the patient's metacarpus in place with one hand.

Instruction: "Bring your fingers back together."

1/0

Starting position: The patient sits. The forearm and hand rest on the examination couch in pronation. The fingers are spread out.

Test procedure: The examiner looks at the movements of the fingers.

Instruction: "Try to bring your fingers back together."

 Problems/comments

- The above description applies to palmar interosseus muscle 1. The remaining palmar interossei muscle are tested accordingly between the middle and ring fingers (fingers III and IV) and between the ring and little fingers (fingers IV and V), respectively.

- The dorsal interossei muscles have an adductor effect relative to an axis running through the middle finger. They also bring about flexion of the metacarpophalangeal joint and extension of the proximal and distal interphalangeal joints. They share this function with the palmar interossei muscles and the lumbricales muscles of the hand.

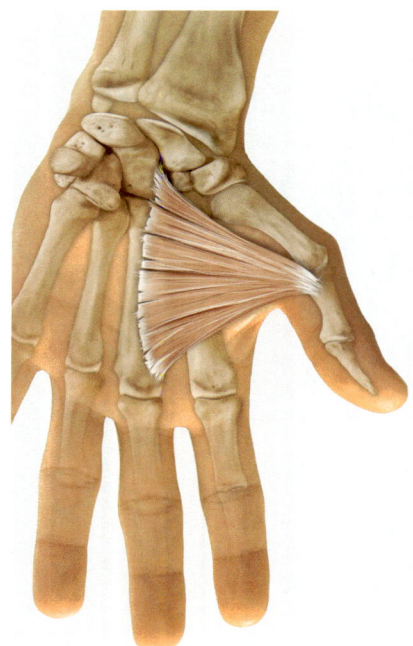

Adductor pollicis muscle

The adductor pollicis muscle (adductor muscle of the thumb) adducts the thumb towards the palm and is jointly responsible for opposition and for the powerful pincer grip between the tips of the thumb and index finger.

Origin	Oblique head: capitate bone (third carpal bone), base of the second and third metacarpal bones, intercarpal ligaments Transverse head: proximal two thirds of the palmar surface of the third metacarpal bone
Insertion	Ulnar sesamoid bone at the carpometacarpal joint of the thumb
Innervation	Ulnar nerve, deep branch, C8–T1

Functions

 Synergists

 Antagonists

Carpometacarpal joint I

Adduction

Dorsal interosseus muscle of the hand I
Opponens pollicis muscle
Extensor pollicis longus muscle
Flexor pollicis brevis muscle (deep head)

Abductor pollicis brevis muscle
Abductor pollicis longus muscle
Flexor pollicis brevis muscle (superficial head)
Extensor pollicis brevis muscle

Opposition

Opponens pollicis muscle
Flexor pollicis longus muscle
Flexor pollicis brevis muscle

Extensor pollicis longus muscle
Extensor pollicis brevis muscle
Abductor pollicis longus muscle

Dip in the skin of the palm over the tightened muscle

Muscle function testing

Muscle power grade	
5/4	
3/2	
1/0	

Starting position: The patient sits. The forearm and hand rest on the examination couch in supination. The fingers are spread out.

Test procedure: The examiner holds the patient's metacarpus in place with one hand and with the other applies pressure to the inner side of the proximal phalanx of the thumb in the direction of abduction.

Instruction: "Bring your thumb towards your index finger against my resistance."

Starting position: The patient sits. The forearm and hand rest on the examination couch in supination. The fingers are spread out.

Test procedure: The examiner holds the patient's metacarpus in place with one hand.

Instruction: "Bring your thumb towards your index finger."

Starting position: The patient sits. The forearm and hand rest on the examination couch in supination. The fingers are spread out.

Test procedure: The examiner palpates the adductor pollicis muscle.

Instruction: "Try to bring your thumb towards your index finger."

 Problems/comments

- The abduction and adduction movements take place at the carpometacarpal joint (base joint of the thumb).

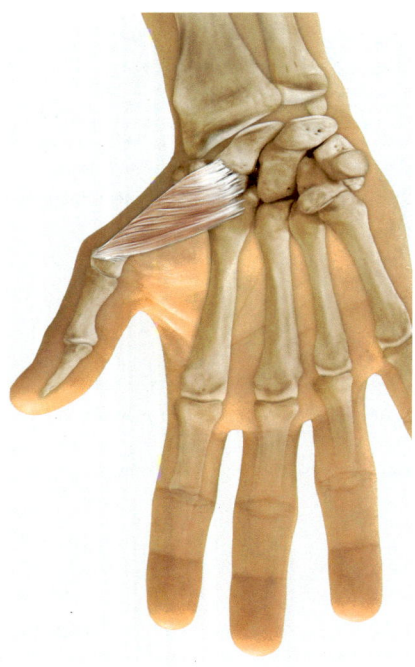

Opponens pollicis muscle

The opponens pollicis muscle (opposing muscle of the thumb) opposes the thumb. This complex movement combines flexion, abduction, rotation and, finally, adduction at the thumb's carpometacarpal joint. Thus, it involves virtually all of the muscles of the thumb, apart from the extensors.

Origin	Trapezium bone (first carpal bone), flexor retinaculum of the hand
Insertion	Radial side of the diaphysis of the first metacarpal bone
Innervation	Median nerve, C7–T1

Functions

 Synergists   Antagonists

Carpometacarpal joint I

Opposition

Adductor pollicis muscle	Extensor pollicis longus muscle
Flexor pollicis longus muscle	Extensor pollicis brevis muscle
Flexor pollicis brevis muscle	Abductor pollicis longus muscle

Flexion

Flexor pollicis longus muscle	Extensor pollicis longus muscle
Flexor pollicis brevis muscle	Extensor pollicis brevis muscle
Abductor pollicis brevis muscle	Abductor pollicis longus muscle
Adductor pollicis muscle	

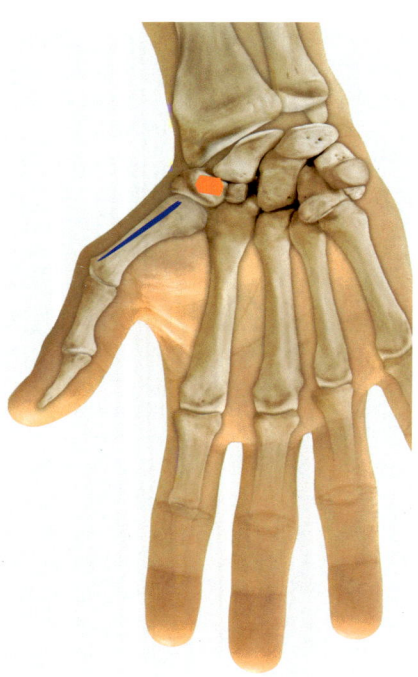

Muscle function testing

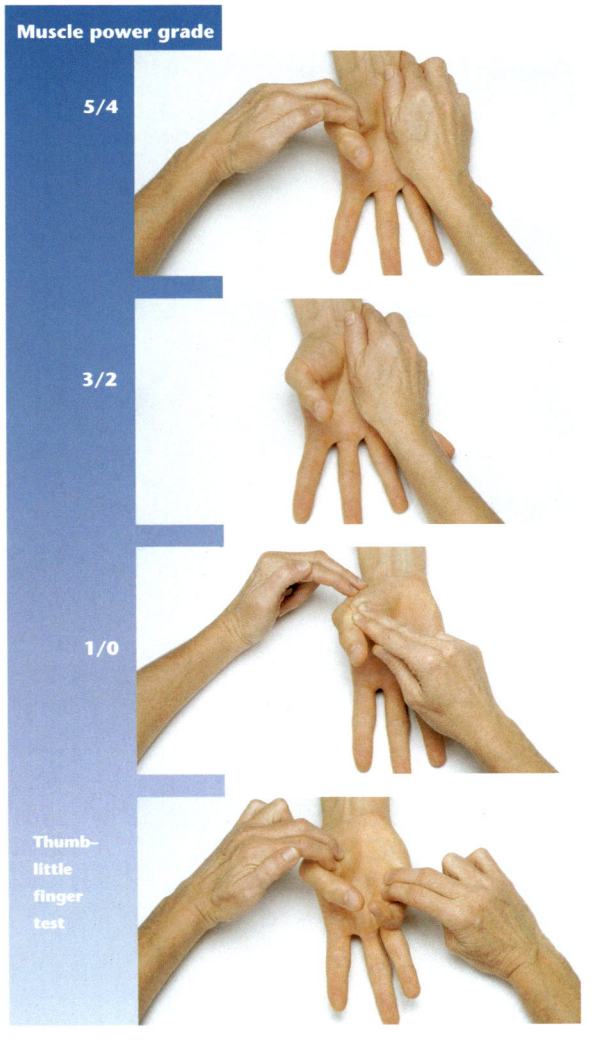

Muscle power grade

5/4

3/2

1/0

Thumb–little finger test

Starting position: The patient sits. The forearm and hand rest on the examination couch in supination. The fingers are spread out.

Test procedure: With one hand, the examiner holds the patient's metacarpus in place from the little finger side up to the middle finger and with the other applies downward pressure to the first metacarpal bone.

Instruction: "Move your thumb towards your little finger against my resistance."

Starting position: The patient sits. The forearm and hand rest on the examination couch in supination. The fingers are spread out.

Test procedure: With one hand, the examiner holds the patient's metacarpus in place from the little finger side up to the middle finger.

Instruction: "Move your thumb towards your little finger."

Starting position: The patient sits. The forearm and hand rest on the examination couch in supination. The fingers are spread out.

Test procedure: The examiner palpates the opponens pollicis muscle.

Instruction: "Try to move your thumb towards your little finger."

Starting position: The patient sits. The forearm and hand rest on the examination couch in supination. The fingers are spread out.

Test procedure: The examiner applies downward pressure to the first and fourth metacarpal bones.

Instruction: "Bring your thumb and little finger together."

⚠️ Problems/comments

- The opponens pollicis performs its function in conjunction with the adductor pollicis and flexor pollicis brevis muscles. .
- The thumb–little finger test assesses the opponens digiti minimi muscle in addition to the muscles of the thumb (movement against resistance, power grade 4 or 5).

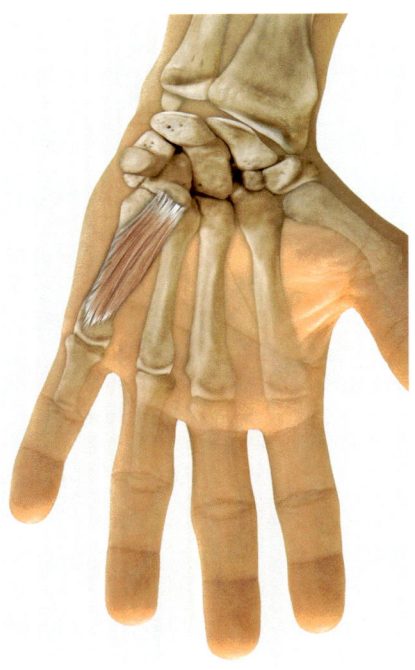

Opponens digiti minimi muscle

The opponens digiti minimi muscle (opposing muscle of the little finger) slightly flexes the fifth metacarpal bone towards the palm at carpometacarpal joint V. It is not inserted at the phalanges of the little finger. Unlike the thumb, the little finger cannot achieve true opposition to the other fingers. Therefore, the name of this muscle is misleading.

Origin	Hamulus of the hamate bone, flexor retinaculum of the hand
Insertion	Ulnar side of the fifth metacarpal bone
Innervation	Ulnar nerve, deep branch, C8–T1

Functions

 Synergists

Antagonists

Metacarpophalangeal joint V

Flexion

Palmar interosseus muscle 3	Extensor digitorum muscle
Flexor digitorum superficialis muscle	Extensor digiti minimi muscle
Flexor digitorum profundus muscle	
Lumbricalis muscle of the hand 4	
Abductor digiti minimi muscle	
Flexor digiti minimi brevis muscle	

Adduction

Palmar interosseus muscle 3	Abductor digiti minimi muscle
Flexor digitorum superficialis muscle	Extensor digiti minimi muscle
Flexor digitorum profundus muscle	

Muscle function testing

Muscle power grade

5/4

Starting position: The patient sits. The forearm and hand rest on the examination couch in supination. The fingers are spread out.

Test procedure: The examiner holds the patient's metacarpus in place with one hand and with the other applies downward pressure to the fifth metacarpal bone.

Instruction: "Move your little finger towards your thumb against my resistance."

3/2

Starting position: The patient sits. The forearm and hand rest on the examination couch in supination. The fingers are spread out.

Test procedure: The examiner holds the patient's metacarpus in place with one hand.

Instruction: "Move your little finger towards your thumb."

1/0

Starting position: The patient sits. The forearm and hand rest on the examination couch in supination. The fingers are spread out.

Test procedure: The examiner looks at the movements of the fingers.

Instruction: "Try to move your little finger towards your thumb."

Thumb–little finger test

Thumb-little finger test

Starting position: The patient sits. The forearm and hand rest on the examination couch in supination. The fingers are spread out.

Test procedure: The examiner applies downward pressure to the first and fourth metacarpal bones.

Instruction: "Bring your thumb and little finger together."

⚠ Problems/comments

- The thumb–little finger test demonstrates the collaboration of the thumb muscles (opponens pollicis, adductor pollicis, flexor pollicis brevis muscles) and the opponens digiti minimi muscle (movement against resistance, power grade 4 or 5).

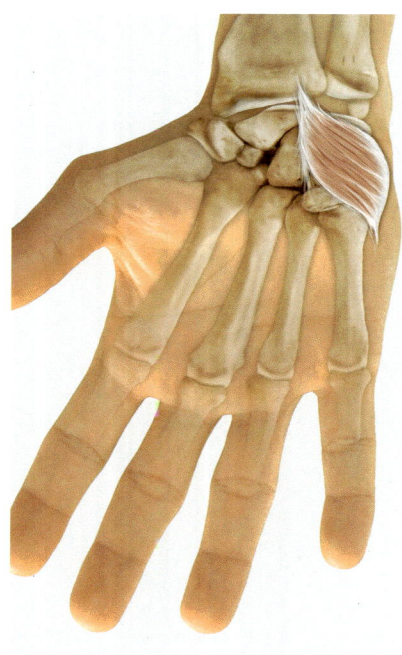

Palmaris brevis muscle

The palmaris brevis muscle protects the ulnar nerve and tightens the palmar aponeurosis slightly. A list of its synergists and antagonists would be superfluous.

Origin	Palmar aponeurosis, flexor retinaculum of the hand
Insertion	Skin of the palm on the ulnar side of the hand
Innervation	Ulnar nerve, superficial branch, C7–T1
Special features	The palmaris brevis is the only muscle innervated by the superficial branch of the ulnar nerve

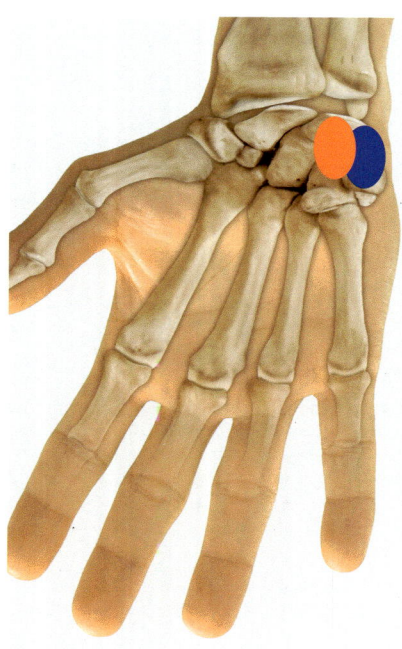

Muscle function testing

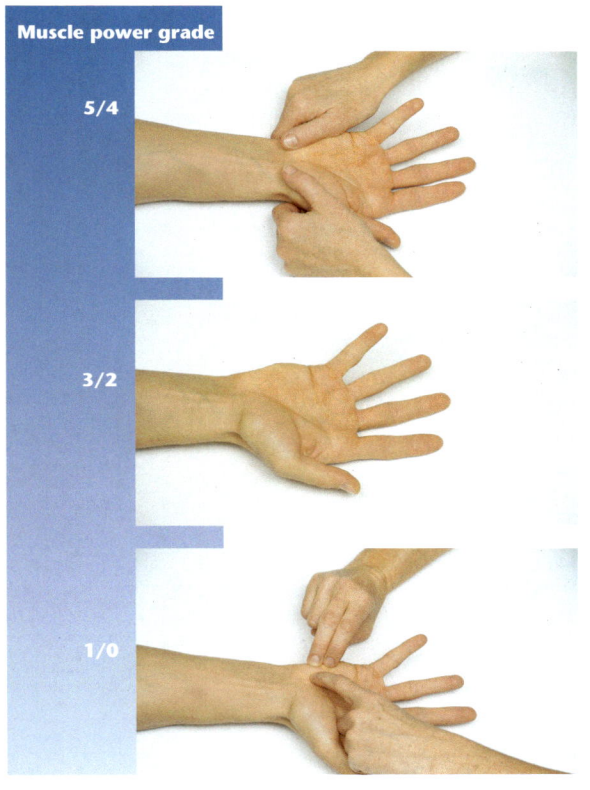

Muscle power grade
5/4
3/2
1/0

Starting position: The patient sits. The forearm and hand rest on the examination couch in supination.

Test procedure: The examiner presses down onto the thenar and hypothenar eminences to flatten out the palm. The examiner looks at the ulnar side of the palm.

Instruction: "Fold up your palm against my resistance."

Starting position: The patient sits. The forearm and hand rest on the examination couch in supination. The fingers are spread out.

Test procedure: The examiner looks at the palm.

Instruction: "Fold up your palm."

Starting position: The patient sits. The forearm and hand rest on the examination couch in supination. The fingers are spread out.

Test procedure: The examiner palpates the palmaris brevis muscle.

Instruction: "Try to fold up your palm."

 Problems/comments

- In evolutionary terms, the palmaris brevis muscle is becoming increasingly vestigial.

Stretch tests

Extensor digitorum, extensor indicis and extensor digiti minimi muscles

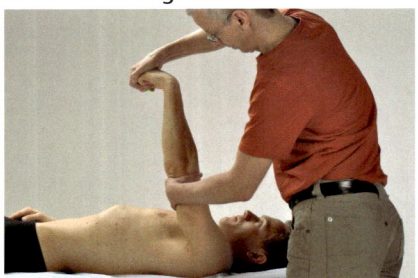

Method
The patient makes a fist, with the metacarpophalangeal and the proximal and distal interphalangeal joints in maximum flexion. The therapist now moves the patient's hand into maximum volar flexion and radial abduction of the wrist.

Finding
Shortening of the relevant muscle is present if the movement cannot be performed to its full extent with the fingers still flexed and if the endpoint feels soft and elastic.
The patient reports a stretching sensation along the course of the muscle.

Flexor digitorum superficialis and flexor digitorum profundus muscles

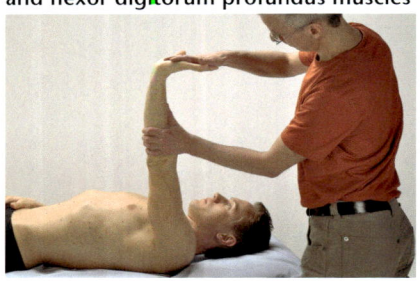

Method
The elbow is in extension, the wrist in dorsal extension. The therapist moves the patient's fingers II to V into maximum extension at the metacarpophalangeal and at the proximal and distal interphalangeal joints. Extension of the distal interphalangeal joints is required only for the stretch test of the flexor digitorum profundus muscle.

Finding
Shortening of the relevant muscle is present if the movement cannot be performed to its full extent and if the endpoint feels soft and elastic.
The patient reports a stretching sensation along the course of the muscle.

Abductor pollicis longus and extensor pollicis brevis muscles

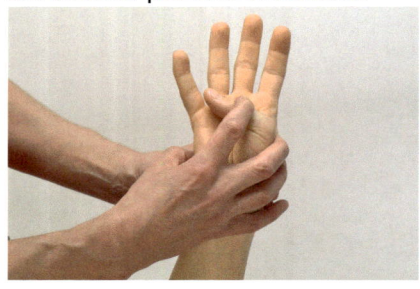

Method
The therapist moves the patient's thumb into maximum flexion at the carpometacarpal and metacarpophalangeal joints. The carpometacarpal joint is also adducted.

Finding
Shortening of the relevant muscle is present if the movement cannot be performed to its full extent and if the endpoint feels soft and elastic.
The patient reports a stretching sensation along the course of the muscle.

Flexor pollicis longus and adductor pollicis muscles

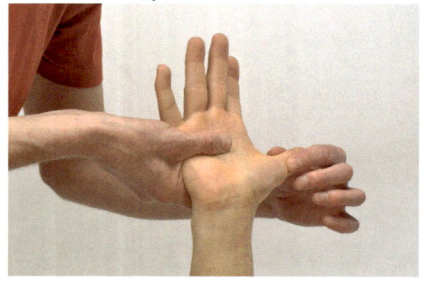

Method
The therapist moves the patient's thumb into abduction at the carpometacarpal joint and into maximum extension at the carpometacarpal, metacarpophalangeal and interphalangeal joints. The wrist is in dorsal extension.

Finding
Shortening of the relevant muscle is present if the movement cannot be performed to its full extent and if the endpoint feels soft and elastic.
The patient reports a stretching sensation along the course of the muscle.

Innervation of the muscles of the upper extremity									
Nerve	**Innervated muscles**	**Segments of origin**							
Accessory nerve and cervical plexus		**Cranial nerve/XI**							
	Trapezius muscle, descending part	**C2 – C4**							
	Trapezius muscle, transverse and ascending parts	**C2 – C4**							
		C3	C4	C5	C6	C7	C8	T1	
Suprascapular nerve			■	■	■				
	Supraspinatus muscle		■	■	■				
	Infraspinatus muscle		■	■	■				
Musculocutaneous nerve				■	■	■			
	Biceps brachii muscle			■	■	■			
	Coracobrachialis muscle			■		■			
	Brachialis muscle			■	■				
Axillary nerve				■	■				
	Deltoideus muscle			■	■				
	Teres minor muscle			■	■				
Lateral and medial pectoral nerves				■	■	■	■	■	
	Pectoralis major muscle			■	■	■	■	■	
	Pectoralis minor muscle				■	■	■	■	
Dorsal nerve of scapula		■	■	■					
	Levator scapulae muscle	■	■	■					
	Rhomboideus major muscle		■	■					
	Rhomboideus minor muscle		■	■					
Subscapular nerve				■	■	■	■		
	Subscapularis muscle			■	■	■			
	Teres major muscle				■	■			
Thoracodorsal nerve					■	■	■		
	Latissimus dorsi muscle				■	■	■		
	Teres major muscle				■	■			
Long thoracic nerve				■	■	■			
	Serratus anterior muscle			■	■	■			
Subclavian nerve				■	■				
	Subclavius muscle			■	■				

Upper extremity – table of nerves						
Nerve	**Innervated muscles**	**Segments of origin**				
		C5	C6	C7	C8	T1
Radial nerve			▒	▒	▒	▒
	Triceps brachii muscle		▒	▒	▒	
	Anconeus muscle		▒	▒	▒	
	Brachialis muscle (lateral and distal parts)	▒	▒			
	Brachioradialis muscle	▒	▒	▒		
	Extensor carpi ulnaris muscle		▒	▒	▒	
	Extensor carpi radialis longus muscle		▒	▒		
	Extensor carpi radialis brevis muscle		▒	▒		
	Supinator muscle	▒	▒	▒		
	Extensor digitorum muscle		▒	▒	▒	
	Extensor digiti minimi muscle		▒	▒	▒	
	Abductor pollicis longus muscle		▒	▒	▒	
	Extensor pollicis brevis muscle		▒	▒	▒	
	Extensor pollicis longus muscle		▒	▒	▒	
	Extensor indicis muscle		▒	▒	▒	
Median nerve			▒	▒	▒	▒
	Pronator teres muscle		▒	▒		
	Pronator quadratus muscle				▒	▒
	Flexor carpi radialis muscle		▒	▒		
	Palmaris longus muscle		▒		▒	▒
	Flexor digitorum superficialis muscle			▒	▒	▒
	Flexor digitorum profundus muscle			▒	▒	▒
	Flexor pollicis longus muscle		▒	▒	▒	▒
	Flexor pollicis brevis muscle		▒	▒	▒	▒
	Lumbricales muscles (I and II)				▒	▒
	Opponens pollicis muscle		▒	▒		
	Abductor pollicis brevis muscle		▒	▒	▒	▒
Ulnar nerve					▒	▒
	Flexor carpi ulnaris muscle			▒	▒	▒
	Flexor digitorum profundus muscle			▒	▒	▒
	Adductor pollicis brevis muscle				▒	▒
	Flexor pollicis brevis muscle				▒	▒
	Abductor digiti minimi muscle				▒	▒
	Flexor digiti minimi muscle				▒	▒
	Opponens digiti minimi muscle				▒	▒
	Lumbricales muscles (III and IV)				▒	▒
	Palmar interossei muscles				▒	▒
	Dorsal interossei muscle of the hand				▒	▒
	Palmaris brevis muscle				▒	▒

3 Lower extremity

Muscles of the hip

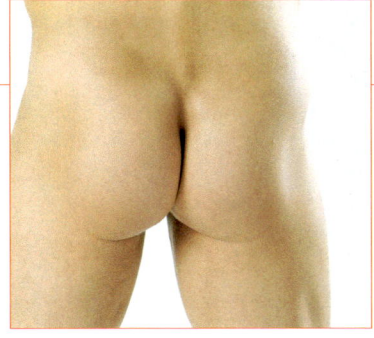

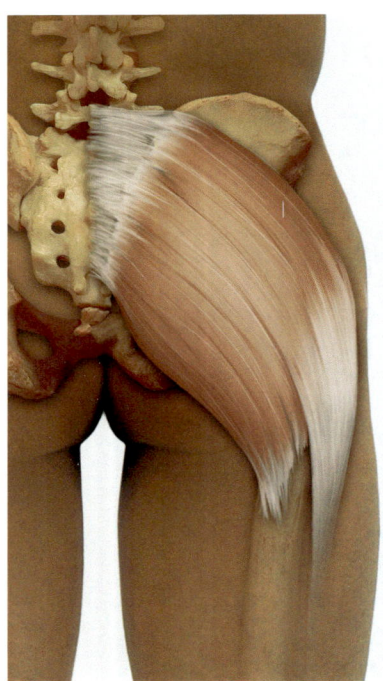

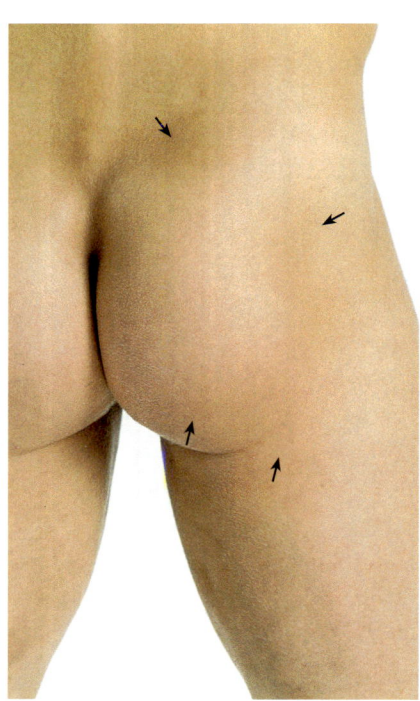

Gluteus maximus muscle

The main function of the gluteus maximus (greatest gluteus) muscle is to right the trunk from a stooped position and to stabilize it if it is at risk of tipping forward from the hips, for example when a heavy load is being carried with the arms outstretched. This muscle is generally not active during gentle walking. It also tilts the pelvis dorsally from the hips, thus flattening out the lordosis of the lumbar spine. When acting via the iliotibial tract, it can secure the extended knee powerfully and act as a tension band to reduce flexion stress on the femur. Occasional contraction of this muscle while sitting leads to pressure distribution, and thus to improved circulation in the soft tissues of the buttocks.

Origin   Dorsal surface of the sacral bone, thoracolumbar fascia, sacrotuberal ligament, dorsal iliac bone, posterior inferior iliac spine

Insertion   Cranial part: iliotibial tract; caudal part: gluteal tuberosity

Innervation   Inferior gluteal nerve, L5–S2

Functions

 Synergists   Antagonists

Hip joint

Extension

Synergists	Antagonists
Semimembranosus muscle, semitendinosus muscle	Iliopsoas muscle, rectus femoris muscle
Biceps femoris muscle (long head)	Tensor fasciae latae muscle
Gluteus medius (dorsal part)	Gluteus medius muscle (ventral part)
Gluteus minimus (dorsal part)	Sartorius muscle, gracilis muscle
Pectineus (from maximum flexion)	Adductor muscles (into the neutral position)
Adductor muscles (into the neutral position)	Pectineus muscle

External rotation

Synergists	Antagonists
Gluteus medius muscle (dorsal part)	Tensor fasciae latae muscle
Gluteus minimus muscle (dorsal part)	Gluteus medius muscle (ventral part)
Quadratus femoris muscle, piriformis muscle	Gluteus minimus muscle (ventral part)
Obturator and gemelli muscles, pectineus muscle	
Sartorius muscle	
Adductor muscles	

Abduction (cranial part)

Synergists	Antagonists
Gluteus medius muscle, gluteus minimus muscle	Adductor muscles, pectineus muscle, gracilis muscle
Tensor fasciae latae muscle	Gluteus maximus muscle (caudal part)
Piriformis muscle (with the hip flexed)	Quadratus femoris muscle (with the hip extended)
Obturator and gemelli muscles (with the hip flexed)	
Quadratus femoris muscle (with the hip flexed)	

Adduction (caudal part)

Synergists	Antagonists
Adductor muscles, pectineus muscle, gracilis muscle	Gluteus medius muscle, gluteus minimus muscle
Quadratus femoris muscle (with the hip extended)	Gluteus maximus muscle (cranial part)
	Tensor fasciae latae muscle
	With the hip flexed: obturator and gemelli muscles, piriformis muscle, quadratus femoris muscle

Hip joint and intervertebral joints (lumbar spine)

Dorsal tilting of the pelvis

Synergists	Antagonists
Rectus abdominis muscle, biceps femoris muscle	Iliopsoas muscle, rectus femoris muscle
Semimembranosus muscle, semitendinosus muscle	Longissimus thoracis muscle
	Iliocostalis muscle

Knee joint

Extension (via the iliotibial tract)

Synergists	Antagonists
Quadriceps femoris muscle	Biceps femoris muscle, semitendinosus muscle
Tensor fasciae latae muscle (via the iliotibial tract)	Semimembranosus muscle, gracilis muscle
	Sartorius muscle, popliteus muscle
	Gastrocnemius muscle

Muscle function testing

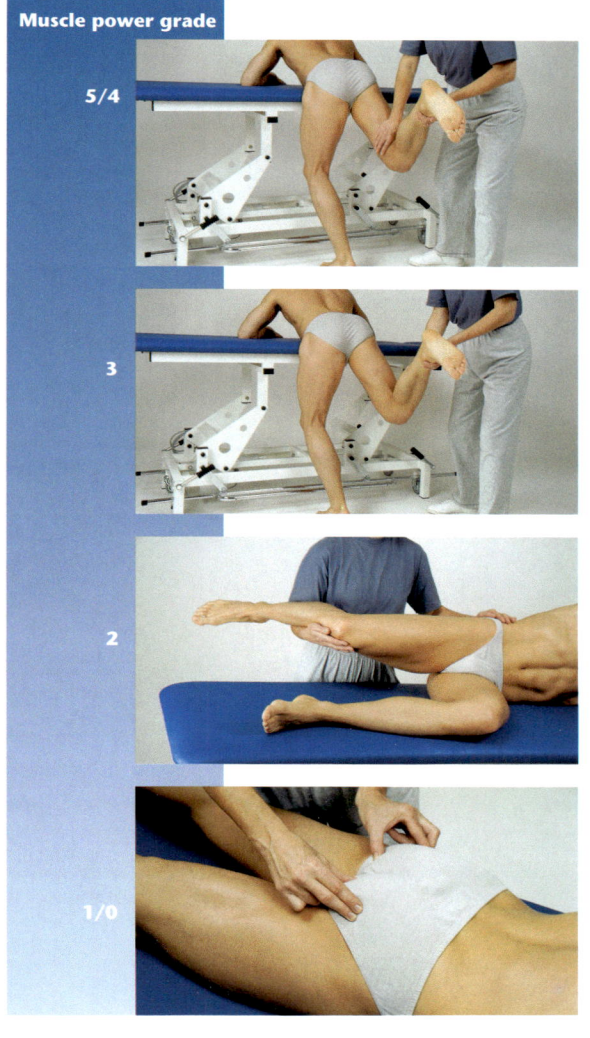

Muscle power grade

5/4

3

2

1/0

Starting position: The patient's rests his trunk, down to the groin and belly side down, on the treatment couch. He stands on one leg with the knee slightly flexed, while the leg being tested is flexed 90° at the knee.

Test procedure: The examiner holds the lower leg on the same side in place with one hand and applies pressure at the distal thigh in the direction of hip flexion.

Instruction: "Extend your leg at the hip against my resistance and hold your final position. Keep the knee flexed."

Starting position: The patient's rests his trunk, down to the groin and belly side down, on the treatment couch. He stands on one leg with the knee slightly flexed, while the leg being tested is flexed 90° at the knee.

Test procedure: The examiner holds the lower leg on the same side in place with one hand.

Instruction: "Extend your leg at the hip, keeping your knee flexed."

Starting position: The patient lies on his side, with both legs flexed 90° at the hip and knee.

Test procedure: The examiner holds the upper side of the pelvis in place and takes the weight of the leg.

Instruction: "Extend your leg at the hip, keeping your knee flexed."

Starting position: The patient lies prone on the treatment couch.

Test procedure: The examiner palpates the gluteus maximus muscle.

Instruction: "Clench your buttocks."

Clinical relevance

• Intramuscular injections must *not* be administered into this muscle!

! Problems/comments

• Carrying out this test with the leg flexed at the knee largely inactivates the ischiocrural musculature, which would otherwise contribute towards the extension of the hip.

• The elasticity of the rectus femoris muscle should be tested before this investigation.

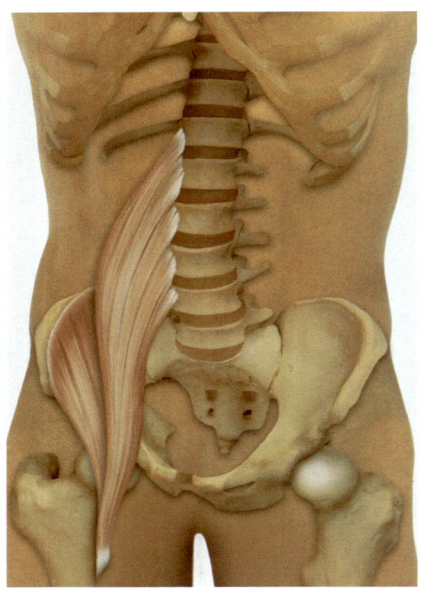

Iliopsoas muscle

The iliopsoas muscle moves the swinging leg at the hip, but only when maximum force is required, for example when flexing the extended hip from a supine position. This muscle is barely active during gentle walking. Its main task is to balance the trunk on the heads of the femurs, that is, when the femurs are fixed. Therefore, it tightens quite noticeably when a person attempts to balance when standing with the trunk bent backwards. The iliopsoas muscle has its origin at the lumbar spine. It acts to enhance lumbar lordosis, particularly during its initial contraction with the hip extended. It also tilts the pelvis ventrally from the hips.

Origin	Iliac muscle: iliac fossa, anterior inferior iliac spine, iliolumbar ligament, anterior sacroiliac ligament Psoas major muscle: lateral surfaces of vertebrae T12–L5, costal processes of vertebrae L1–L5
Insertion	Just distally of the greater trochanter
Innervation	Iliac muscle: femoral nerve, L2–L3 Psoas major muscle: ventral branches, L2–L4

Functions

 Synergists   Antagonists

Hip joint

Extension

Rectus femoris muscle	Gluteus maximus muscle
Tensor fasciae latae muscle	Semimembranosus muscle
Gluteus medius muscle (ventral part)	Semitendinosus muscle
Sartorius muscle	Biceps femoris muscle
Gracilis muscle	Gluteus medius (dorsal part)
Pectineus muscle	Gluteus minimus (dorsal part)
Adductor muscles (from maximum extension)	Adductor muscles (into the neutral position)
	Pectineus muscle (from maximum flexion)

Hip joint and intervertebral joints (lumbar spine)

Lordosis of the lumbar spine and ventral tilting of the pelvis

Longissimus thoracis muscle	Gluteus maximus muscle (dorsal part)
Iliocostalis lumborum muscle	Biceps femoris muscle (long head)
Rectus femoris muscle	Semimembranosus muscle
Quadratus lumborum muscle (lordosis only)	Semitendinosus muscle
Sartorius muscle	Rectus abdominis muscle
Tensor fasciae latae muscle	

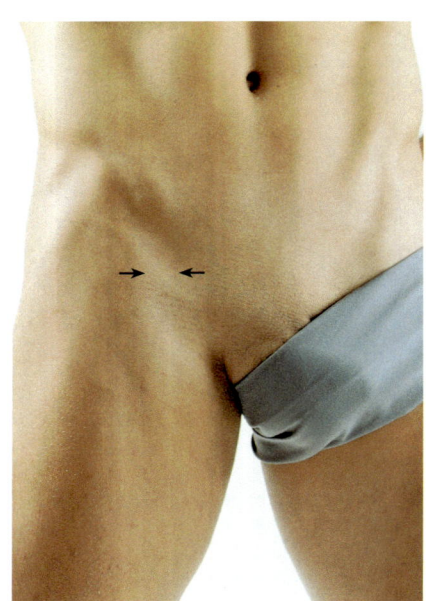

Muscle function testing

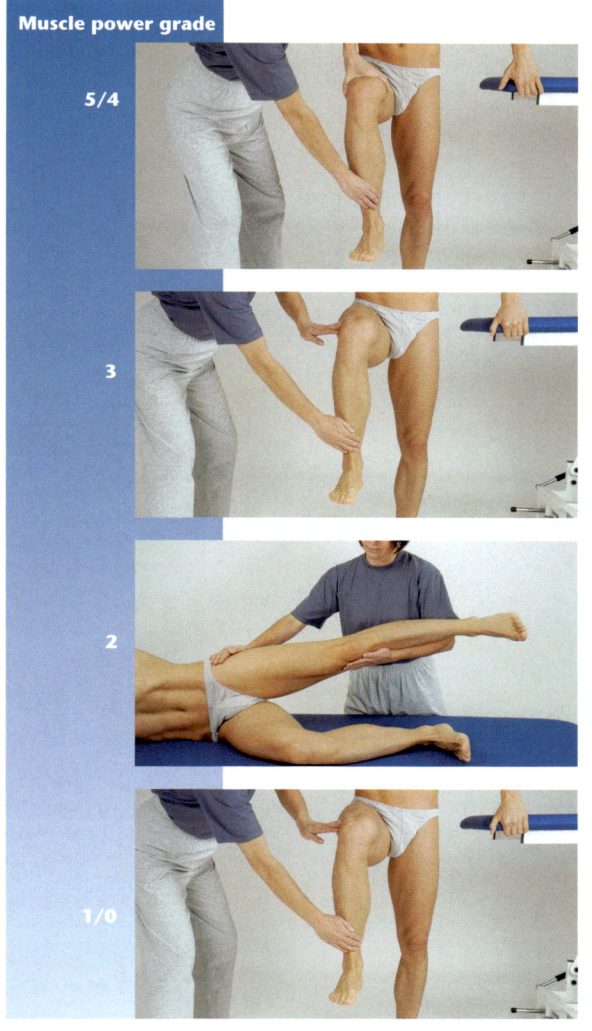

Muscle power grade

5/4

3

2

1/0

Starting position: The patient stands and holds onto the treatment couch.

Test procedure: The examiner applies pressure to the patient's distal thigh in the direction of hip extension.

Instruction: "Pull your knee up towards your nose against my resistance and hold your final position while standing upright."

Starting position: The patient stands and holds onto the treatment couch.

Test procedure: The examiner looks at the movement of the leg and, if necessary, supports the patient to ensure the leg moves in a straight line.

Instruction: "Pull your knee up towards your nose while standing upright."

Starting position: The patient lies on his side. The uppermost leg is extended at the hip.

Test procedure: The examiner looks at the movement of the leg resting on the couch.

Instruction: "Pull your knee up towards your nose."

Starting position: The patient stands and holds onto the treatment couch.

Test procedure: The examiner looks at the movement of the leg.

Instruction: "Try to pull your knee up towards your nose."

Clinical relevance

- Contracture of the iliopsoas muscle permanently enhances lumbar lordosis, which can result in damage to the intervertebral disks in this region.
- Pain on abrupt extension of the flexed right hip may be an indication of appendicitis (psoas sign)

Problems/comments

- If the patient seems too unsteady when standing in the starting position, some of the movement may also be tested with the patient sitting down.
- If both legs are fixed, the iliopsoas can help to lift the trunk from a supine position.

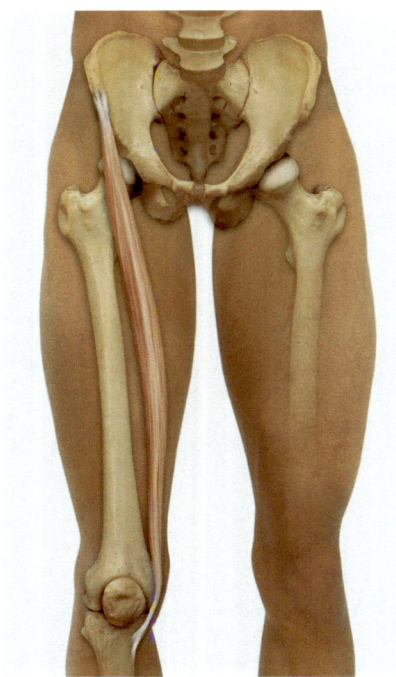

Sartorius muscle

The sartorius (tailor's) muscle flexes the hip and knee joints and causes external rotation of the leg at the hip. Its two functions are united when a person sits cross-legged (like a tailor, hence its name); another example is when the legs are pulled back towards the body when swimming the breast stroke.

Origin Anterior superior iliac spine

Insertion Proximal, medial surface of the tibia via the pes anserinus

Innervation Femoral nerve, L2–L3

Special features The fascia lata holds the sartorius muscle in place along its winding course

Functions

Synergists

Antagonists

Hip joint

Flexion
Iliopsoas muscle
Rectus femoris muscle
Tensor fasciae latae muscle
Gluteus medius muscle (ventral part)
Gracilis muscle
Pectineus muscle
Adductor muscles (bringing the leg back from maximum extension)

Gluteus maximus muscle
Semimembranosus muscle
Semitendinosus muscle
Biceps femoris muscle
Gluteus medius (dorsal part)
Gluteus minimus (dorsal part)
Pectineus (from maximum flexion)
Adductor muscles (bringing the leg back from maximum flexion)

External rotation
Gluteus maximus muscle
Gluteus medius muscle (dorsal part)
Gluteus minimus muscle (dorsal part)
Quadratus femoris muscle
Piriformis muscle
Obturator and gemelli muscles
Pectineus muscle
Adductor muscles (bringing the leg back from maximum internal rotation)

Tensor fasciae latae muscle
Gluteus medius muscle (ventral part)
Gluteus minimus muscle (ventral part)
Adductor muscles (bringing the leg back from maximum external rotation)

Knee joint

Flexion
Biceps femoris muscle
Semitendinosus muscle
Semimembranosus muscle
Gracilis muscle
Gastrocnemius muscle (not with the foot in plantar flexion)

Quadriceps femoris muscle
Gluteus maximus muscle (via the iliotibial tract)
Tensor fasciae latae muscle (via the iliotibial tract)

Internal rotation
Semimembranosus muscle
Semitendinosus muscle
Popliteus muscle
Gracilis muscle
Vastus medialis muscle

Biceps femoris muscle
Gluteus maximus muscle (via the iliotibial tract)
Tensor fasciae latae muscle (via the iliotibial tract)
Vastus lateralis muscle

Muscle function testing

Muscle power grade	
5/4	
3	
2	
1/0	

Starting position: The patient sits on the treatment couch.

Test procedure: With one hand, the examiner applies pressure to the distal thigh in the direction of hip joint extension and adduction. The other hand applies pressure to the medial side of the distal lower leg in the lateral direction and in the direction of knee joint extension.

Instruction: "Force your knee upwards and outwards against my resistance and hold your final position."

Starting position: The patient sits on the treatment couch.

Test procedure: The examiner looks at the movement of the leg.

Instruction: "Move your knee upwards and outwards."

Starting position: The patient lies on his back, with the leg flexed slightly at the hip and knee.

Test procedure: The examiner takes the weight of the leg.

Instruction: "Pull your knee up towards the shoulder on the same side."

Starting position: The patient lies on his back, with the leg flexed slightly at the hip and knee.

Test procedure: The examiner takes the weight of the leg. The examiner palpates the sartorius muscle.

Instruction: "Try to pull your knee up towards the shoulder on the same side."

 Problems/comments

- The sartorius muscle should only be tested in conjunction with the other flexors of the hip joint.

Gluteus medius muscle

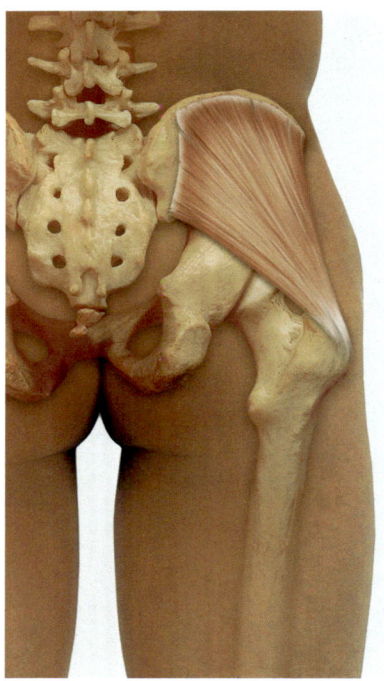

The gluteus medius (middle gluteus) muscle abducts the leg at the hip. During walking, however, it contracts on the supporting leg side, stops the pelvis from tilting on the swinging leg side and even inclines the pelvis slightly towards the supporting leg side, making it easier to for the foot to be lifted off the ground. At the same time, it also has an internal rotatory component, which when it contracts on the side of the fixed supporting leg, brings the swinging leg forward very slightly.

Origin	Wing of the ilium between the anterior and posterior gluteal line
Insertion	Greater trochanter
Innervation	Superior gluteal nerve, L4–S1

Functions

 Synergists   Antagonists

Hip joint

Abduction

Gluteus minimus muscle	Adductor muscles
Tensor fasciae latae muscle	Pectineus muscle
Piriformis muscle (with the hip flexed)	Gracilis muscle
Quadratus femoris muscle (with the hip flexed)	Gluteus maximus muscle (caudal part)
Gluteus maximus muscle (cranial part)	Quadratus femoris muscle (with the hip extended)

Internal rotation (ventral part only)

Tensor fasciae latae muscle	Gluteus maximus muscle
Gluteus minimus muscle (ventral part)	Gluteus minimus muscle (dorsal part)
Adductor muscles (from maximum external rotation)	Quadratus femoris muscle, piriformis muscle
	Pectineus muscle, sartorius muscle
	Adductor muscles (from maximum internal rotation)
	Obturator and gemelli muscles

External rotation (dorsal part only)

Gluteus maximus muscle	Tensor fasciae latae muscle
Gluteus minimus muscle (dorsal part)	Gluteus medius muscle (ventral part)
Quadratus femoris muscle, piriformis muscle	Gluteus minimus muscle (ventral part)
Pectineus muscle, sartorius muscle	Adductor muscles (from maximum external rotation)
Adductor muscles (from maximum internal rotation)	
Obturator and gemelli muscles	

Flexion (ventral part only)

Iliopsoas muscle, rectus femoris muscle	Gluteus maximus muscle, semimembranosus muscle
Tensor fasciae latae muscle	Semitendinosus muscle, biceps femoris muscle
Sartorius muscle, gracilis muscle	Gluteus medius muscle (dorsal part)
Pectineus muscle	Gluteus minimus muscle (dorsal part)
Adductor muscles (from maximum extension)	Adductor muscles (from maximum flexion)
	Pectineus muscle (from maximum flexion)

Extension (dorsal part only)

Gluteus maximus muscle	liopsoas muscle, rectus femoris muscle
Semimembranosus muscle	Tensor fasciae latae muscle
Semitendinosus muscle	Gluteus medius muscle (ventral part)
Biceps femoris muscle	Sartorius muscle, gracilis muscle, pectineus muscle
Gluteus minimus muscle (dorsal part)	
Adductor muscles (from maximum flexion)	Adductor muscles (from maximum extension)
Pectineus muscle (from maximum flexion)	

Hip joint and intervertebral joints (lumbar spine)

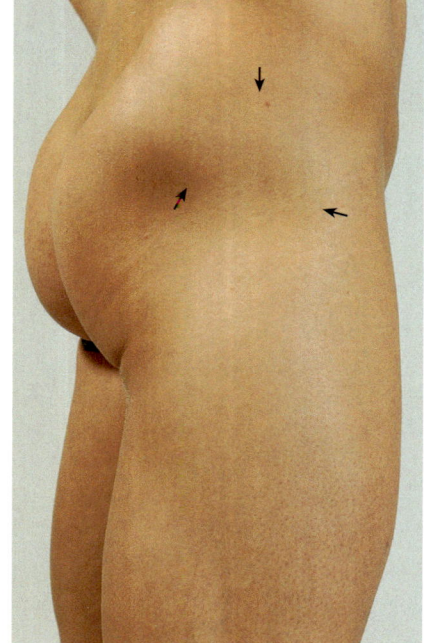

Preventing the pelvis from subsiding on the swinging leg side

Gluteus minimus muscle, tensor fasciae latae muscle	Adductor muscles, pectineus muscle
	Gracilis muscle, quadratus femoris muscle

Muscle function testing

Muscle power grade

5/4

Starting position: The patient lies on his side. The legs are extended at the hip and lie on top of one another, with the uppermost leg flexed 90° at the knee.

Test procedure: The examiner holds the patient's pelvis in place with one hand and with the other applies pressure at the distal thigh of the uppermost leg in the direction of hip joint adduction.

Instruction: "Pull your uppermost leg apart from the other one against my resistance and hold your final position. Do not allow the lower half of the leg to droop."

3

Starting position: The patient lies on his side. The legs are extended at the hip and lie on top of one another, with the uppermost leg flexed 90° at the knee.

Test procedure: The examiner holds the patient's pelvis in place with one hand.

Instruction: "Pull your uppermost leg apart from the other one. Do not allow the lower half of the leg to droop."

2

Starting position: The patient lies on his back, his thighs resting on the treatment couch. The lower legs are flexed 90° and hang down.

Test procedure: The examiner holds the distal thigh in place and takes the weight of the leg.

Instruction: "Pull your leg apart from the other one."

1/0

Starting position: The patient lies on his side. The legs are extended at the hip and lie on top of one another, with the uppermost leg flexed 90° at the knee.

Test procedure: The examiner palpates the gluteus medius muscle.

Instruction: "Try to pull your uppermost leg apart from the other one."

🔖 Clinical relevance

- Intramuscular injections can be administered into this muscle.
- During walking, weakness of the lesser gluteal muscles leads to subsidence of the pelvis on the swinging leg side. As a result, the swinging leg needs to be flexed more powerfully at the hip and knee joints (Trendelenburg's sign).

Problems/comments

- The gluteus medius cannot be tested in isolation from the gluteus minimus muscle.
- In contrast to the test for the tensor fasciae latae muscle, the single-joint hip abductors are tested with the knee flexed. This reduces the leverage of the tensor fasciae latae to a great extent, so that it has only a minimal added effect on the adduction of the hip.
- The elasticity of the rectus femoris muscle should be tested before this investigation.

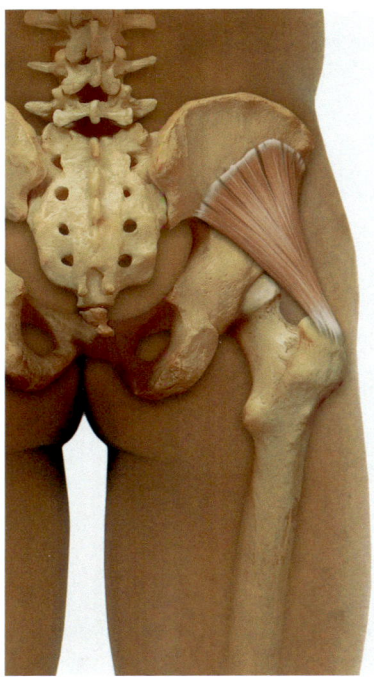

Gluteus minimus muscle

Like the gluteus medius, the gluteus minimus (least gluteus) muscle abducts the leg at the hip. During walking, it also contracts on the supporting leg side, stops the pelvis from tilting on the swinging leg side, inclines the pelvis slightly towards the supporting leg side, and brings the swinging leg forward by means of its internal rotatory component.

Origin   Wing of the ilium between the anterior and posterior gluteal line
Insertion   Greater trochanter
Innervation   Superior gluteal nerve, L4–S1

Functions

Synergists   Antagonists

Hip joint

Abduction

Synergists	Antagonists
Gluteus medius muscle	Adductor muscles
Tensor fasciae latae muscle	Pectineus muscle
Piriformis muscle (with the hip flexed)	Gracilis muscle
Obturator and gemelli muscles (with the hip flexed)	Gluteus maximus muscle (caudal part)
Quadratus femoris muscle (with the hip flexed)	Quadratus femoris muscle (with the hip extended)
Gluteus maximus muscle (cranial part)	

Internal rotation (ventral part only)

Synergists	Antagonists
Tensor fasciae latae muscle	Gluteus maximus muscle
Gluteus medius muscle (ventral part)	Gluteus minimus muscle (dorsal part)
Adductor muscles (from maximum external rotation)	Gluteus medius muscle (dorsal part)
	Quadratus femoris muscle
	Piriformis muscle
	Pectineus muscle
	Sartorius muscle
	Adductor muscles (from maximum internal rotation)
	Obturator and gemelli muscles

External rotation (dorsal part only)

Synergists	Antagonists
Gluteus maximus muscle	Tensor fasciae latae muscle
Gluteus medius muscle (dorsal part)	Gluteus medius muscle (ventral part)
Quadratus femoris muscle	Gluteus minimus muscle (ventral part)
Piriformis muscle	Adductor muscles (from maximum external rotation)
Pectineus muscle	
Sartorius muscle	
Adductor muscles (from maximum internal rotation)	
Obturator and gemelli muscles	

Extension (dorsal part only)

Synergists	Antagonists
Gluteus maximus muscle	Iliopsoas muscle
Semimembranosus muscle	Rectus femoris muscle
Semitendinosus muscle	Tensor fasciae latae muscle
Biceps femoris muscle	Gluteus medius muscle (ventral part)
Gluteus medius (dorsal part)	Sartorius muscle
Adductor muscles (from maximum flexion)	Gracilis muscle
Pectineus muscle (from maximum flexion)	Pectineus muscle
	Adductor muscles (from maximum extension)

Hip joint and intervertebral joints (lumbar spine)

Preventing the pelvis from subsiding on the swinging leg side

Synergists	Antagonists
Gluteus medius muscle	Adductor muscles
Tensor fasciae latae muscle	Pectineus muscle
	Gracilis muscle
	Quadratus femoris muscle

Muscle power grade

5/4

3

2

1/0

Muscle function testing

Starting position: The patient lies on his side. The legs are extended at the hip and lie on top of one another, with the uppermost leg flexed 90° at the knee.

Test procedure: The examiner holds the patient's pelvis in place with one hand and with the other applies pressure at the distal thigh of the uppermost leg in the direction of hip joint adduction.

Instruction: ""Pull your uppermost leg apart from the other one against my resistance and hold your final position. Do not allow the lower half of the leg to droop."

Starting position: The patient lies on his side. The legs are extended at the hip and lie on top of one another, with the uppermost leg flexed 90° at the knee.

Test procedure: The examiner holds the patient's pelvis in place with one hand.

Instruction: ""Pull your uppermost leg apart from the other one. Do not allow the lower half of the leg to droop."

Starting position: The patient lies on his back, his thighs resting on the treatment couch. The lower legs are flexed 90° and hang down.

Test procedure: The examiner holds the distal thigh in place and takes the weight of the leg.

Instruction: "Pull your leg apart from the other one."

Starting position: The patient lies on his side. The legs are extended at the hip and lie on top of one another, with the uppermost leg flexed 90° at the knee.

Test procedure: The examiner holds the patient's pelvis in place with one hand.

Instruction: "Try to pull your uppermost leg apart from the other one. Do not allow the lower half of the leg to droop."

Clinical relevance

- During walking, weakness of the lesser gluteal muscles leads to subsidence of the pelvis on the swinging leg side. As a result, the swinging leg needs to be flexed more powerfully at the hip and knee joints (Trendelenburg's sign).

! Problems/comments

- The gluteus minimus cannot be tested in isolation from the gluteus medius muscle.

- In contrast to the test for the tensor fasciae latae muscle, the single-joint hip abductors are tested with the knee flexed. This reduces the leverage of the tensor fasciae latae to a great extent, so that it has only a minimal added effect on the adduction of the hip.

- The elasticity of the rectus femoris muscle should be tested before this investigation.

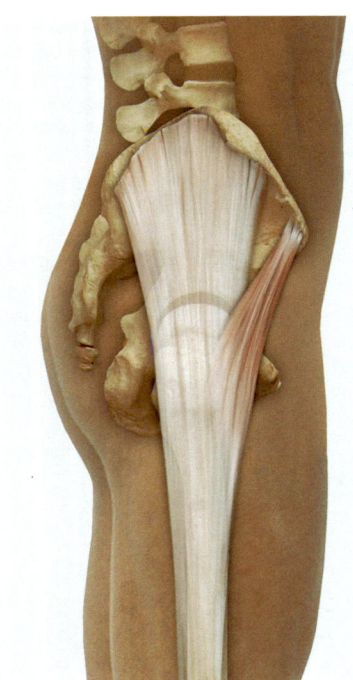

Tensor fasciae latae muscle

The tensor fasciae latae muscle (tensor muscle of the fascia lata) flexes and abducts the leg at the hip. However, an adequate number of far more powerful synergists are available for both these functions. However, it is also important as a powerful internal rotator of the hip, in which capacity it can even counter the effect of the gluteus maximus on the iliotibial tract. At the knee, its extending effect (conducted via the iliotibial tract) is powerful enough to replace – at least in part – that of a paralyzed quadriceps femoris muscle, if necessary. Tightening of the iliotibial tract produces a tension band effect which counteracts flexion stress on the femur of the supporting leg. In this respect, it acts in conjunction with the adductor magnus muscle.

Origin   Iliac crest, near the anterior superior iliac spine
Insertion   Iliotibial tract in the middle third of the femur
Innervation   Superior gluteal nerve, L4–L5

Functions

 Synergists   Antagonists

Hip joint

Flexion
Iliopsoas muscle   Gluteus maximus muscle
Rectus femoris muscle   Semimembranosus muscle
Gluteus medius muscle (ventral part)   Semitendinosus muscle
Sartorius muscle   Biceps femoris muscle
Gracilis muscle   Gluteus medius muscle (dorsal part)
Adductor muscles (from maximum extension)   Gluteus minimus muscle (dorsal part)
Pectineus muscle   Pectineus muscle (from maximum flexion)
                                                Adductor muscles (from maximum flexion)

**Preventing the pelvis from subsiding
on the swinging leg side**
Gluteus medius muscle (supporting leg side)   Adductor muscles (supporting leg side)
Gluteus minimus muscle (supporting leg side)   Pectineus muscle (supporting leg side)
Quadratus lumborum muscle (swinging leg side)   Gracilis muscle (supporting leg side)
Iliocostalis lumborum muscle (swinging leg side) Quadratus femoris muscle (supporting leg side)
Longissimus lumborum muscle (swinging leg side)

Internal rotation
Gluteus minimus muscle (ventral part)   Gluteus maximus muscle
Gluteus medius muscle (ventral part)   Gluteus minimus muscle (dorsal part)
Adductor muscles (from maximum external   Gluteus medius muscle (dorsal part)
rotation)   Quadratus femoris muscle
                                                Piriformis muscle
                                                Pectineus muscle
                                                Sartorius muscle
                                                Adductor muscles (from maximum internal rotation)
                                                Obturator and gemelli muscles

Abduction
Gluteus medius muscle   Adductor muscles
Gluteus minimus muscle   Pectineus muscle
Piriformis muscle (with the hip flexed)   Gracilis muscle
Gluteus maximus muscle (cranial part)   Quadratus femoris muscle (with the hip extended)
Obturator and gemelli muscles (with the hip flexed) Gluteus maximus muscle (caudal part)
Quadratus femoris muscle (with the hip flexed) Rectus femoris muscle

Knee joint

Extension (via the iliotibial tract)
Quadriceps femoris muscle   Biceps femoris muscle
Gluteus maximus muscle (via the iliotibial tract) Semitendinosus muscle
                                                Semimembranosus muscle
                                                Gracilis muscle
                                                Sartorius muscle
                                                Gastrocnemius muscle
                                                Popliteus muscle

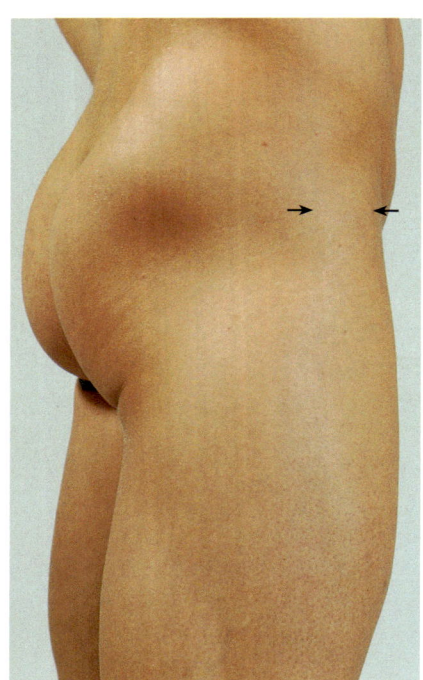

Muscle function testing

Muscle power grade	
5/4	

Starting position: The patient lies on his side. The legs are extended and lie on top of one another.

Test procedure: The examiner holds the patient's pelvis in place with one hand and with the other applies pressure at the distal thigh of the uppermost leg in the direction of hip joint adduction.

Instruction: "Pull your uppermost leg apart from the other one against my resistance and hold your final position. Do not allow the lower half of the leg to droop."

3

Starting position: The patient lies on his side. The legs are extended and lie on top of one another.

Test procedure: The examiner holds the patient's pelvis in place with one hand.

Instruction: "Pull your uppermost leg apart from the other one. Do not allow the lower half of the leg to droop."

2

Starting position: The patient lies on his back.

Test procedure: The examiner takes the weight of the leg by supporting the lower leg.

Instruction: "Pull your leg apart from the other one."

1/0

Starting position: The patient lies on his side. The legs are extended and lie on top of one another.

Test procedure: The examiner palpates the tensor fasciae latae muscle.

Instruction: "Try to pull your uppermost leg apart from the other one."

 Problems/comments

- Abduction at the hip is produced not only by the tensor fasciae latae but also by the gluteus medius and gluteus minimus muscles.

145

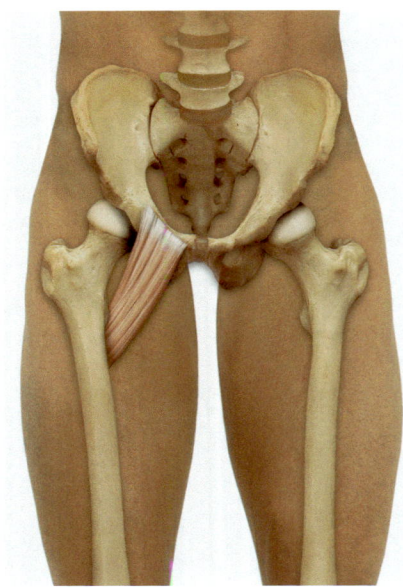

Pectineus muscle

The pectineus (pectineal) muscle adducts the thigh from any position of the hip joint. When the hip joint is extended, this muscle flexes; however, when the hip is powerfully flexed, it becomes an extensor, for example when a person begins to stand up from a very deep sitting position. This muscle also has an external rotatory component, as when the legs are crossed while sitting.

Origin	Pecten of pubis
Insertion	Pectineal line of the femur, distally of the lesser trochanter
Innervation	Femoral nerve, L2–L3

Functions

Synergists

Antagonists

Hip joint

Adduction

Adductor muscles
Gracilis muscle
Gluteus maximus muscle (caudal part)
Quadratus femoris muscle (with the hip extended)

Gluteus medius muscle
Gluteus minimus muscle
Piriformis muscle (with the hip flexed)
Tensor fasciae latae muscle
Gluteus maximus muscle (cranial part)
Quadratus femoris muscle (with the hip flexed)
Obturator and gemelli muscles (with the hip flexed)

External rotation

Gluteus maximus muscle
Gluteus minimus muscle (dorsal part)
Gluteus medius muscle (dorsal part)
Piriformis muscle
Quadratus femoris muscle
Obturator and gemelli muscles
Sartorius muscle
Adductor muscles (from maximum internal rotation)

Tensor fasciae latae muscle
Gluteus minimus muscle (ventral part)
Gluteus medius muscle (ventral part)
Adductor muscles (from maximum external rotation)

Flexion

Iliopsoas muscle
Rectus femoris muscle
Tensor fasciae latae muscle
Gluteus medius muscle (ventral part)
Sartorius muscle
Gracilis muscle
Adductor muscles (from maximum extension)

Gluteus maximus muscle
Semimembranosus muscle
Semitendinosus muscle
Biceps femoris muscle
Gluteus medius muscle (dorsal part)
Adductor muscles (from maximum flexion)

Extension (from maximum flexion)

Gluteus maximus muscle
Semimembranosus muscle
Semitendinosus muscle
Biceps femoris muscle
Gluteus medius muscle (dorsal part)
Gluteus minimus muscle (dorsal part)
Adductor muscles (from maximum flexion)

liopsoas muscle
Rectus femoris muscle
Tensor fasciae latae muscle
Gluteus medius muscle (ventral part)
Sartorius muscle
Gracilis muscle
Adductor muscles (from maximum extension)

Muscle function testing

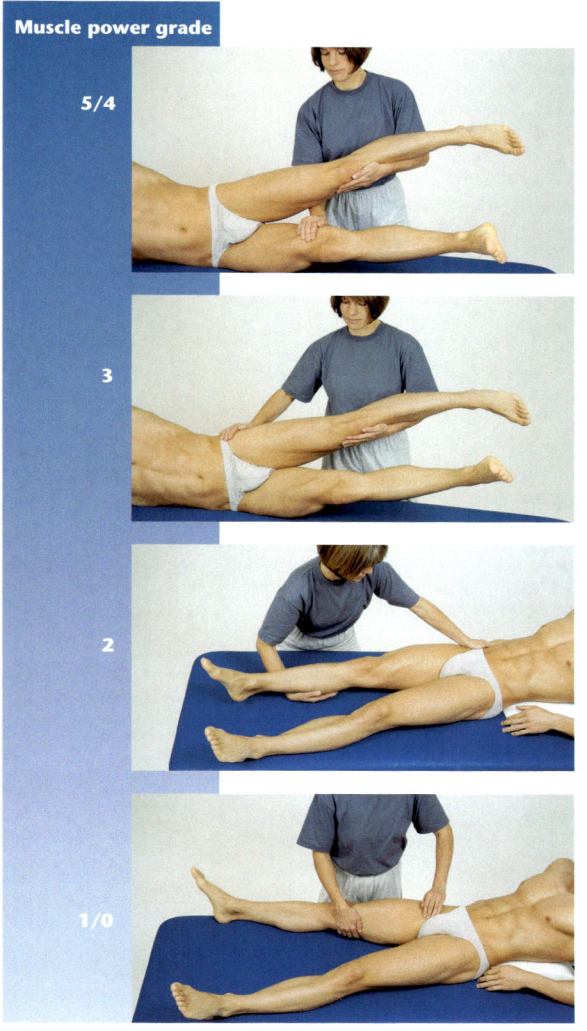

Muscle power grade

5/4

3

2

1/0

Starting position: The patient lies on his side. The leg being tested rests on the couch, the other is abducted at the hip.

Test procedure: The examiner holds the patient's uppermost leg with one hand and with the other applies downward pressure onto the distal thigh of the leg which is resting on the couch.

Instruction: "Working against my resistance, lift your leg off the couch in front of the other leg. Hold your final position."

Starting position: The patient lies on his side. The leg being tested rests on the couch, the other is abducted at the hip.

Test procedure: The examiner holds the patient's uppermost leg with one hand.

Instruction: "Lift your leg off the couch in front of the other leg."

Starting position: The patient lies on his back. He lifts his upper body slightly, supporting himself in that position. Both legs are abducted at the hip.

Test procedure: The examiner holds the lower leg in place and takes the weight of the leg.

Instruction: "Pull your leg towards the other one."

Starting position: The patient lies on his back. He lifts his upper body slightly, supporting himself in that position. Both legs are abducted at the hip.

Test procedure: The examiner palpates the pectineus muscle.

Instruction: "Try to pull your leg towards the other one."

⚠️ Problems/comments

- The adductors, which also have a hip flexor function, are at the forefront in this test.
- The pectineus muscle should not be tested in isolation, but in conjunction with the other adductors.

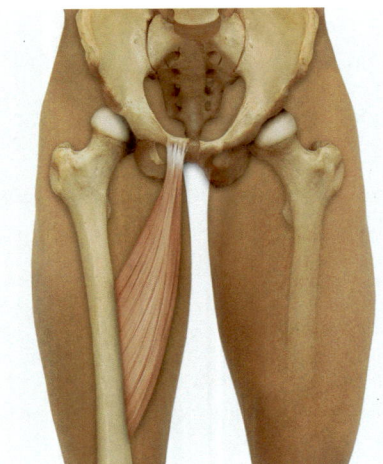

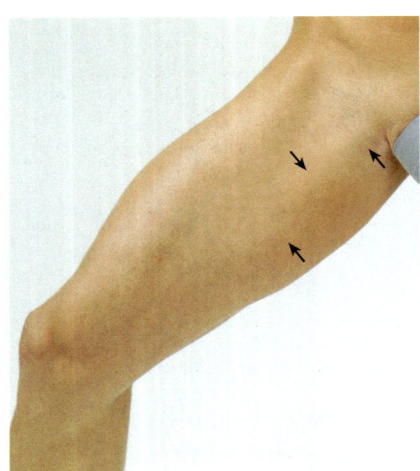

Adductor longus muscle

The action of the adductor longus (long adductor) muscle is equivalent to that of the adductor magnus. It brings the leg back to its neutral position from flexion or marked extension at the hip, and also from extreme rotation in both directions.

Origin   Superior pubic ramus
Insertion   Pectineal line
              Linea aspera (femoral crest)
Innervation   Obturator nerve, anterior branch, L2–L4

Functions

 Synergists   Antagonists

Hip joint

Synergists	Antagonists
Adduction	
Adductor brevis muscle	Gluteus medius muscle
Adductor magnus muscle	Gluteus minimus muscle
Pectineus muscle	Piriformis muscle (with the hip flexed)
Gracilis muscle	Tensor fasciae latae muscle
Gluteus maximus muscle (caudal part)	Gluteus maximus muscle (cranial part)
Rectus femoris muscle	Obturator and gemelli muscles (with the hip flexed)
Quadratus femoris muscle (with the hip extended)	Quadratus femoris muscle (with the hip flexed)
External rotation (from maximum internal rotation)	
Gluteus maximus muscle	Tensor fasciae latae muscle
Gluteus minimus muscle (dorsal part)	Gluteus minimus muscle (ventral part)
Gluteus medius muscle (dorsal part)	Gluteus medius muscle (ventral part)
Quadratus femoris muscle	Adductor muscles (from maximum external rotation)
Piriformis muscle	
Pectineus muscle	
Sartorius muscle	
Adductor brevis muscle	
Adductor magnus muscle	
Obturator and gemelli muscles	
Internal rotation (from maximum external rotation)	
Tensor fasciae latae muscle	Gluteus maximus muscle
Gluteus minimus muscle (ventral part)	Gluteus minimus muscle (dorsal part)
Gluteus medius muscle (ventral part)	Gluteus medius muscle (dorsal part)
Adductor brevis muscle	Quadratus femoris muscle, piriformis muscle
Adductor magnus muscle	Pectineus muscle, sartorius muscle
	Adductor muscles (from maximum internal rotation)
	Obturator and gemelli muscles
Extension (from maximum flexion)	
Gluteus maximus muscle, semimembranosus muscle	Iliopsoas muscle, rectus femoris muscle
Semitendinosus muscle, biceps femoris muscle	Tensor fasciae latae muscle
Gluteus medius muscle (dorsal part)	Gluteus medius muscle (ventral part)
Gluteus minimus muscle (dorsal part)	Sartorius muscle, gracilis muscle, pectineus muscle
Adductor brevis muscle	Adductor muscles (from maximum extension)
Adductor magnus muscle	
Pectineus muscle	
Flexion (from maximum extension)	
Iliopsoas muscle, rectus femoris muscle	Gluteus maximus muscle, semimembranosus muscle
Tensor fasciae latae muscle	Semitendinosus muscle, biceps femoris muscle
Gluteus medius muscle (ventral part)	Gluteus medius muscle (dorsal part)
Sartorius muscle, gracilis muscle, pectineus muscle	Gluteus minimus muscle (dorsal part)
Adductor brevis muscle	Adductor muscles (from maximum flexion)
	Pectineus muscle (from maximum flexion)

Muscle function testing

Muscle power grade

5/4

Starting position: The patient lies on his side. The leg being tested rests on the couch, the other is abducted at the hip.

Test procedure: The examiner holds the patient's uppermost leg with one hand and with the other applies downward pressure onto the distal thigh of the leg which is resting on the couch.

Instruction: "Working against my resistance, lift your leg off the couch in front of the other leg. Hold your final position."

3

Starting position: The patient lies on his side. The leg being tested rests on the couch, the other is abducted at the hip.

Test procedure: The examiner holds the patient's uppermost leg with one hand.

Instruction: "Lift your leg off the couch in front of the other leg."

2

Starting position: The patient lies on his back. He lifts his upper body slightly, supporting himself in that position. Both legs are abducted at the hip.

Test procedure: The examiner holds the lower leg in place and takes the weight of the leg.

Instruction: "Pull your leg towards the other one."

1/0

Starting position: The patient lies on his back. He lifts his upper body slightly, supporting himself in that position. Both legs are abducted at the hip.

Test procedure: The examiner palpates the adductor longus muscle.

Instruction: "Try to pull your leg towards the other one."

Clinical relevance

- Damage to the adductors is a common sports injury. A classic example of this is groin strain (adductor rupture) in soccer.

Problems/comments

- The adductors, which also have a hip flexor function, are at the forefront in this test.
- The adductor longus muscle should not be tested in isolation, but in conjunction with the other adductors.

149

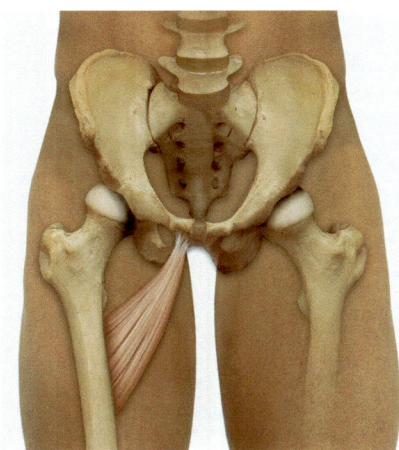

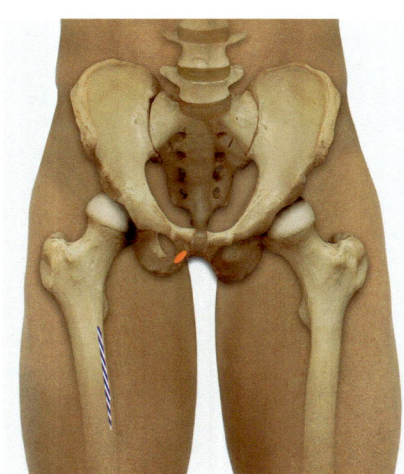

Adductor brevis muscle

The adductor brevis (short adductor) muscle adducts the leg at the hip. It brings the leg back to its neutral position from flexion or marked extension at the hip, and also from extreme rotation in both directions.

Origin   Inferior pubic ramus
Insertion   Cranial part of the linea aspera (femoral crest), laterally of its medial lip
Innervation   Obturator nerve, anterior branch, L2–L4

Functions

 Synergists

Antagonists

Hip joint

Adduction

Adductor longus muscle	Gluteus medius muscle
Adductor magnus muscle	Gluteus minimus muscle
Pectineus muscle	Piriformis muscle (with the hip flexed)
Gracilis muscle	Tensor fasciae latae muscle
Gluteus maximus muscle (caudal part)	Gluteus maximus muscle (cranial part)
Rectus femoris muscle	Obturator and gemelli muscles (with the hip flexed)
Quadratus femoris muscle (with the hip extended)	Quadratus femoris muscle (with the hip flexed)

External rotation (from maximum internal rotation)

Gluteus maximus muscle	Tensor fasciae latae muscle
Gluteus minimus muscle (dorsal part)	Gluteus medius muscle (ventral part)
Gluteus medius muscle (dorsal part)	Gluteus minimus muscle (ventral part)
Quadratus femoris muscle	Adductor muscles (from maximum external rotation)
Piriformis muscle	
Pectineus muscle	
Sartorius muscle	
Adductor longus muscle	
Adductor magnus muscle	
Obturator and gemelli muscles	

Internal rotation (from maximum external rotation)

Tensor fasciae latae muscle	Gluteus maximus muscle
Gluteus medius muscle (ventral part)	Gluteus minimus muscle (dorsal part)
Gluteus minimus muscle (ventral part)	Gluteus medius muscle (dorsal part)
Adductor longus muscle	Quadratus femoris muscle, piriformis muscle
Adductor magnus muscle	Pectineus muscle, sartorius muscle
	Adductor muscles (from maximum internal rotation)
	Obturator and gemelli muscles

Extension (from maximum flexion)

Gluteus maximus muscle, semimembranosus muscle	Iliopsoas muscle, rectus femoris muscle
Semitendinosus muscle, biceps femoris muscle	Tensor fasciae latae muscle
Gluteus medius muscle (dorsal part)	Gluteus medius muscle (ventral part)
Gluteus minimus muscle (dorsal part)	Sartorius muscle, gracilis muscle, pectineus muscle
Adductor longus muscle	Adductor muscles (from maximum extension)
Adductor magnus muscle	
Pectineus muscle	

Flexion (from maximum extension)

Iliopsoas muscle, rectus femoris muscle	Gluteus maximus muscle, semimembranosus muscle
Tensor fasciae latae muscle	Semitendinosus muscle, biceps femoris muscle
Gluteus medius muscle (ventral part)	Gluteus medius muscle (dorsal part)
Sartorius muscle, gracilis muscle, pectineus muscle	Gluteus minimus muscle (dorsal part)
Adductor longus muscle	Adductor muscles (from maximum flexion)
Adductor magnus muscle	Pectineus muscle (from maximum flexion)

Muscle function testing

Muscle power grade

5/4

3

2

1/0

Starting position: The patient lies on his side. The leg being tested rests on the couch, the other is abducted at the hip.

Test procedure: The examiner holds the patient's uppermost leg with one hand and with the other applies downward pressure onto the distal thigh of the leg which is resting on the couch.

Instruction: "Working against my resistance, lift your leg off the couch in front of the other leg. Hold your final position."

Starting position: The patient lies on his side. The leg being tested rests on the couch, the other is abducted at the hip.

Test procedure: The examiner holds the patient's uppermost leg with one hand.

Instruction: "Lift your leg off the couch in front of the other leg."

Starting position: The patient lies on his back. He lifts his upper body slightly, supporting himself in that position. Both legs are abducted at the hip.

Test procedure: The examiner holds the lower leg in place and takes the weight of the leg.

Instruction: "Pull your leg towards the other one."

Starting position: The patient lies on his back. He lifts his upper body slightly, supporting himself in that position. Both legs are abducted at the hip.

Test procedure: The examiner holds the uppermost leg with one hand.

Instruction: "Try to pull your leg towards the other one."

 Clinical relevance

• Damage to the adductors is a common sports injury. A classic example of this is groin strain (adductor rupture) in soccer.

Problems/comments

• The adductor brevis muscle cannot be palpated.

• The adductors, which also have a hip flexor function, are at the forefront in this test.

• The adductor brevis muscle should not be tested in isolation, but in conjunction with the other adductors.

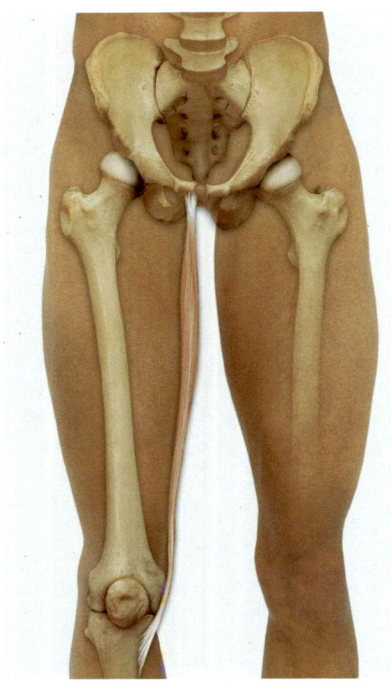

Gracilis muscle

The gracilis muscle flexes the leg at the hip and the knee, for example at the initiation of the swing phase of the lead leg during walking. It also has an adductor effect on the thigh. When the knee is flexed, this muscle functions as an internal rotator.

Origin	Inferior pubic ramus
Insertion	Pes anserinus: proximal end of the tibia, just below the medial epicondyle
Innervation	Obturator nerve, anterior branch, L2–L4

Functions

 Synergists   Antagonists

Hip joint

Adduction

Synergists	Antagonists
Adductor muscles	Gluteus medius muscle
Pectineus muscle	Gluteus minimus muscle
Gluteus maximus muscle (caudal part)	Piriformis muscle (with the hip flexed)
Rectus femoris muscle	Tensor fasciae latae muscle
Quadratus femoris muscle (with the hip extended)	Gluteus maximus muscle (cranial part)
	Obturator and gemelli muscles (with the hip flexed)
	Quadratus femoris muscle (with the hip flexed)

Flexion (from maximum extension)

Synergists	Antagonists
Iliopsoas muscle	Gluteus maximus muscle
Rectus femoris muscle	Semimembranosus muscle
Tensor fasciae latae muscle	Semitendinosus muscle
Gluteus medius muscle (ventral part)	Biceps femoris muscle
Sartorius muscle	Gluteus medius muscle (dorsal part)
Pectineus muscle	Gluteus minimus muscle (dorsal part)
Adductor muscles	Adductor muscles (from maximum flexion)
	Pectineus muscle (from maximum flexion)

Knee joint

Flexion

Synergists	Antagonists
Biceps femoris muscle	Quadriceps femoris muscle
Semitendinosus muscle	Gluteus maximus muscle (via the iliotibial tract)
Sartorius muscle	Tensor fasciae latae muscle (via the iliotibial tract)
Gastrocnemius muscle (not with the foot in plantar flexion)	

Internal rotation

Synergists	Antagonists
Semimembranosus muscle	Biceps femoris muscle
Semitendinosus muscle	Gluteus maximus muscle (via the iliotibial tract)
Sartorius muscle	Tensor fasciae latae muscle (via the iliotibial tract)
Popliteus muscle	Vastus lateralis muscle
Vastus medialis muscle	

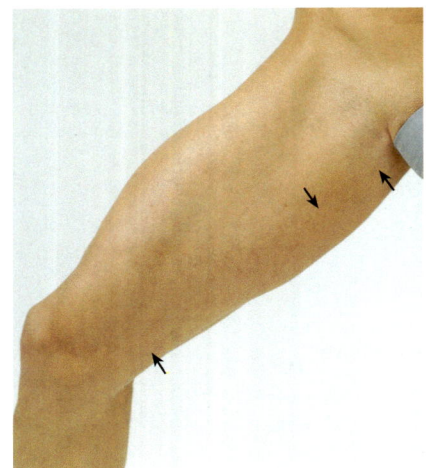

Muscle function testing

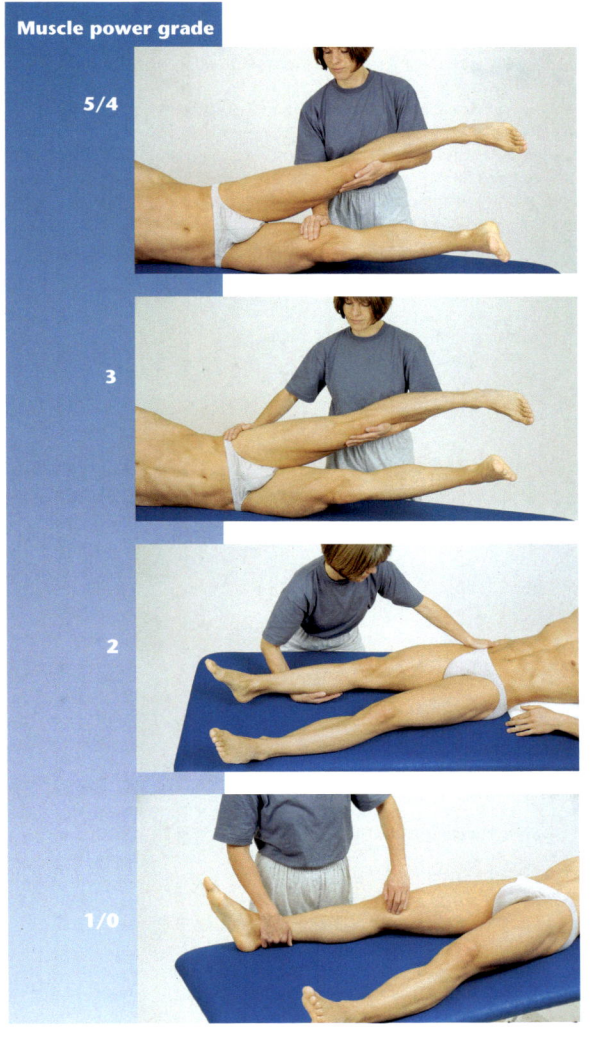

Muscle power grade

5/4

3

2

1/0

Starting position: The patient lies on his side. The leg being tested rests on the couch, the other is abducted at the hip.

Test procedure: The examiner holds the patient's uppermost leg with one hand and with the other applies downward pressure onto the distal thigh of the leg which is resting on the couch.

Instruction: "Working against my resistance, lift your leg off the couch in front of the other leg. Hold your final position."

Starting position: The patient lies on his side. The leg being tested rests on the couch, the other is abducted at the hip.

Test procedure: The examiner holds the patient's uppermost leg with one hand.

Instruction: "Lift your leg off the couch in front of the other leg."

Starting position: The patient lies on his back. He lifts his upper body slightly, supporting himself in that position. Both legs are abducted at the hip.

Test procedure: The examiner holds the lower leg in place and takes the weight of the leg.

Instruction: "Pull your leg towards the other one."

Starting position: The patient lies on his back. He lifts his upper body slightly, supporting himself in that position. Both legs are abducted at the hip.

Test procedure: The examiner palpates the gracilis muscle close to its insertion at the pes anserinus.

Instruction: "Try to pull your leg towards the other one."

 Clinical relevance

- The gracilis muscle is frequently damaged during sports.

Problems/comments

- The adductors, which also have a hip flexor function, are at the forefront in this test.
- The gracilis muscle should not be tested in isolation, but in conjunction with the other adductors.

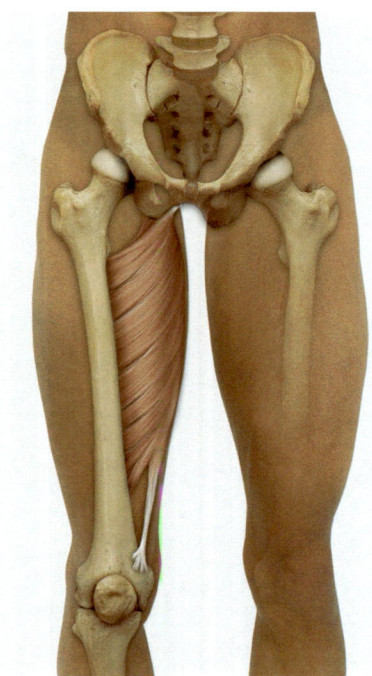

Adductor magnus muscle

The adductor magnus (great adductor) muscle acts as an adductor on the swinging leg. When a person stands with legs wide apart, it stops the legs from splaying out as a result of the weight of the body. Its rotational components depend on the flexion of the hip and are insignificant. During walking, the muscle contracts in the supporting leg and, together with the lesser gluteal muscles, balances the pelvis on the head of the femur, aligning the body's centre of gravity. It also has an important action on the femur, balancing out the lateral flexion stress on this bone, acting in conjunction with the tensor fasciae latae muscle. It brings the leg back to its neutral position from flexion or marked extension at the hip, and also from extreme rotation in both directions.

Origin	Inferior pubic ramus, ischial ramus, ischial tuberosity
Insertion	Ventral part: linea aspera (femoral crest) Dorsal part: adductor tubercle of the femur
Innervation	Ventral part: obturator nerve, L2–L4 Dorsal part: sciatic nerve, L4–S1
Special features	An opening formed between the two parts of the muscle, the adductor hiatus, allows passage of the femoral vessels to the popliteal fossa.

Functions

 Synergists   Antagonists

Hip joint

Adduction

Adductor longus muscle	Gluteus medius muscle
Adductor brevis muscle	Gluteus minimus muscle
Pectineus muscle	Piriformis muscle (with the hip flexed)
Gracilis muscle	Tensor fasciae latae muscle
Gluteus maximus muscle (caudal part)	Gluteus maximus muscle (cranial part)
Rectus femoris muscle	Obturator and gemelli muscles (with the
Quadratus femoris muscle (with the hip extended)	hip flexed)
	Quadratus femoris muscle (with the hip flexed)

Extension (from maximum flexion)

Gluteus maximus muscle	Iliopsoas muscle
Semimembranosus muscle	Rectus femoris muscle
Semitendinosus muscle	Tensor fasciae latae muscle
Biceps femoris muscle	Gluteus medius muscle (ventral part)
Gluteus medius muscle (dorsal part)	Sartorius muscle
Gluteus minimus muscle (dorsal part)	Gracilis muscle
Adductor longus muscle	Pectineus muscle
Adductor brevis muscle	Adductor muscles (from maximum extension)
Pectineus muscle	

Flexion (from maximum extension)

Iliopsoas muscle	Gluteus maximus muscle
Rectus femoris muscle	Semimembranosus muscle
Tensor fasciae latae muscle	Semitendinosus muscle
Gluteus medius muscle (ventral part)	Biceps femoris muscle
Sartorius muscle	Gluteus medius muscle (dorsal part)
Gracilis muscle	Gluteus minimus muscle (dorsal part)
Pectineus muscle	Adductor muscles (from maximum flexion)
Adductor longus muscle	Pectineus muscle (from maximum flexion)
Adductor brevis muscle	

Reduction of flexion stress on the femur

Tensor fasciae latae muscle
Gluteus maximus muscle
Adductor muscles

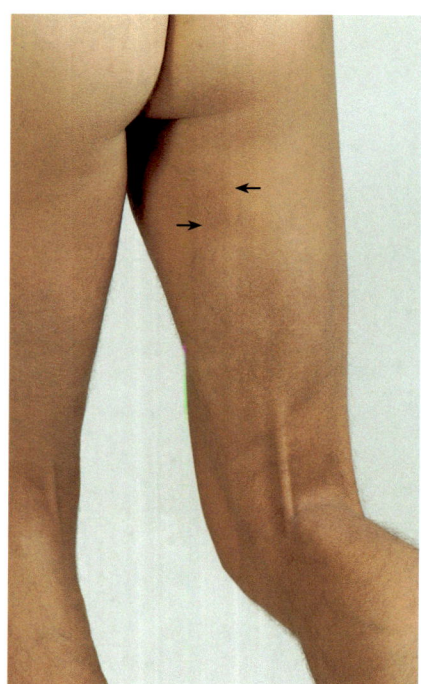

Muscle function testing

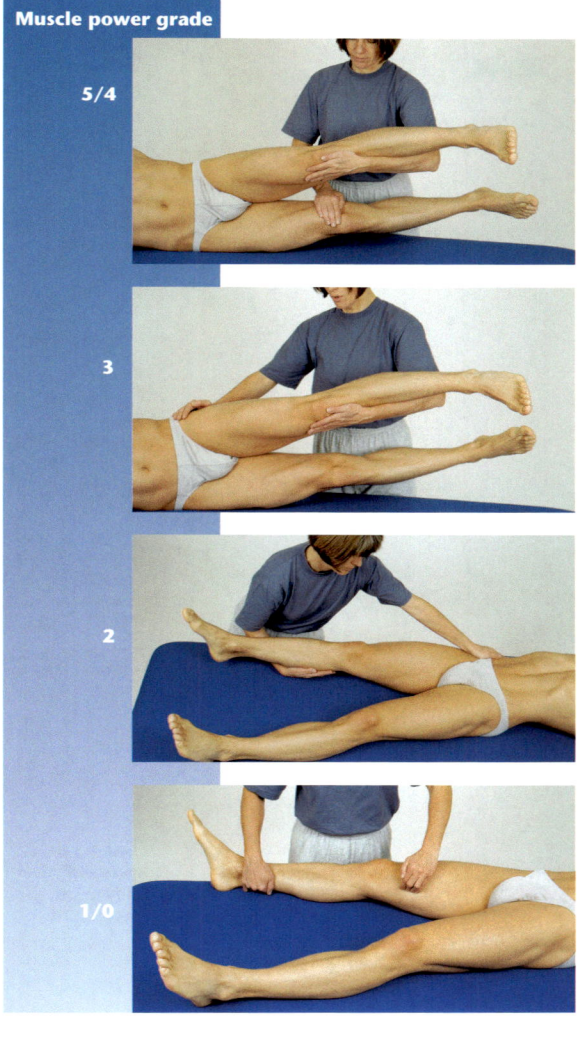

Muscle power grade

5/4

3

2

1/0

Starting position: The patient lies on his side. Both legs are extended, with the leg being tested resting on the couch and the other abducted at the hip.

Test procedure: The examiner holds the patient's uppermost leg with one hand and with the other applies downward pressure onto the distal thigh of the leg which is resting on the couch.

Instruction: "Working against my resistance, lift your leg off the couch and move it behind the other leg. Hold your final position."

Starting position: The patient lies on his side. Both legs are extended, with the leg being tested resting on the couch and the other abducted at the hip.

Test procedure: The examiner holds the patient's uppermost leg with one hand.

Instruction: "Lift your leg off the couch and move it behind of the other leg."

Starting position: The patient lies on his back. He lifts his upper body slightly, supporting himself in that position. Both legs are abducted at the hip.

Test procedure: The examiner holds the lower leg in place and takes the weight of the leg.

Instruction: "Pull your leg towards the other one."

Starting position: The patient lies on his back. He lifts his upper body slightly, supporting himself in that position. Both legs are abducted at the hip.

Test procedure: The examiner palpates the adductor magnus muscle.

Instruction: "Try to pull your leg towards the other one."

Clinical relevance

- Damage to the adductors is a common sports injury. A classic example of this is groin strain (adductor rupture) in soccer.

! Problems/comments

- The adductors, which also have a hip flexor function, are at the forefront in this test.
- The adductor brevis muscle should not be tested in isolation, but in conjunction with the other adductors.

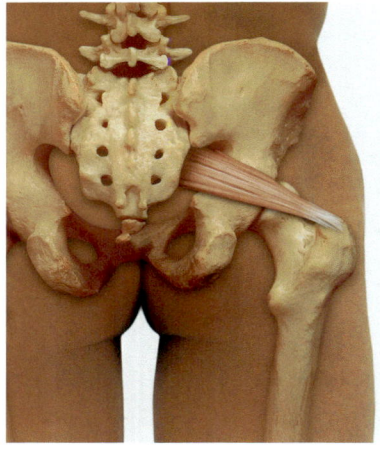

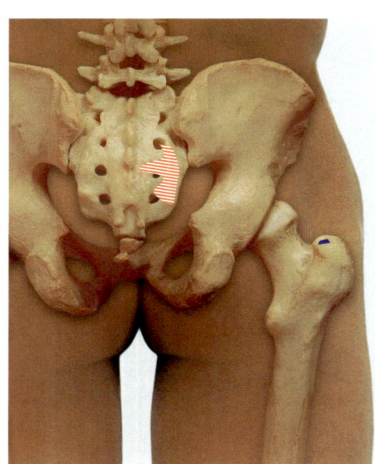

Piriformis muscle

Like all the "short external rotators," the piriformis (piriform) muscle is an external rotator of the thigh when the hip is extended, and an abductor when it is flexed.

Origin	Pelvic surface of the sacral bone
Insertion	Upper boundary of the greater trochanter
Innervation	Sciatic nerve or direct branches of the sacral plexus, L5–S2
Special features	In a small percentage of individuals, parts of the sciatic nerve run through the piriformis muscle

Functions

 Synergists   Antagonists

Hip joint

External rotation (with the hip extended)

Synergists	Antagonists
Gluteus maximus muscle	Tensor fasciae latae muscle
Gluteus medius muscle (dorsal part)	Gluteus minimus muscle (ventral part)
Gluteus minimus muscle (dorsal part)	Gluteus medius muscle (ventral part)
Quadratus femoris muscle	Adductor muscles (from maximum external rotation)
Obturator and gemelli muscles	
Pectineus muscle	
Sartorius muscle	
Adductor muscles (from maximum internal rotation)	

Abduction (with the hip flexed)

Synergists	Antagonists
Gluteus medius muscle	Adductor muscles
Gluteus minimus muscle	Pectineus muscle
Tensor fasciae latae muscle	Gracilis muscle
Obturator and gemelli muscles	Gluteus maximus muscle (caudal part)
Quadratus femoris muscle	Quadratus femoris muscle (with the hip extended)

Gemellus superior muscle

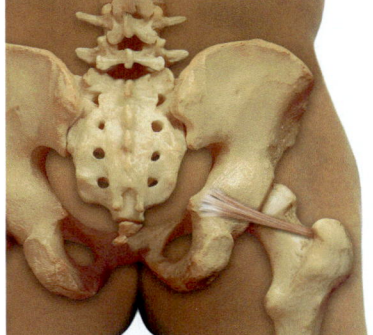

The gemellus superior muscle, acting in conjunction with the gemellus inferior, can aid the obturatorius internus muscle in producing external rotation. In doing so, the power of these muscles is not reduced by a fulcrum, as with the obturatorius internus. The gemellus superior also acts as an abductor when the hip joint is flexed. Its adductor effect on the extended hip is negligible.

Origin	Sciatic spine
Insertion	Trochanteric fossa
Innervation	Direct branches of the sacral plexus, L5–S2

Functions

 Synergists   Antagonists

Hip joint

External rotation (with the hip extended)

Gluteus maximus muscle
Gluteus medius muscle (dorsal part)
Gluteus minimus muscle (dorsal part)
Quadratus femoris muscle
Piriformis muscle
Obturator muscles
Gemellus inferior muscle
Pectineus muscle
Sartorius muscle
Adductor muscles (from maximum internal rotation)

Tensor fasciae latae muscle
Gluteus minimus muscle (ventral part)
Gluteus medius muscle (ventral part)
Adductor muscles (from maximum external rotation)

Abduction (with the hip flexed)

Gluteus medius muscle
Gluteus minimus muscle
Piriformis muscle
Tensor fasciae latae muscle
Gluteus maximum (cranial part)
Obturator muscles
Gemellus inferior muscle
Quadratus femoris muscle (with the hip flexed)

Adductor muscles
Pectineus muscle
Gracilis muscle
Gluteus maximus muscle (caudal part)
Quadratus femoris muscle (with the hip extended)

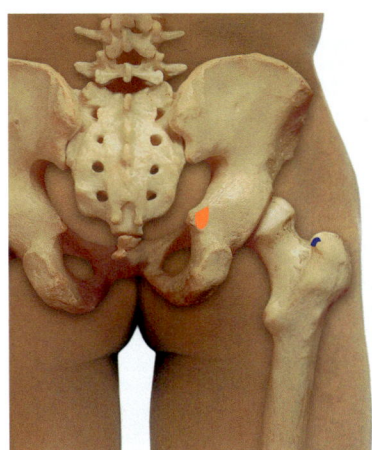

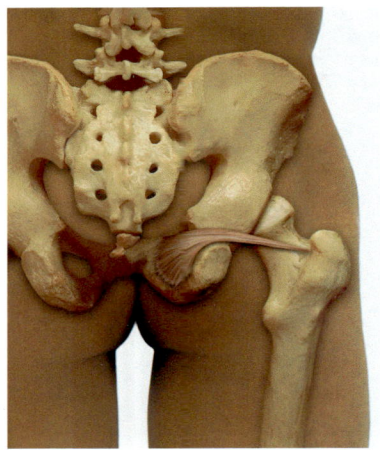

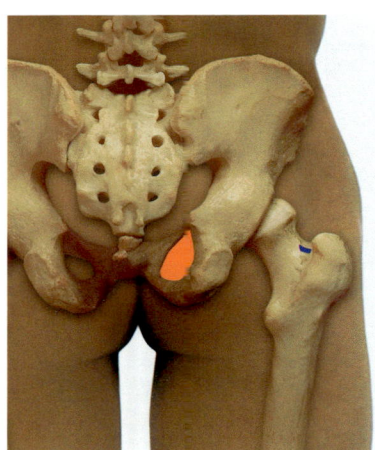

Obturatorius internus muscle

When the hip is extended, the obturatorius internus (internal obturator) muscle rotates the thigh outwards. However, when the hip is flexed at a right angle, as when sitting, this muscle becomes an abductor of the femur. Its adductor effect on the extended hip is negligible.

Origin	Margin of the obturator foramen (medial surface), Obturator membrane
Insertion	Trochanteric fossa
Innervation	Internal obturator nerve, L5–S2
Special features	The lesser ischial notch acts as a fulcrum for this muscle

Functions

 Synergists

 Antagonists

Hip joint

External rotation (with the hip extended)

Gluteus maximus muscle
Gluteus medius muscle (dorsal part)
Gluteus minimus muscle (dorsal part)
Quadratus femoris muscle
Piriformis muscle
Obturatorius externus muscle
Gemelli muscles
Pectineus muscle
Sartorius muscle
Adductor muscles (from maximum internal rotation)

Tensor fasciae latae muscle
Gluteus minimus muscle (ventral part)
Gluteus medius muscle (ventral part)
Adductor muscles (from maximum external rotation)

Abduction (with the hip flexed)

Gluteus medius muscle
Gluteus minimus muscle
Piriformis muscle
Tensor fasciae latae muscle
Gluteus maximum (cranial part)
Obturatorius externus muscle (with the hip flexed)
Gemelli muscles (with the hip flexed)
Quadratus femoris muscle (with the hip flexed)

Adductor muscles
Pectineus muscle
Gracilis muscle
Gluteus maximus muscle (caudal part)
Quadratus femoris muscle (with the hip extended)

Gemellus inferior muscle

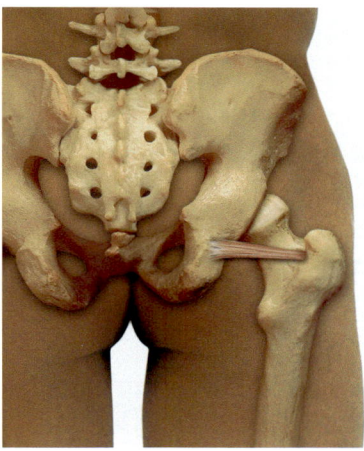

The gemellus inferior muscle, acting in conjunction with the gemellus superior, can aid the obturatorius internus muscle in producing external rotation. In doing so, the power of these muscles is not reduced by a fulcrum, as with the obturatorius internus. It also acts as an abductor when the hip joint is flexed. Its adductor effect on the extended hip is negligible.

Origin	Ischial tuberosity
Insertion	Trochanteric fossa
Innervation	Internal obturator nerve or quadrate femoral nerve or pudendal nerve or sciatic nerve, L5–S2

Functions

 Synergists

 Antagonists

Hip joint

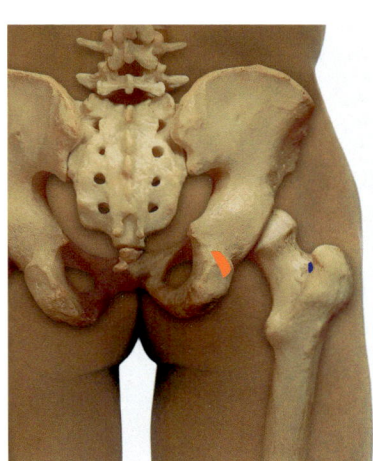

External rotation (with the hip extended)

Synergists	Antagonists
Gluteus maximus muscle	Tensor fasciae latae muscle
Gluteus medius muscle (dorsal part)	Gluteus minimus muscle (ventral part)
Gluteus minimus muscle (dorsal part)	Gluteus medius muscle (ventral part)
Quadratus femoris muscle	Adductor muscles (from maximum external rotation)
Piriformis muscle	
Obturator muscles	
Gemellus superior muscle	
Pectineus muscle	
Sartorius muscle	
Adductor muscles (from maximum internal rotation)	

Abduction (with the hip flexed)

Synergists	Antagonists
Gluteus medius muscle	Adductor muscles
Gluteus minimus muscle	Pectineus muscle
Piriformis muscle	Gracilis muscle
Tensor fasciae latae muscle	Gluteus maximus muscle (caudal part)
Gluteus maximum (cranial part)	Quadratus femoris muscle (with the hip extended)
Obturator muscles (with the hip flexed)	
Gemellus superior muscle(with the hip flexed)	
Quadratus femoris muscle (with the hip flexed)	

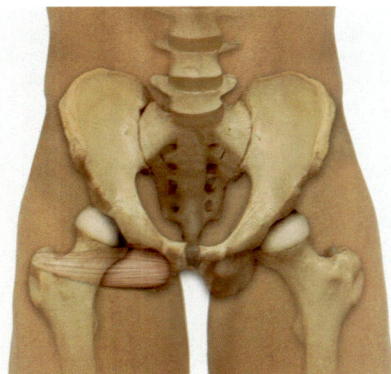

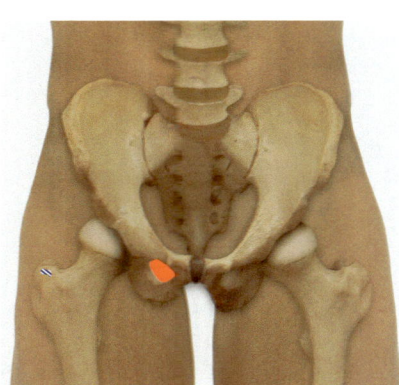

Obturatorius externus muscle

Like the obturatorius internus, the obturatorius externus (external obturator) muscle also rotates the thigh outwards when the hip joint is extended. When the hip is flexed, it abducts the femur. Its adductor effect on the extended hip is negligible. Acting in conjunction with the quadratus femoris, it fixes the neck of the femur and reduces its tendency to subside under the weight of the body.

Origin   Lateral margin of the obturator foramen (medial surface)
                  Obturator membrane

Insertion   Trochanteric fossa

Innervation   Obturator nerve, anterior branch, L2–L4

Functions

 Synergists   Antagonists

Hip joint

External rotation (with the hip extended)
Gluteus maximus muscle   Tensor fasciae latae muscle
Gluteus medius muscle (dorsal part)   Gluteus minimus muscle (ventral part)
Gluteus minimus muscle (dorsal part)   Gluteus medius muscle (ventral part)
Quadratus femoris muscle   Adductor muscles (from maximum external
Piriformis muscle   rotation)
Obturatorius internus muscle
Gemelli muscles
Pectineus muscle
Sartorius muscle
Adductor muscles (from maximum internal
rotation)

Abduction (with the hip flexed)
Gluteus medius muscle   Adductor muscles
Gluteus minimus muscle   Pectineus muscle
Piriformis muscle   Gracilis muscle
Tensor fasciae latae muscle   Gluteus maximus muscle (caudal part)
Gluteus maximum (cranial part)   Quadratus femoris muscle (with the hip
Obturatorius internus muscle (with the hip   extended)
flexed)
Gemelli muscles (with the hip flexed)
Quadratus femoris muscle (with the hip flexed)

Quadratus femoris muscle

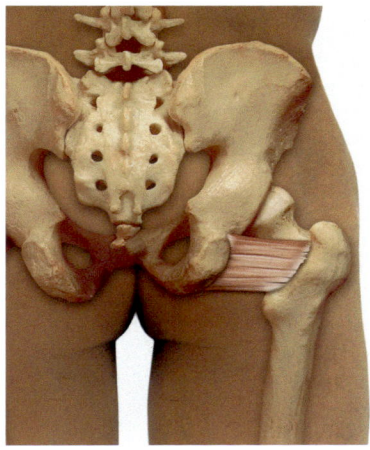

The quadratus femoris muscle (quadrate muscle of thigh) is a very powerful and efficient external rotator of the thigh when the knee joint is extended. When the knee is flexed, it abducts the swinging leg. Its contraction reduces the flexion stress on the neck of the femur, thus slowing the tendency of the head of the femur to subside with age.

Origin	Lateral margin of the ischial tuberosity
Insertion	Intertrochanteric crest
Innervation	Quadrate femoral nerve, rarely also the sciatic nerve, L5–S2

Functions

 Synergists Antagonists

Hip joint

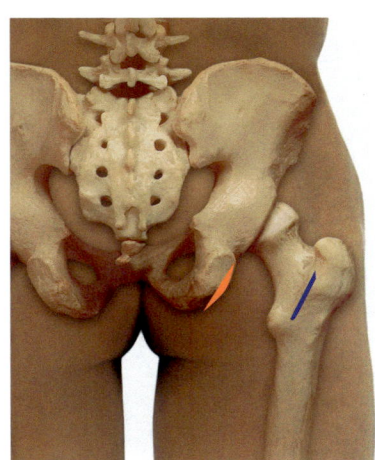

External rotation (with the thigh extended)

Synergists	Antagonists
Gluteus maximus muscle	Tensor fasciae latae muscle
Gluteus medius muscle (dorsal part)	Gluteus minimus muscle (ventral part)
Gluteus minimus muscle (dorsal part)	Gluteus medius muscle (ventral part)
Piriformis muscle	Adductor muscles (from maximum external rotation)
Obturator and gemelli muscles	
Pectineus muscle	
Sartorius muscle	
Adductor muscles (from maximum internal rotation)	

Abduction (with the hip flexed)

Synergists	Antagonists
Obturator and gemelli muscles	Adductor muscles
Piriformis muscle	Pectineus muscle
Tensor fasciae latae muscle	Gracilis muscle
Gluteus minimus muscle	Gluteus maximus muscle (caudal part)
Gluteus medius muscle	

Adduction (with the hip extended)

Synergists	Antagonists
Adductor muscles	Obturator and gemelli muscles
Pectineus muscle	Piriformis muscle
Gracilis muscle	Tensor fasciae latae muscle
Gluteus maximus muscle (caudal part)	Gluteus minimus muscle
	Gluteus medius muscle

The following muscles are tested jointly:

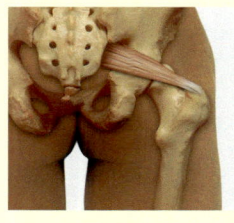

Piriformis muscle, p. 156

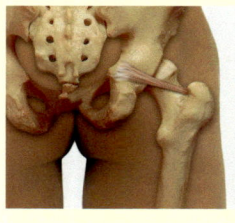

Gemellus superior muscle, p. 157

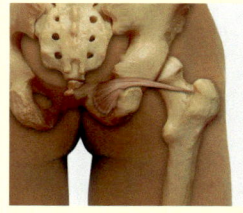

Obturatorius internus muscle, p. 158

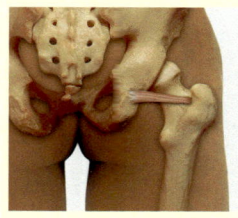

Gemellus inferior muscle, p. 159

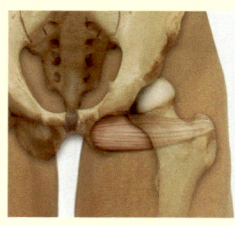

Obturatorius externus muscle, p. 160

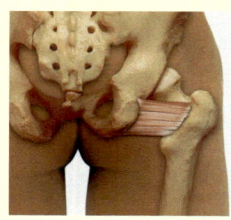

Quadratus femoris muscle, p. 161

Muscle function testing

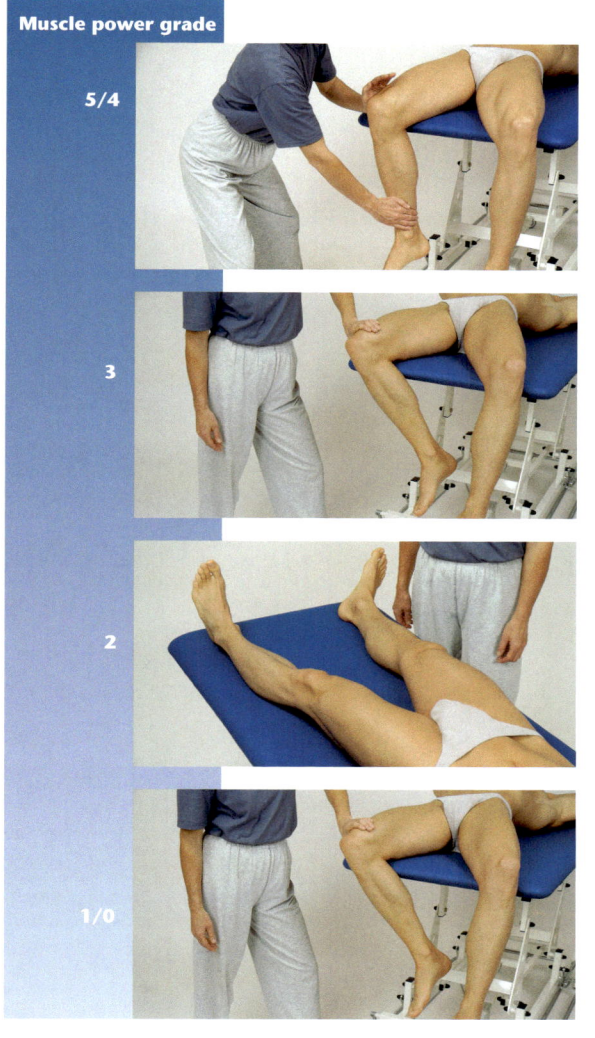

Muscle power grade

5/4

Starting position: The patient lies on his back, with both lower legs hanging over the edge of the couch.

Test procedure: The examiner holds the patient's distal thigh in place with one hand and with the other applies pressure to the medial side of the distal lower leg in the direction of internal rotation at the hip.

Instruction: "Rotate your lower leg inwards against my resistance and hold your final position, without allowing your thigh to shift."

3

Starting position: The patient lies on his back, with both lower legs hanging over the edge of the couch.

Test procedure: The examiner holds the patient's distal thigh in place with one hand.

Instruction: "Rotate your lower leg inwards without allowing your thigh to shift."

2

Starting position: The patient lies on his back.

Test procedure: The examiner looks at the movement of the leg.

Instruction: "Rotate your leg outwards."

1/0

Starting position: The patient lies on his back, with both lower legs hanging over the edge of the couch.

Test procedure: The examiner holds the patient's distal thigh in place with one hand and looks at the movement of the leg.

Instruction: "Try to rotate your lower leg inwards."

 Problems/comments

• Since the test for muscle power 4 and 5 involves a lever action across the knee, ensure that resistance is applied carefully.

Stretch tests

Tensor fasciae latae muscle

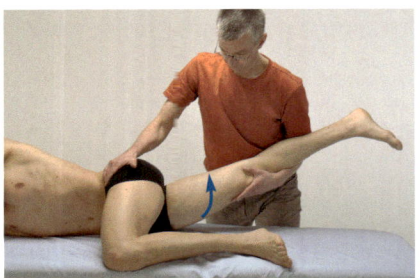

Method
The therapist moves the patient's thigh to achieve maximum extension, adduction and external rotation of the hip.

Finding
Muscle shortening is present if the movement cannot be performed to its full extent and if the endpoint feels soft and elastic.
The patient reports a stretching sensation along the course of the muscle.

Piriformis muscle

Method
The hip joint of the side being tested is flexed 45°. The therapist moves the patient's thigh to achieve maximum internal rotation and adduction of the hip.

Finding
Muscle shortening is present if the movement cannot be performed to its full extent and if the endpoint feels soft and elastic.
The patient reports a stretching sensation along the course of the muscle.

Adductor muscles

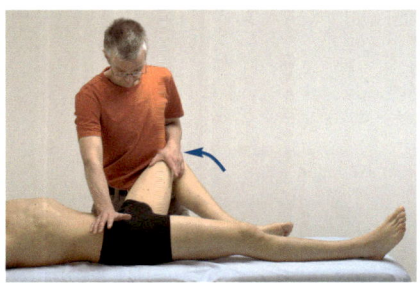

Method
The therapist holds the opposite side of the patient's pelvis in place from the ventral side. The hip joint of the side being tested is flexed 45°. The therapist moves the thigh to achieve maximum abduction of the hip.

Finding
Muscle shortening is present if the movement cannot be performed to its full extent and if the endpoint feels soft and elastic.
The patient reports a stretching sensation along the course of the muscle.
To perform a stretch test of the gracilis muscle, the knee should be extended.

Iliopsoas muscle

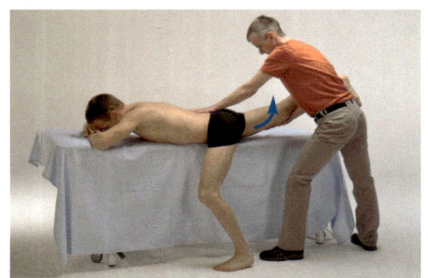

Method
The leg not being tested is flexed at the hip to the maximum extend and stands on the floor. The therapist moves the patient's thigh to achieve maximum extension and internal rotation of the hip.

Finding
Muscle shortening is present if the movement cannot be performed to its full extent and if the endpoint feels soft and elastic.
The patient reports a stretching sensation along the course of the muscle.

3 Lower extremity

Muscles of the knee

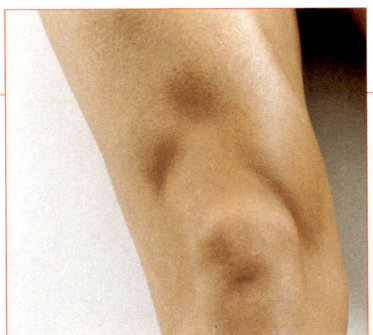

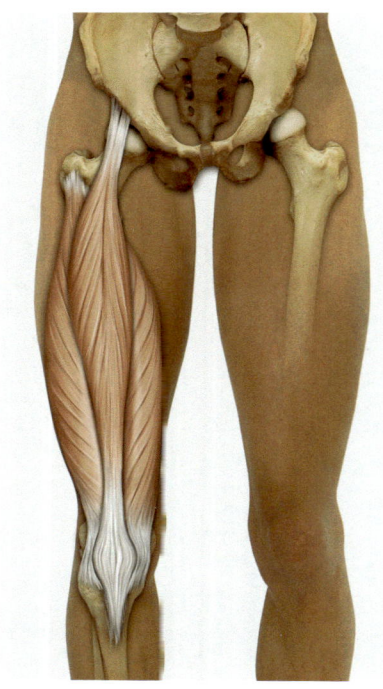

Quadriceps femoris muscle

With the two-joint rectus femoris muscle, the quadriceps femoris muscle (quadriceps muscle of the thigh) flexes the leg at the hip and extends the knee with all of its heads. In doing so, it also holds the patella in place over the patellar surface of the femur. The rotational components of its lateral parts balance out to a great extent, so that the muscle as a whole does not exert any rotational forces on the knee during walking.

Origin	Vastus lateralis muscle: linea aspera (femoral crest), greater trochanter, intertrochanteric line
	Vastus medialis muscle: linea aspera (femoral crest), intertrochanteric line, tendons of the adductor magnus and longus muscle
	Vastus intermedius muscle: upper two thirds of the shaft of the femur
	Rectus femoris muscle, straight head: anterior inferior iliac spine
	Rectus femoris muscle, reflexed head: supra-acetabular groove
Insertion	Jointly via the patellar ligament at the tibial tuberosity
Innervation	Femoral nerve, L2–L4
Special features	The superficial fibers of these muscles are pinnate, the deep ones are parallel
	The patella is a sesamoid bone in the quadriceps femoris tendon and acts as a fulcrum when the knee is flexed. It improves the leverage or torque of the quadriceps femoris muscle by guiding the tendon at some distance to the axis of movement
	The patellar ligament forms part of the tendon of this muscle

Functions

 Synergists    Antagonists

Hip joint

Flexion (rectus femoris muscle only)

Iliopsoas muscle	Gluteus maximus muscle
Tensor fasciae latae muscle	Semimembranosus muscle
Sartorius muscle	Semitendinosus muscle
Gluteus medius muscle (ventral part)	Biceps femoris muscle
Gracilis muscle	Gluteus medius muscle (dorsal part)
Pectineus muscle	Gluteus minimus muscle (dorsal part)
Adductor muscles (from maximum extension)	Pectineus muscle (from maximum flexion)
	Adductor muscles (from maximum flexion)

Knee joint

Extension

Gluteus maximus muscle (via the iliotibial tract)	Biceps femoris muscle
Tensor fasciae latae (via the iliotibial tract)	Semitendinosus muscle
	Semimembranosus muscle
	Sartorius muscle
	Gracilis muscle
	Gastrocnemius muscle (not with the foot in plantar flexion)

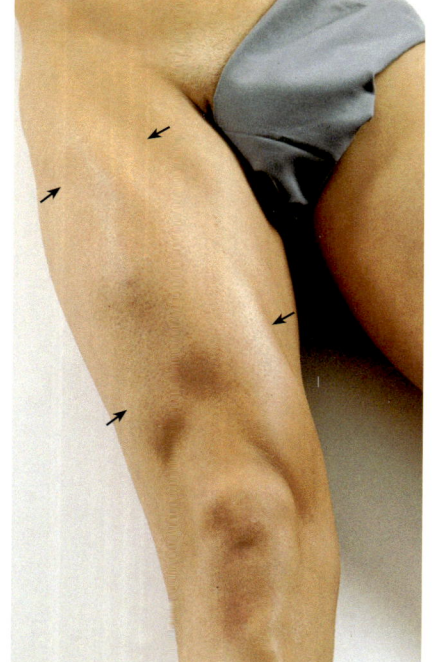

Muscle function testing

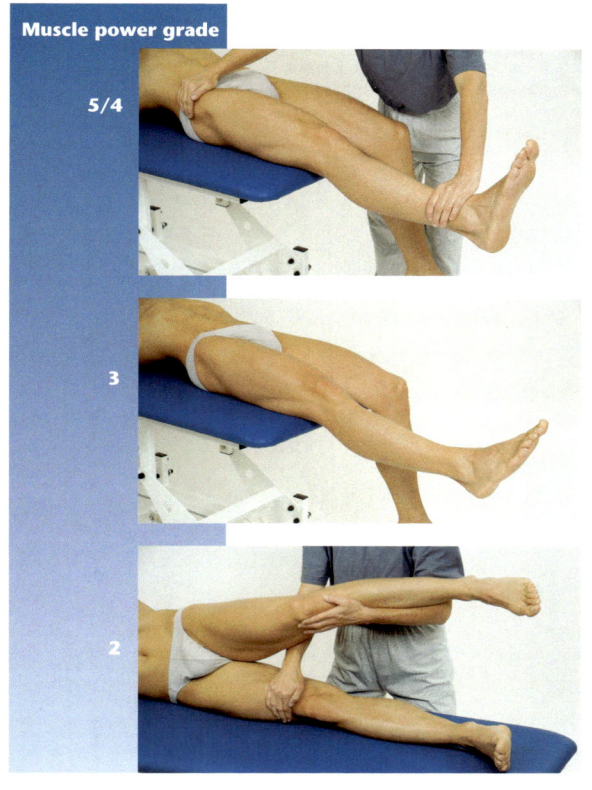

Muscle power grade

5/4

3

2

Starting position: The patient lies on his back, with the lower legs hanging over the narrow end of the treatment couch.

Test procedure: The examiner holds the patient's pelvis in place with one hand and with the other applies pressure to the distal lower leg in the direction of flexion at the knee.

Instruction: "Extend your leg at the knee against my resistance and hold your final position."

Starting position: The patient lies on his back, with the lower legs hanging over the narrow end of the treatment couch.

Test procedure: The examiner looks at the movement of the leg.

Instruction: "Extend your leg at the knee."

Starting position: The patient lies on his side. Both legs are extended at the hip. The knee of the leg being tested, which is resting on the couch, is flexed 90°.

Test procedure: The examiner holds the uppermost leg and fixes the thigh of the other leg in place.

Instruction: "Extend your leg at the knee."

Clinical relevance

- The term "jumper's knee" refers to an insertion tendinopathy of the quadriceps femoris tendon at the patella. It is also known as "patellar tendinopathy."

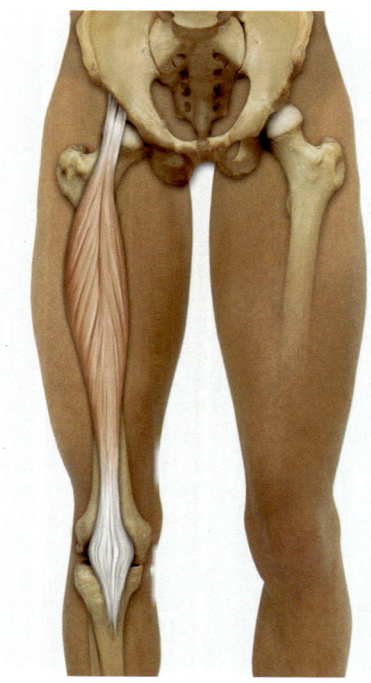

Rectus femoris muscle

The rectus femoris muscle flexes the hip joint powerfully and extends the knee. Both functions are used when the swinging leg is brought forward. The rectus femoris is the main hip flexor during slow walking. In contrast, it shows virtually no tightening when a person stands still on both legs. This muscle has a negligibly weak action as an adductor of the hip.

Origin	Rectus femoris muscle, straight head: anterior inferior iliac spine
	Rectus femoris muscle, reflexed head: supra-acetabular groove
Insertion	Via the patellar ligament at the tibial tuberosity
Innervation	Femoral nerve, L2–L4

Functions

 Synergists

Antagonists

Hip joint

Flexion (rectus femoris muscle only)

Iliopsoas muscle	Gluteus maximus muscle
Tensor fasciae latae muscle	Semimembranosus muscle
Gluteus medius muscle (ventral part)	Semitendinosus muscle
Sartorius muscle	Biceps femoris muscle
Gracilis muscle	Gluteus medius muscle (dorsal part)
Pectineus muscle	Gluteus minimus muscle (dorsal part)
Adductor muscles (from maximum extension)	Adductor muscles (from maximum flexion)

Knee joint

Extension

Vastus lateralis muscle	Biceps femoris muscle
Vastus intermedius muscle	Semitendinosus muscle
Vastus medialis muscle	Semimembranosus muscle
Gluteus maximus muscle (via the iliotibial tract)	Sartorius muscle
Tensor fasciae latae (via the iliotibial tract)	Gracilis muscle
	Gastrocnemius muscle (not with the foot in plantar flexion)
	Popliteus muscle

Hip joint and intervertebral joints (lumbar spine)

Ventral tilting of the pelvis

Iliopsoas muscle	Gluteus maximus muscle
Longissimus thoracis muscle	Rectus abdominis muscle
Iliocostalis lumborum muscle	Biceps femoris muscle
Quadratus lumborum muscle (lordosis only)	Semitendinosus muscle
Sartorius muscle	Semimembranosus muscle
Tensor fasciae latae muscle	

Muscle function testing

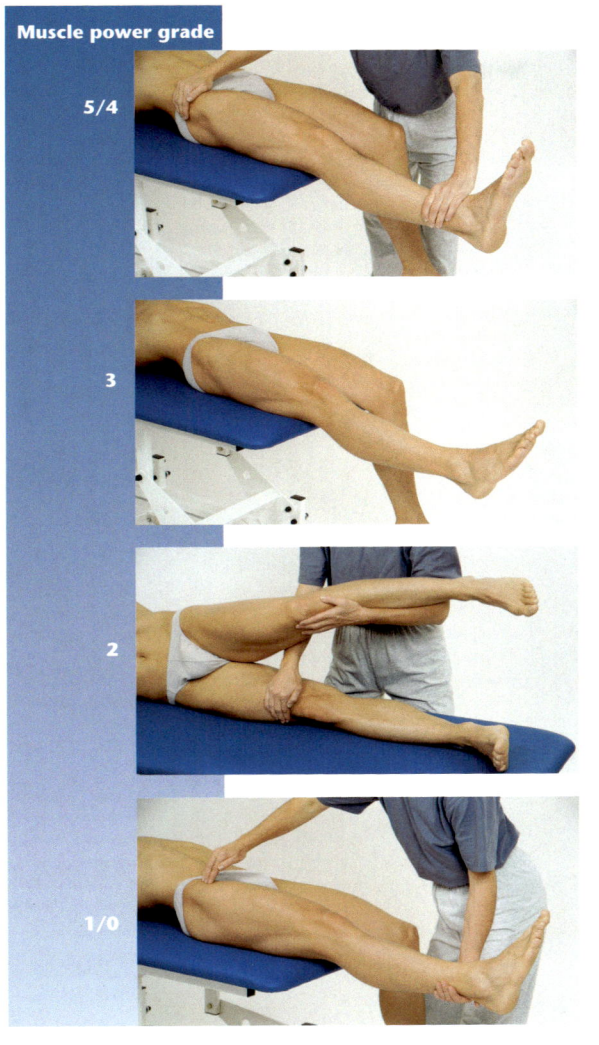

5/4

Starting position: The patient lies on his back, with the lower legs hanging over the narrow end of the treatment couch.

Test procedure: The examiner holds the patient's pelvis in place with one hand and with the other applies pressure to the distal lower leg in the direction of flexion at the knee.

Instruction: "Extend your leg at the knee against my resistance and hold your final position."

3

Starting position: The patient lies on his back, with the lower legs hanging over the narrow end of the treatment couch.

Test procedure: The examiner looks at the movement of the leg.

Instruction: "Extend your leg at the knee."

2

Starting position: The patient lies on his side. Both legs are extended at the hip. The knee of the leg being tested, which is resting on the couch, is flexed 90°.

Test procedure: The examiner holds the uppermost leg and fixes the thigh of the other leg in place.

Instruction: "Extend your leg at the knee."

1/0

Starting position: The patient lies on his back, with the lower legs hanging over the narrow end of the treatment couch.

Test procedure: The examiner palpates the tendon of origin of the rectus femoris muscle.

Instruction: "Try to extend your leg."

 Clinical relevance

- The term "jumper's knee" refers to an insertion tendinopathy of the quadriceps femoris tendon at the patella. It is also known as "patellar tendinopathy."

 Problems/comments

- When performing its hip-flexing and knee-extending functions simultaneously, the rectus femoris muscle becomes actively insufficient; that is, it can no longer deploy its full power.

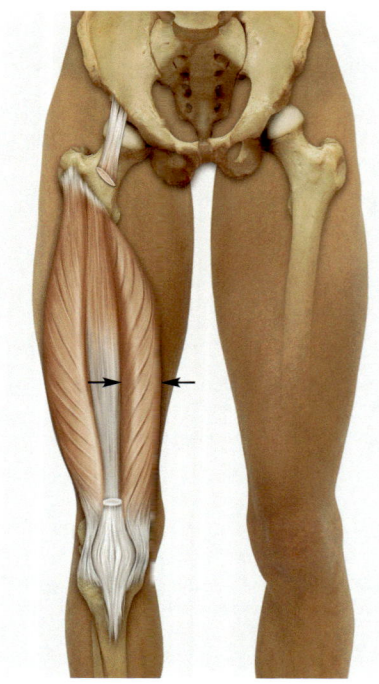

Vastus medialis muscle

Like the other vasti muscles, the vastus medialis muscle extends the leg at the knee, tightening particularly in the final phase of extension, stopping the patella from deviating and laterally. It antagonizes the external rotatory components of the vastus lateralis muscle.

Origin	Medial lip of the linea aspera (femoral crest)
	Intertrochanteric line
Insertion	Via the patellar ligament at the tibial tuberosity
Innervation	Femoral nerve, L2–L4

Functions

 Synergists

 Antagonists

Knee joint

Extension

Synergists	Antagonists
Rectus femoris muscle	Biceps femoris muscle
Vastus lateralis muscle	Semitendinosus muscle
Vastus intermedius muscle	Semimembranosus muscle
Gluteus maximus muscle (via the iliotibial tract)	Sartorius muscle
Tensor fasciae latae (via the iliotibial tract)	Gracilis muscle
	Gastrocnemius muscle (not with the foot in plantar flexion)
	Popliteus muscle

Internal rotation (with the knee flexed)

Synergists	Antagonists
Semimembranosus muscle	Biceps femoris muscle
Semitendinosus muscle	Vastus lateralis muscle
Sartorius muscle	Gluteus maximus muscle (via the iliotibial tract)
Popliteus muscle	Tensor fasciae latae (via the iliotibial tract)
Gracilis muscle	

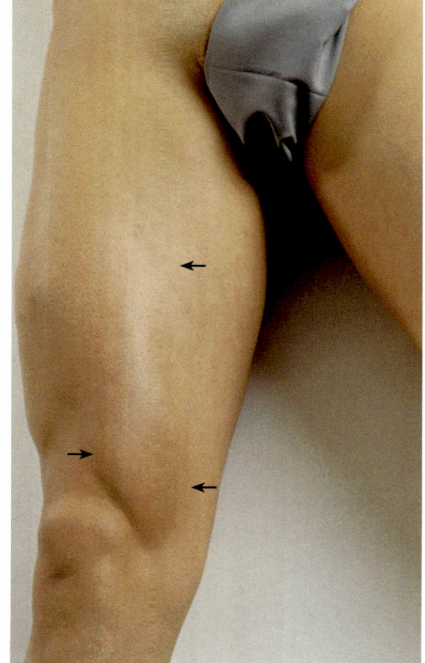

Muscle function testing

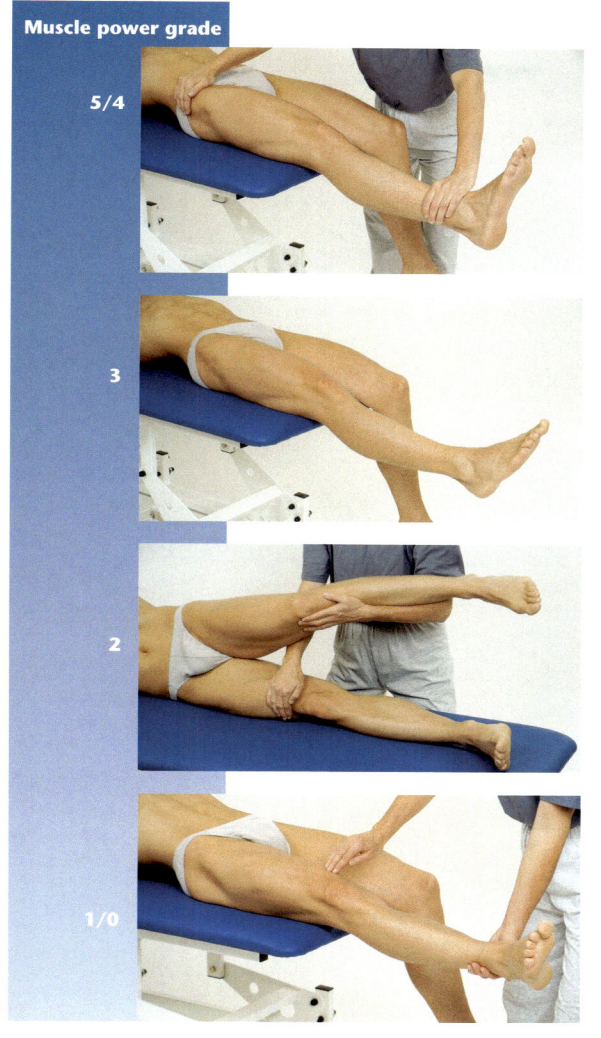

5/4

3

2

1/0

Starting position: The patient lies on his back, with the lower legs hanging over the narrow end of the treatment couch.

Test procedure: The examiner holds the patient's pelvis in place with one hand and with the other applies pressure to the distal lower leg in the direction of flexion at the knee.

Instruction: "Rotate your leg outwards and extend it at the knee against my resistance; hold your final position."

Starting position: The patient lies on his back, with the lower legs hanging over the narrow end of the treatment couch.

Test procedure: The examiner looks at the movement of the leg.

Instruction: "Rotate your leg outwards and extend it at the knee."

Starting position: The patient lies on his side. Both legs are extended at the hip. The knee of the leg being tested, which is resting on the couch, is flexed 90°.

Test procedure: The examiner holds the uppermost leg and fixes the thigh of the other leg in place.

Instruction: "Rotate your leg outwards and extend it at the knee."

Starting position: The patient lies on his back, with the lower legs hanging over the narrow end of the treatment couch.

Test procedure: The examiner palpates the vastus medialis muscle.

Instruction: "Rotate your leg outwards and try to extend it at the knee."

 Problems/comments

• External rotation at the hip causes more pronounced tightening of the vastus medialis muscle.

Vastus intermedius muscle

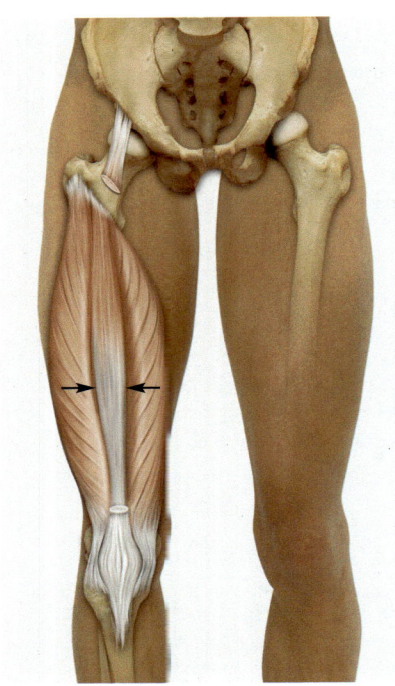

| The vastus intermedius muscle also causes extension at the knee.

Origin

Upper two thirds of the ventral shaft of the femur, Intertrochanteric line

Insertion

Jointly via the patellar ligament at the tibial tuberosity

Innervation

Femoral nerve, L2–L4

Functions

 Synergists

Antagonists

Knee joint

Extension

Rectus femoris muscle	Biceps femoris muscle
Vastus lateralis muscle	Semitendinosus muscle
Vastus medialis muscle	Semimembranosus muscle
Gluteus maximus muscle (via the iliotibial tract)	Sartorius muscle
Tensor fasciae latae (via the iliotibial tract)	Gracilis muscle
	Gastrocnemius muscle (not with the foot in plantar flexion)
	Popliteus muscle

Muscle function testing

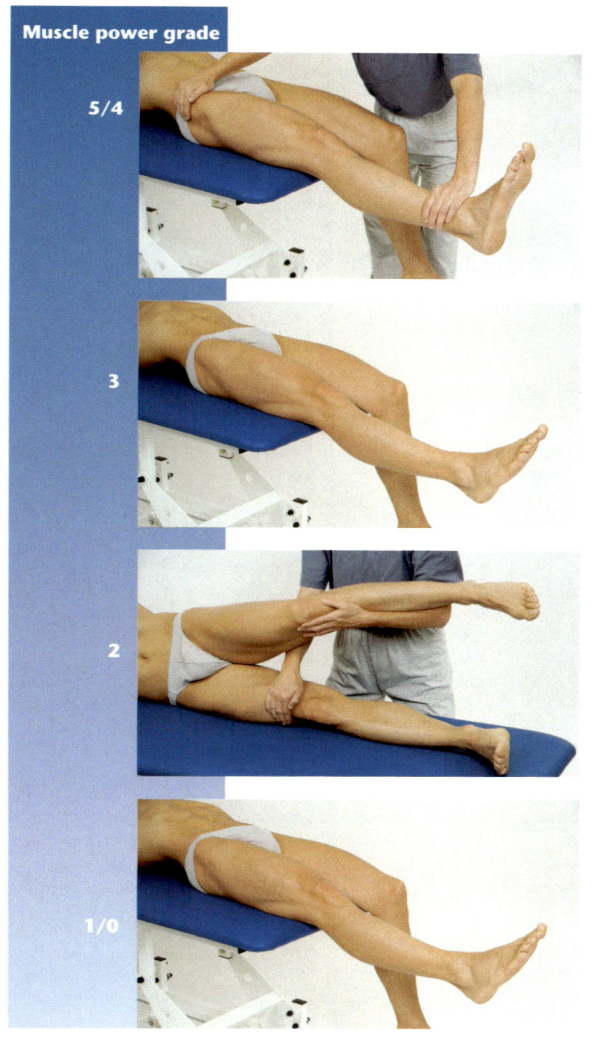

Muscle power grade

5/4

3

2

1/0

Starting position: The patient lies on his back, with the lower legs hanging over the narrow end of the treatment couch.

Test procedure: The examiner holds the patient's pelvis in place with one hand and with the other applies pressure to the distal lower leg in the direction of flexion at the knee.

Instruction: "Extend your leg at the knee against my resistance and hold your final position."

Starting position: The patient lies on his back, with the lower legs hanging over the narrow end of the treatment couch.

Test procedure: The examiner looks at the movement of the leg.

Instruction: "Extend your leg at the knee."

Starting position: The patient lies on his side. Both legs are extended at the hip. The knee of the leg being tested, which is resting on the couch, is flexed 90°.

Test procedure: The examiner holds the uppermost leg and fixes the thigh of the other leg in place.

Instruction: "Extend your leg at the knee."

Starting position: The patient lies on his back, with the lower legs hanging over the narrow end of the treatment couch.

Test procedure: The examiner looks at the movement of the leg.

Instruction: "Try to extend your leg at the knee."

 Problems/comments

- The vastus intermedius muscle cannot be palpated.
- The functions of the vastus intermedius and rectus femoris muscles cannot be distinguished during knee extension.

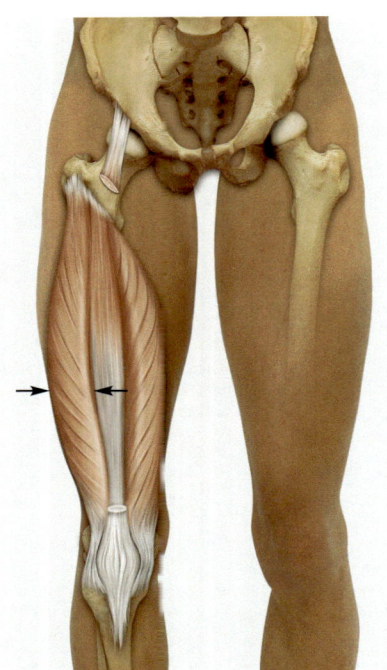

Vastus lateralis muscle

The vastus lateralis muscle extends the knee joint. In doing so, it antagonizes the internal rotatory components of the vastus medialis muscle.

Origin   Linea aspera (femoral crest), lateral lip, Basal greater trochanter

Insertion   Via the patellar ligament at the tibial tuberosity

Innervation   Femoral nerve, L2–L4

Functions

Synergists   Antagonists

Knee joint

Extension

Rectus femoris muscle	Biceps femoris muscle
Vastus medialis muscle	Semitendinosus muscle
Vastus intermedius muscle	Semimembranosus muscle
Gluteus maximus muscle (via the iliotibial tract)	Sartorius muscle
Tensor fasciae latae (via the iliotibial tract)	Gracilis muscle
	Gastrocnemius muscle (not with the foot in plantar flexion)
	Popliteus muscle

External rotation (with the knee flexed)

Biceps femoris muscle	Semimembranosus muscle
Gluteus maximus muscle (via the iliotibial tract)	Semitendinosus muscle
Tensor fasciae latae (via the iliotibial tract)	Sartorius muscle
	Popliteus muscle
	Gracilis muscle
	Vastus medialis muscle

Muscle function testing

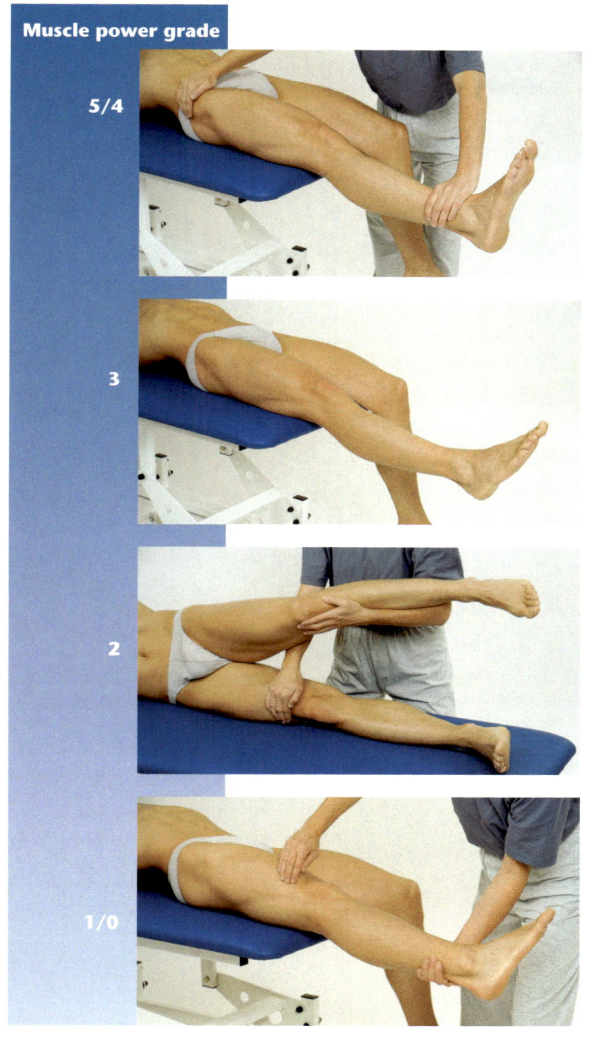

Muscle power grade

5/4

3

2

1/0

Starting position: The patient lies on his back, with the lower legs hanging over the narrow end of the treatment couch.

Test procedure: The examiner holds the patient's pelvis in place with one hand and with the other applies pressure to the distal lower leg in the direction of flexion at the knee.

Instruction: "Rotate your leg inwards and extend it at the knee against my resistance; hold your final position."

Starting position: The patient lies on his back, with the lower legs hanging over the narrow end of the treatment couch.

Test procedure: The examiner looks at the movement of the leg.

Instruction: "Extend your leg at the knee."

Starting position: The patient lies on his side. Both legs are extended at the hip. The knee of the leg being tested, which is resting on the couch, is flexed 90°.

Test procedure: The examiner holds the uppermost leg and fixes the thigh of the other leg in place.

Instruction: "Extend your leg at the knee."

Starting position: The patient lies on his back, with the lower legs hanging over the narrow end of the treatment couch.

Test procedure: The examiner palpates the vastus medialis muscle.

Instruction: "Rotate your leg inwards and try to extend it at the knee."

 Problems/comments

- Internal rotation at the hip causes more pronounced tightening of the vastus lateralis muscle.

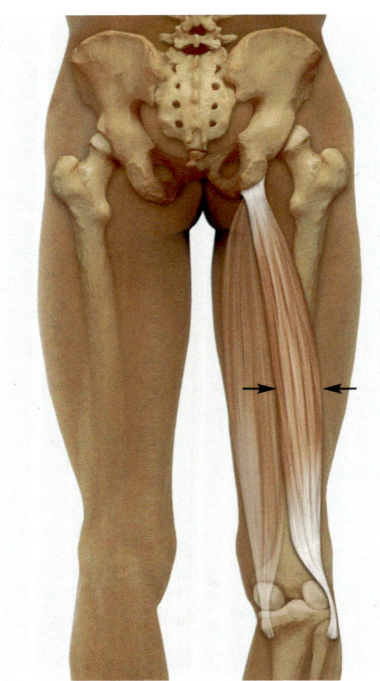

Biceps femoris muscle

The biceps femoris muscle (biceps muscle of the thigh) extends the leg at the hip joint and rotates the femur outwards. It can flex the extended knee powerfully and rotate the lower leg outwards with the knee flexed. When the knee is flexed and the lower leg fixed in place (f.e., when a person who is seated on a bench, with the feet placed firmly on the floor, shifts his or her buttocks sideways), this muscle rotates the femur inwards on the lower leg. When the trunk is righted from a stooped posture, its pulls the pelvis upright, thus indirectly flattening out the lordosis of the lumbar spine.

Origin	Long head: ischial tuberosity and sacrotuberal ligament
	Short head: linea aspera (femoral crest), lateral intermuscular septum
Insertion	Lateral surface of the head of the fibula, lateral condyle of the tibia
Innervation	Long head: sciatic nerve, tibial part, L5–S2
	Short head: common peroneal nerve, L5–S2
Special features	The muscle's insertion tendon forms the upper lateral boundary of the popliteal cavity

Functions

 Synergists Antagonists

Hip joint (long head only)

Extension

Gluteus maximus muscle	Iliopsoas muscle
Semimembranosus muscle	Rectus femoris muscle
Semitendinosus muscle	Tensor fasciae latae muscle
Gluteus medius muscle (dorsal part)	Gluteus medius muscle (ventral part)
Gluteus minimus muscle (dorsal part)	Sartorius muscle
Pectineus muscle (from maximum flexion)	Gracilis muscle
Adductor muscles (into the neutral position)	Adductor muscles (into the neutral position)

External rotation

Gluteus maximus muscle	Tensor fasciae latae muscle
Gluteus medius muscle (dorsal part)	Gluteus minimus muscle (ventral part)
Gluteus minimus muscle (dorsal part)	Gluteus medius muscle (ventral part)
Piriformis muscle	Adductor muscles (from maximum external rotation)
Quadratus femoris muscle	
Obturator and gemelli muscles	
Pectineus muscle	
Sartorius muscle	
Adductor muscles (from maximum internal rotation)	

Knee joint (both heads)

Flexion

Semitendinosus muscle, semimembranosus muscle	Quadriceps femoris muscle
Sartorius muscle, gracilis muscle	Gluteus maximus muscle (via the iliotibial tract)
Gastrocnemius muscle (not with the foot in plantar flexion)	Tensor fasciae latae
	Popliteus muscle

External rotation

Gluteus maximus muscle (via the iliotibial tract)	Semitendinosus muscle, semimembranosus muscle
Tensor fasciae latae (via the iliotibial tract)	Sartorius muscle, popliteus muscle, gracilis muscle
Vastus lateralis muscle	Vastus medialis muscle

Hip joint and intervertebral joints (lumbar spine)

Dorsal tilting of the pelvis

Gluteus maximus muscle	Iliopsoas muscle
Semimembranosus muscle	
Semitendinosus muscle	

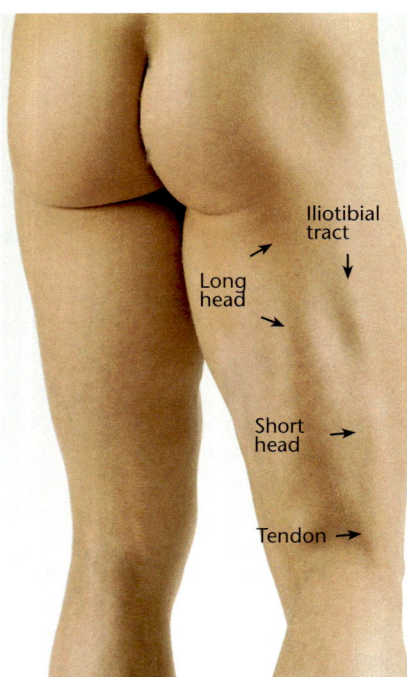

Iliotibial tract

Long head

Short head

Tendon

Muscle function testing

Muscle power grade

5/4

3

2

1/0

Starting position: The patient rests his upper body, stomach down, on the treatment couch, with both hips flexed. One leg, flexed slightly at the knee, stands on the floor by the couch, while the other is tested.

Test procedure: The examiner holds the distal thigh of the leg being tested in place with one hand and, with the other, applies pressure to the distal lower leg in the direction of extension at the knee.

Instruction: "Pull your heel up towards your buttocks against my resistance and hold your final position."

Starting position: The patient rests his upper body, stomach down, on the treatment couch, with both hips flexed. One leg, flexed slightly at the knee, stands on the floor by the couch, while the other is tested.

Test procedure: The examiner holds the distal thigh of the leg being tested in place with one hand.

Instruction: "Pull your heel up towards your buttocks."

Starting position: The patient lies on his side. The uppermost leg is extended, the other – the one being tested – rests on couch with the hip flexed 90° and the knee slightly flexed.

Test procedure: The examiner holds the uppermost leg.

Instruction: "Pull your heel up towards your buttocks."

Starting position: The patient sits. The legs hang over the narrow end of the treatment couch.

Test procedure: The examiner palpates the biceps femoris muscle.

Instruction: "Try to pull your heel under the treatment couch, turning the tip of your foot outwards."

Clinical relevance

- A deficit of the ischiocrural muscles has virtually no effect on everyday functions such as walking, standing up and climbing stairs, as the gluteus maximus can compensate for the defect. However, this does lead to hyperextension of the knee.

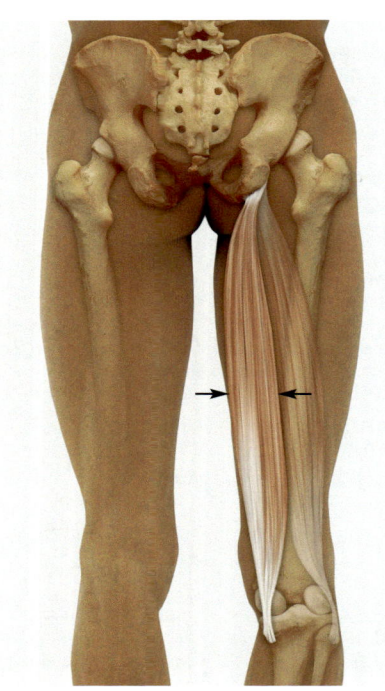

Semimembranosus muscle

During walking, the semimembranosus (semimembranous) muscle, together with the other ischiocrural muscles, causes powerful extension of the supporting leg at the hip, thus generating propulsion. In the swinging leg, this muscle also causes flexion of the knee joint. When it tightens in isolation with the knee flexed, it produces internal rotation of the lower leg at the knee.

Origin	Ischial tuberosity, proximally and laterally of the joint origin of the semitendinosus and the long head of the biceps femoris (the "caput commune")
Insertion	Posteromedial part of the medial condyle of the tibia via the pes anserinus
Innervation	Sciatic nerve, tibial part, L5–S2

Functions

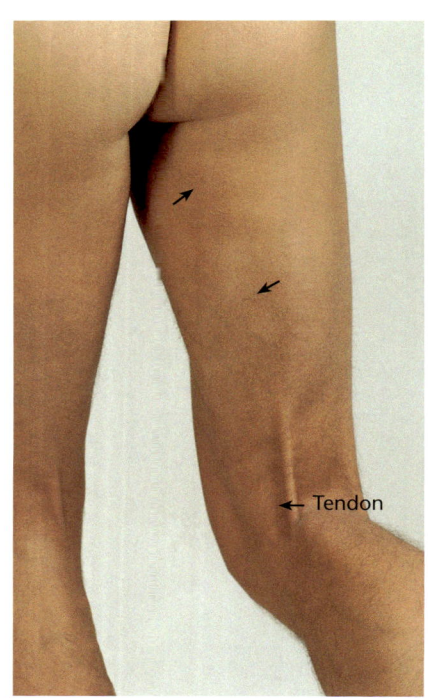

← Tendon

Synergists   Antagonists

Hip joint

Extension

Gluteus maximus muscle	Iliopsoas muscle
Semitendinosus muscle	Rectus femoris muscle
Biceps femoris muscle	Tensor fasciae latae muscle
Gluteus medius muscle (dorsal part)	Gluteus medius muscle (ventral part)
Gluteus minimus muscle (dorsal part)	Sartorius muscle
Pectineus muscle (from maximum flexion)	Gracilis muscle
Adductor muscles (from maximum flexion)	Adductor muscles (from maximum extension)
	Pectineus muscle

Knee joint

Flexion

Biceps femoris muscle	Quadriceps femoris muscle
Semitendinosus muscle	Gluteus maximus muscle (via the iliotibial tract)
Sartorius muscle	Tensor fasciae latae (via the iliotibial tract)
Gracilis muscle	
Gastrocnemius muscle (not with the foot in plantar flexion)	
Popliteus muscle	

Internal rotation

Semitendinosus muscle	Biceps femoris muscle
Sartorius muscle	Vastus lateralis muscle
Popliteus muscle	Gluteus maximus muscle (via the iliotibial tract)
Gracilis muscle	Tensor fasciae latae (via the iliotibial tract)
Vastus medialis muscle	

Hip joint and intervertebral joints (lumbar spine)

Dorsal tilting of the pelvis

Gluteus maximus muscle	Iliopsoas muscle
Biceps femoris muscle, long head	Sartorius muscle
Semitendinosus muscle	Tensor fasciae latae muscle
	Rectus femoris muscle

Muscle function testing

Muscle power grade	
5/4	

Starting position: The patient rests his upper body, stomach down, on the treatment couch, with both hips flexed. One leg, flexed slightly at the knee, stands on the floor by the couch, while the other is tested.

Test procedure: The examiner holds the distal thigh of the leg being tested in place with one hand and with the other applies pressure to the distal lower leg in the direction of extension at the knee.

Instruction: "Pull your heel up towards your buttocks against my resistance and hold your final position."

Starting position: The patient rests his upper body, stomach down, on the treatment couch, with both hips flexed. One leg, flexed slightly at the knee, stands on the floor by the couch, while the other is tested.

Test procedure: The examiner holds the distal thigh of the leg being tested in place with one hand.

Instruction: "Pull your heel up towards your buttocks."

Starting position: The patient lies on his side. The uppermost leg is extended, the other – the one being tested – rests on the couch with the hip flexed 90° and the knee slightly flexed.

Test procedure: The examiner holds the uppermost leg.

Instruction: "Pull your heel up towards your buttocks."

Starting position: The patient sits. The legs hang over the narrow end of the treatment couch.

Test procedure: The examiner palpates the semimembranosus and semitendinosus muscles.

Instruction: "Try to pull your heel under the treatment couch, rotating the tip of your foot outwards."

Clinical relevance

- A deficit of the ischiocrural muscles has virtually no effect on everyday functions such as walking, standing up and climbing stairs, as the gluteus maximus can compensate for the defect. However, this does lead to hyperextension of the knee.

Problems/comments

- The semimembranosus muscle is tested together with the semitendinosus and popliteus muscles.

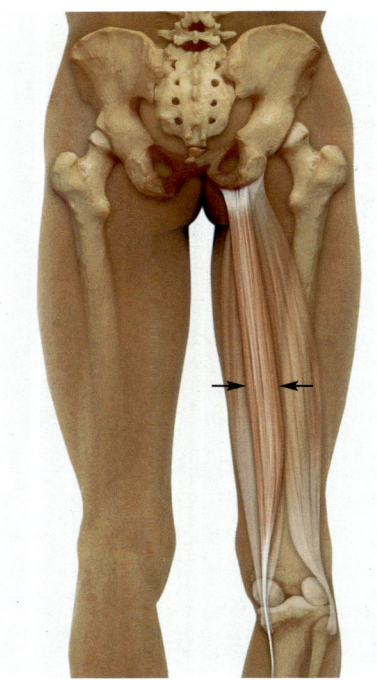

Semitendinous muscle

During walking, the semitendinosus (semitendinous) muscle, together with the other is-chiocrural muscles, causes powerful extension of the supporting leg at the hip, thus gener-ating propulsion. In the swinging leg, it prevents extension of the knee, as might occur dur-ing the swing of the lower leg. It also controls the forward inclination of the trunk over the hips and is a powerful aid in righting the trunk from a forward bend. In doing so, its con-traction indirectly flattens lumbar lordosis, thus antagonizing the effect of the iliopsoas mus-cle. In the swinging leg, this muscle also has a flexor effect on the knee joint. When it tight-ens in isolation with the knee flexed, it produces internal rotation of the lower leg at the knee.

Origin	Ischial tuberosity (common tendon, common origin with the long head of the biceps femoris, the "caput commune")
Insertion	Tibial tuberosity via the pes anserinus
Innervation	Sciatic nerve, tibial part, L5–S2

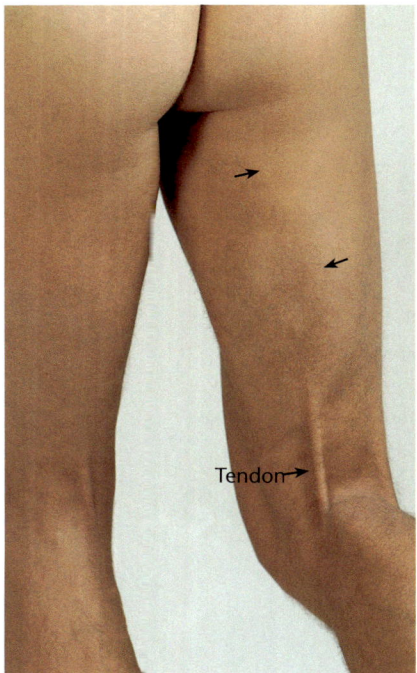

Functions

 Synergists

 Antagonists

Hip joint

Extension

Gluteus maximus muscle	Iliopsoas muscle
Semimembranosus muscle	Rectus femoris muscle
Biceps femoris muscle	Tensor fasciae latae muscle
Gluteus medius muscle (dorsal part)	Gluteus medius muscle (ventral part)
Gluteus minimus muscle (dorsal part)	Sartorius muscle
Pectineus muscle (from maximum flexion)	Gracilis muscle
Adductor muscles (back to neutral from maxi-mum flexion or extension)	Adductor muscles (from maximum extension)
	Pectineus muscle

Knee joint

Flexion

Biceps femoris muscle	Quadriceps femoris muscle
Semimembranosus muscle	Gluteus maximus muscle (via the iliotibial tract)
Sartorius muscle	Tensor fasciae latae (via the iliotibial tract)
Gracilis muscle	
Gastrocnemius muscle (not with the foot in plantar flexion)	
Popliteus muscle	

Internal rotation

Semitendinosus muscle	Biceps femoris muscle
Sartorius muscle	Gluteus maximus muscle (via the iliotibial tract)
Popliteus muscle	Tensor fasciae latae (via the iliotibial tract)
Gracilis muscle	Vastus lateralis muscle
Vastus medialis muscle	

Hip joint and intervertebral joints (lumbar spine)

Dorsal tilting of the pelvis

Gluteus maximus muscle	Iliopsoas muscle
Biceps femoris muscle, long head	
Semimembranosus muscle	

Muscle function testing

Muscle power grade

5/4

Starting position: The patient rests his upper body, stomach down, on the treatment couch, with both hips flexed slightly at the knee, stands on the floor by the couch, while the other is tested.

Test procedure: The examiner holds the distal thigh of the leg being tested in place with one hand and with the other applies pressure to the distal lower leg in the direction of extension at the knee.

Instruction: "Pull your heel up towards your buttocks against my resistance and hold your final position."

3

Starting position: The patient rests his upper body, stomach down, on the treatment couch, with both hips flexed. One leg, flexed slightly at the knee, stands on the floor by the couch, while the other is tested.

Test procedure: The examiner holds the distal thigh of the leg being tested in place with one hand.

Instruction: "Pull your heel up towards your buttocks."

2

Starting position: The patient lies on his side. The uppermost leg is extended, the other – the one being tested – rests on the couch with the hip flexed 90° and the knee slightly flexed.

Test procedure: The examiner holds the uppermost leg.

Instruction: "Pull your heel up towards your buttocks."

1/0

Starting position: The patient sits. The legs hang over the narrow end of the treatment couch.

Test procedure: The examiner palpates the semimembranosus and semitendinosus muscles.

Instruction: "Try to pull your heel under the treatment couch, turning the tip of your foot outwards."

🔱 Clinical relevance

- A deficit of the ischiocrural muscles has virtually no effect on everyday functions such as walking, standing up and climbing stairs, as the gluteus maximus can compensate for the defect. However, this does lead to hyperextension of the knee.
- The semitendinosus muscle's tendon may be used as a homologous graft for the cruciate ligaments.

⚠ Problems/comments

- The semitendinosus muscle is tested together with the semimembranosus and popliteus muscles.

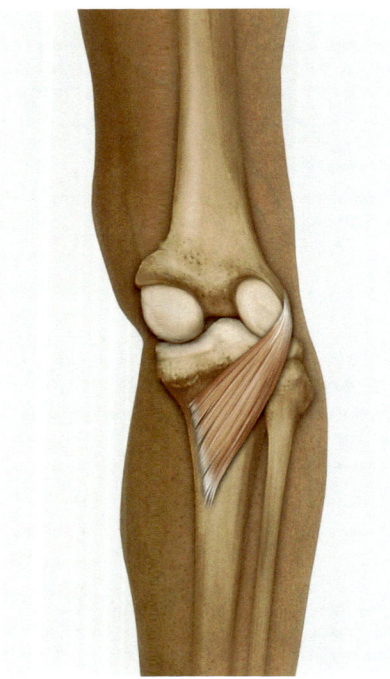

Popliteus muscle

The popliteus (popliteal) muscle carries out two important tasks: in the firmly extended knee joint (e.g. in the supporting leg), its contraction, together with a simultaneous reduction in extensor tone, causes external rotation of the femur on the tibia. This loosens the cruciate ligaments, thus allowing already initiated flexion of the knee to proceed. As flexion progresses, this muscle pulls the lateral meniscus dorsally on the tibia, stopping it from becoming impacted. Thus, the lateral meniscus is not only passively more mobile than the medial one but is also pulled actively in the dorsal direction during flexion.

Origin	Lateral condyle of the femur
	Lateral meniscus
	Arcuate ligament, part of the joint capsule of the knee
Insertion	Posterior surface of the proximal third of the tibia
Innervation	Tibial nerve, L5–S1

Functions

 Synergists   Antagonists

Knee joint

Internal rotation
Semimembranosus muscle
Semitendinosus muscle
Sartorius muscle
Gracilis muscle

Biceps femoris muscle
Gluteus maximus muscle (via the iliotibial tract)
Tensor fasciae latae (via the iliotibial tract)

Flexion
Biceps femoris muscle
Semitendinosus muscle
Semimembranosus muscle
Sartorius muscle
Gracilis muscle
Gastrocnemius muscle (not with the foot in plantar flexion)

Quadriceps femoris muscle
Gluteus maximus muscle (via the iliotibial tract)
Tensor fasciae latae (via the iliotibial tract)

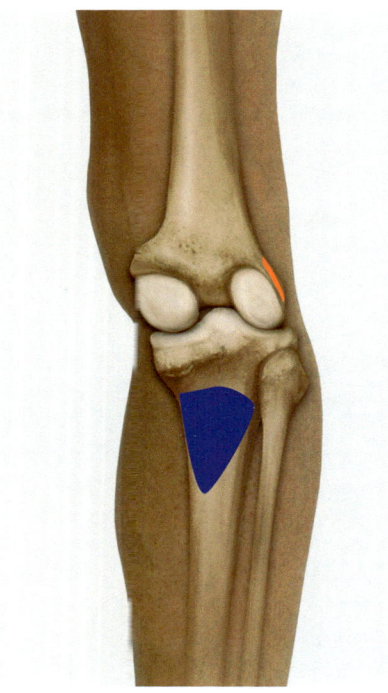

Muscle function testing

Muscle power grade

5/4

3

2

1/0

Starting position: The patient rests his upper body, stomach down, on the treatment couch, with both hips flexed. One leg, flexed slightly at the knee, stands on the floor by the couch, while the other is tested.

Test procedure: The examiner holds the distal thigh of the leg being tested in place with one hand and with the other applies pressure to the distal lower leg in the direction of extension at the knee.

Instruction: "Pull your heel up towards your buttocks against my resistance and hold your final position."

Starting position: The patient rests his upper body, stomach down, on the treatment couch, with both hips flexed. One leg, flexed slightly at the knee, stands on the floor by the couch, while the other is tested.

Test procedure: The examiner holds the distal thigh of the leg being tested in place with one hand.

Instruction: "Pull your heel up towards your buttocks."

Starting position: The patient lies on his side. The uppermost leg is extended, the other – the one being tested – rests on couch with the hip flexed 90° and the knee slightly flexed.

Test procedure: The examiner holds the uppermost leg.

Instruction: "Pull your heel up towards your buttocks."

Starting position: The patient rests his upper body, stomach down, on the treatment couch, with both hips flexed. One leg, flexed slightly at the knee, stands on the floor by the couch while the other is tested.

Test procedure: The examiner holds the distal thigh of the leg being tested in place with one hand and looks at the movement of the lower leg.

Instruction: "Pull your heel up towards your buttocks, turning the tip of your foot inwards."

! Problems/comments

- The popliteus muscle cannot be palpated.
- The popliteus muscle is tested together with the semimembranosus and semitendinosus muscles.

Stretch tests

Biceps femoris, semitendinosus and semimembranosus muscles

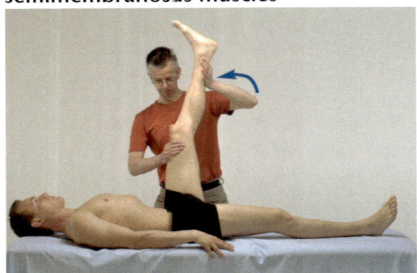

Method
The patient's hip is flexed 90°. The therapist moves the patient's lower leg to achieve maximum extension of the knee.

Finding
Muscle shortening is present if the movement cannot be performed to its full extent and if the endpoint feels soft and elastic.
The patient reports a stretching sensation along the course of the muscle.

Rectus femoris muscle

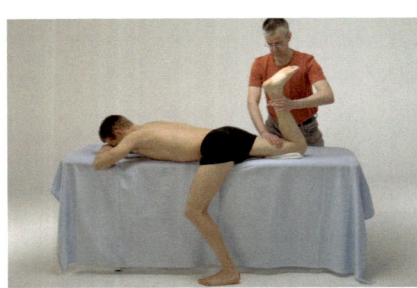

Method
The leg not being tested is flexed to the maximum at the hip and stands on the floor. The therapist moves the patient's lower leg to achieve maximum flexion of the knee.

Finding
Muscle shortening is present if the movement cannot be performed to its full extent and if the endpoint feels soft and elastic.
The patient reports a stretching sensation along the course of the muscle.

3 Lower extremity

Muscles of the ankle

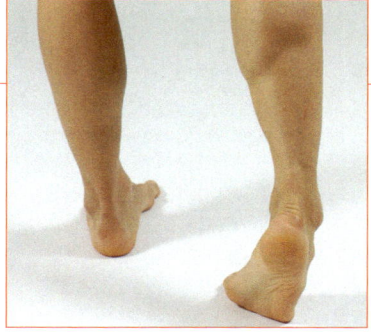

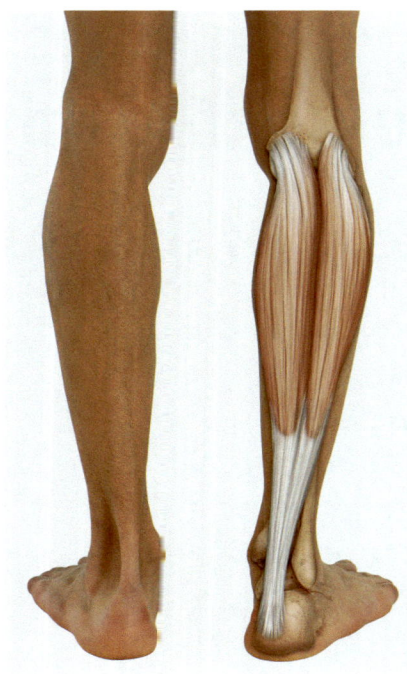

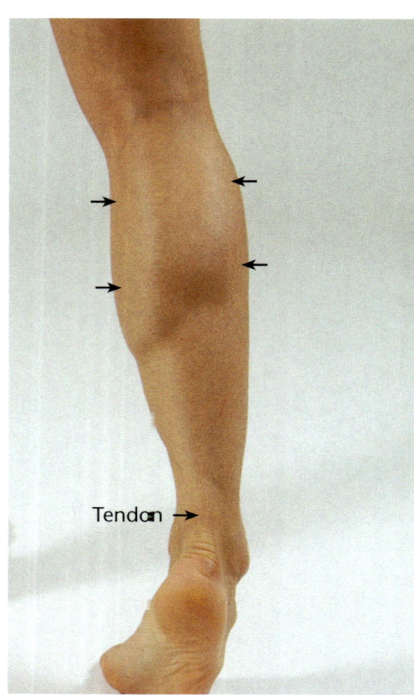

Tendon →

Gastrocnemius muscle

The gastrocnemius muscle flexes the knee and ankle joints powerfully. This means it has an important task in lifting off the supporting leg, with both components of its function causing propulsion during walking. Its other task is to halt extension of the knee in the swinging leg. In the subtalar and talocalcaneonavicular joints, this muscle acts as a supinator: that is, it elevates the medial edge of the foot during flexion.

Origin	Medial and lateral condyles of the femur
Insertion	Cranial and medial parts of the calcaneal tuberosity
Innervation	Tibial nerve, S1–S2
Special features	The gastrocnemius muscle forms the triceps surae muscle together with the soleus and plantaris muscles

Functions

 Synergists

 Antagonists

Knee joint

Flexion

Biceps femoris muscle
Semitendinosus muscle
Semimembranosus muscle
Sartorius muscle
Gracilis muscle

Quadriceps femoris muscle
Gluteus maximus muscle (via the iliotibial tract)
Tensor fasciae latae muscle (via the iliotibial tract)

Talocrural joint

Flexion

Soleus muscle
Flexor hallucis longus muscle
Peronei muscles
Tibialis posterior muscle
Flexor digitorum longus muscle

Tibialis anterior muscle
Extensor digitorum longus muscle
Extensor hallucis longus muscle
Peroneus tertius muscle

Subtalar and talocalcaneonavicular joints

Supination (inversion and flexion)

Soleus muscle
Tibialis posterior muscle
Flexor digitorum longus muscle
Flexor hallucis longus muscle
Tibialis anterior muscle

Peronei muscles
Extensor digitorum longus muscle

Muscle function testing

Muscle power grade

5/4

Starting position: The patient stands on one leg. He may hold on to something to keep his balance, if necessary.

Test procedure: The examiner observes.

Instruction: "Stand high up on tiptoes and then go down slowly until your heel almost touches the floor. Repeat this movement five times without pushing yourself up with your hands."

3

Starting position: The patient stands on one leg. He may hold on to something to keep his balance, if necessary.

Test procedure: The examiner observes.

Instruction: "Stand on tiptoes as high as you can and then go down again slowly."

2

Starting position: The patient lies on his side.

Test procedure: The examiner holds the lower leg in place with one hand from the ventral side, above the ankles.

Instruction: "Push the tip of your foot downwards."

1/0

Starting position: The patient lies on his side.

Test procedure: The examiner palpates the gastrocnemius muscle.

Instruction: "Try to push the tip of your foot downwards."

Clinical reference

- The term achillodynia refers to pain in the Achilles tendon associated with load-bearing stress.
- An Achilles tendon rupture is usually preceded by degenerative processes in that tendon.

Problems/comments

- Muscle power grade 4 is achieved if the patient can carry out the full movement sequence only two to three times.
- The patient needs to be able to stabilize the forefoot if he is to stand on his toes.

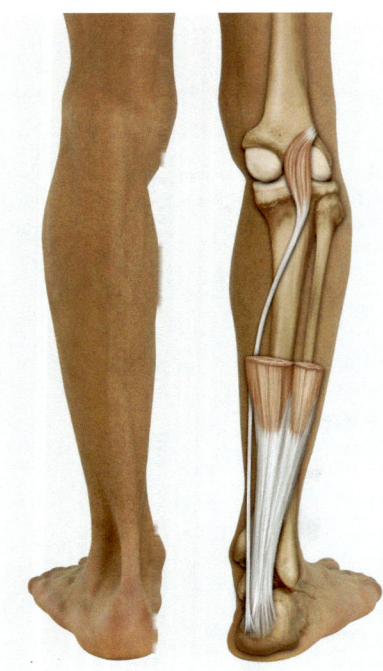

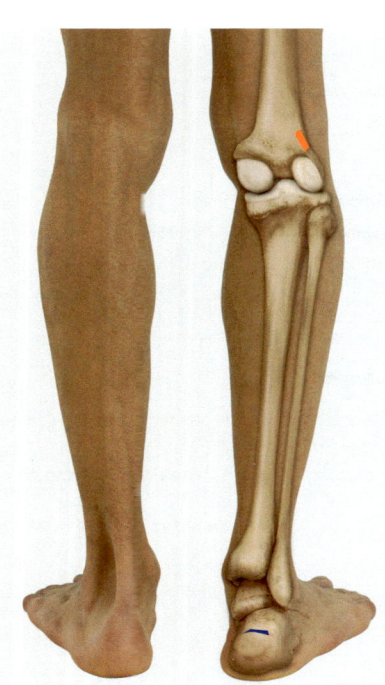

Plantaris muscle

The plantaris (plantar) muscle produces negligibly weak flexion at the knee and ankle. Its main effect during flexion lies in keeping the posterior tibial vessels open: the plantaris muscle is linked to the tunica adventitia of these vessels via the connective tissue of the popliteal cavity. Therefore, it has no synergists or antagonists with respect to its functions in producing joint movement.

Origin	Popliteal surface and lateral epicondyle of the femur
Insertion	Cranial and medial part of the calcaneal tuberosity
Innervation	Tibial nerve, S1–S2
Special features	The plantaris muscle forms the triceps surae muscle together with the gastrocnemius and soleus muscles

Muscle function testing

5/4

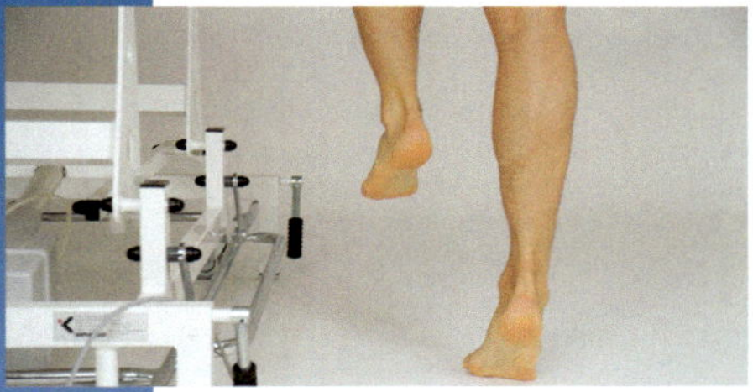

Starting position: The patient stands on one leg. He may hold on to something to keep his balance, if necessary.

Test procedure: The examiner observes.

Instruction: "Stand high up on tiptoes and then go down slowly until your heel almost touches the floor. Repeat this movement five times without pushing yourself up with your hands."

3

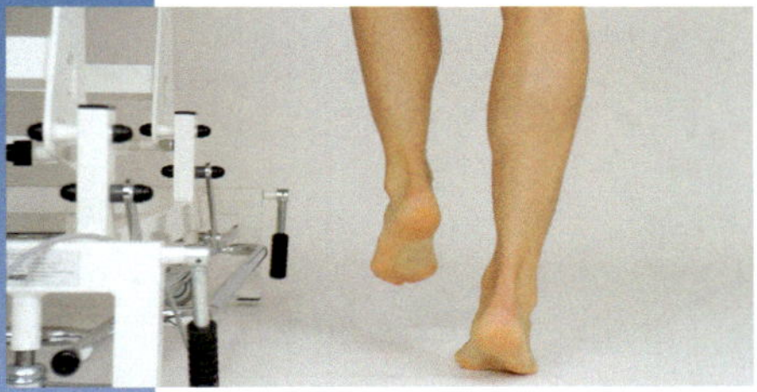

Starting position: The patient stands on one leg. He may hold on to something to keep his balance, if necessary.

Test procedure: The examiner observes.

Instruction: "Stand on tiptoes as high as you can and then go down again slowly."

2

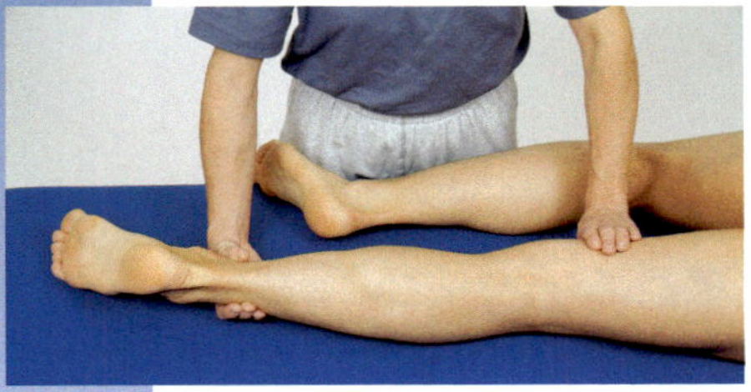

Starting position: The patient lies on his side.

Test procedure: The examiner holds the lower leg in place with one hand from the ventral side, above the ankles.

Instruction: "Push the tip of your foot downwards."

1/0

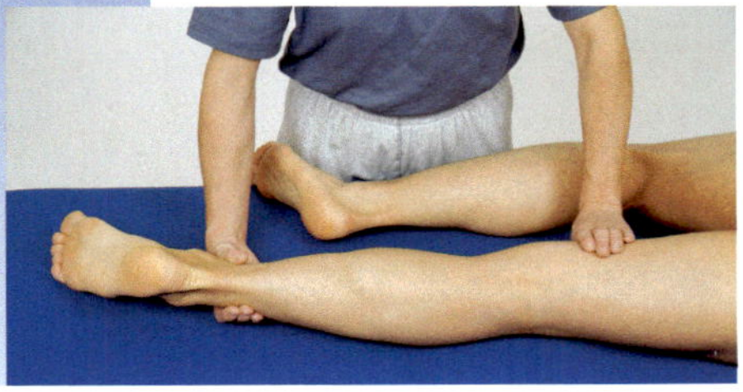

Starting position: The patient lies on his side.

Test procedure: The examiner looks at the movement of the foot.

Instruction: "Try to push the tip of your foot downwards."

⚠ Problems/comments

- The plantaris muscle can only be tested together with the gastrocnemius and soleus muscles.
- The patient needs to be able to stabilize the forefoot if he is to stand on his toes.
- It is not possible to palpate the plantaris muscle.

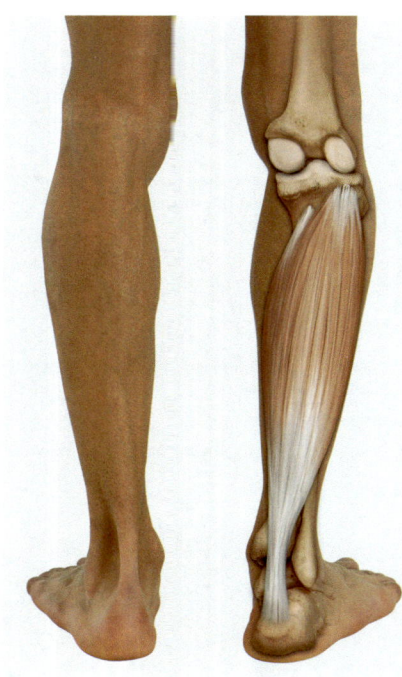

Soleus muscle

The soleus muscle is an important flexor of the talocrural joint and a supinator of the subtalar and talocalcaneonavicular joints. However, it gains particular significance in the standing posture, when it balances the leg on the joints of the ankle.

Origin	Dorsal, proximal third of the fibula Middle third of the tibia Tendinous arch of the soleus muscle
Insertion	Cranial and medial part of the calcaneal tuberosity
Innervation	Tibial nerve, S1–S2
Special features	The soleus muscle forms the triceps surae muscle together with the gastrocnemius and plantaris muscles

Functions

 Synergists

Antagonists

Talocrural joint

Flexion

Gastrocnemius muscle	Tibialis anterior muscle
Flexor hallucis longus muscle	Extensor digitorum longus muscle
Peronei muscles	Extensor hallucis longus muscle
Tibialis posterior muscle	Peroneus tertius muscle
Flexor digitorum longus muscle	

Subtalar and talocalcaneonavicular joints

Supination (inversion and flexion)

Gastrocnemius muscle	Peronei muscles
Tibialis posterior muscle	Extensor digitorum longus muscle
Flexor digitorum longus muscle	
Flexor hallucis longus muscle	

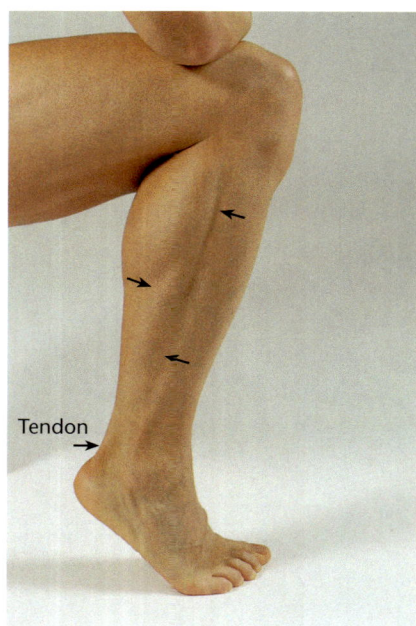

Tendon

Muscle function testing

5/4

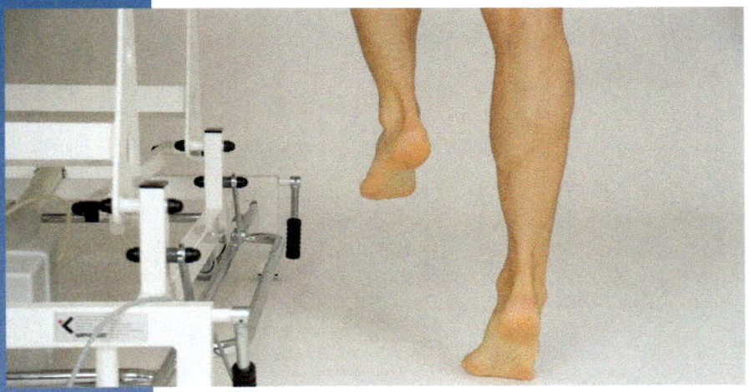

Starting position: The patient stands on one leg. He may hold on to something to keep his balance, if necessary.

Test procedure: The examiner observes.

Instruction: "Stand high up on tiptoes and then go down slowly until your heel almost touches the floor. Repeat this movement five times without pushing yourself up with your hands."

3

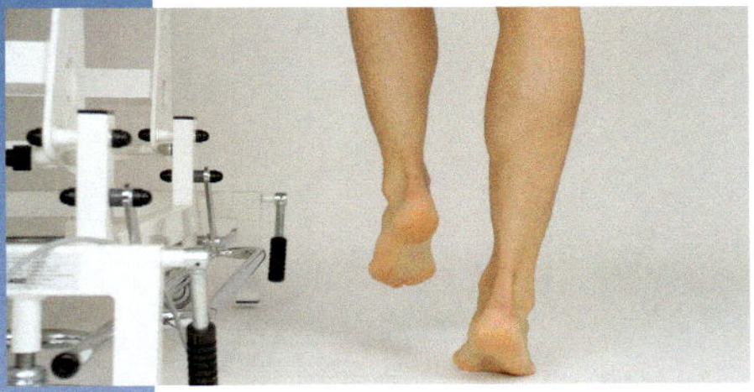

Starting position: The patient stands on one leg. He may hold on to something to keep his balance, if necessary.

Test procedure: The examiner observes.

Instruction: "Stand on tiptoes as high as you can and then go down again slowly."

2

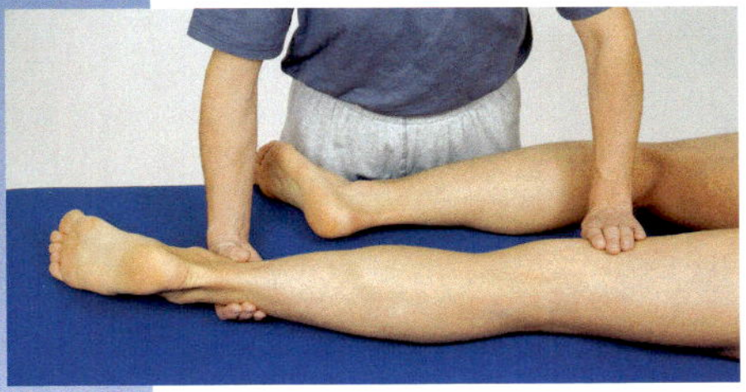

Starting position: The patient lies on his side.

Test procedure: The examiner holds the lower leg in place with one hand from the ventral side, above the ankles.

Instruction: "Push the tip of your foot downwards."

1/0

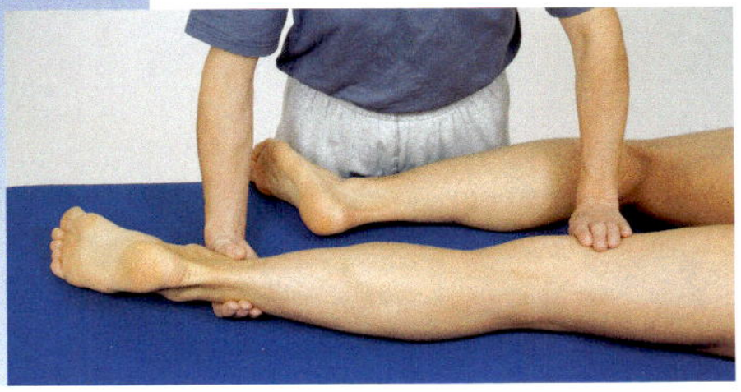

Starting position: The patient lies on his side.

Test procedure: The examiner looks at the movement of the foot.

Instruction: "Try to push the tip of your foot downwards."

[!] **Problems/comments**

- The soleus muscle can only be tested together with the gastrocnemius and plantaris muscles.
- The patient needs to be able to stabilize the forefoot if he is to stand on his toes.

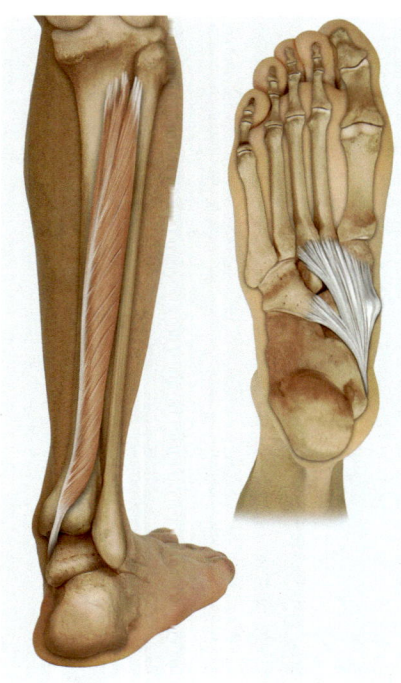

Tibialis posterior muscle

The tibialis posterior (posterior tibial) muscle supinates and flexes the foot. Since it runs along the sole, it supports the arch of the foot, together with the peroneus longus muscle.

Origin	Proximal two thirds of the posterior surface of the tibia Proximal two thirds of the medial surface of the fibula Interosseous membrane of the leg
Insertion	Navicular bone Cuneiform bones Cuboid bone Second to fourth metatarsal bones
Innervation	Tibial nerve, L5–S1

Functions

 Synergists Antagonists

Talocrural joint

Flexion

Gastrocnemius muscle	Tibialis anterior muscle
Soleus muscle	Extensor digitorum longus muscle
Flexor hallucis longus muscle	Extensor hallucis longus muscle
Peronei muscles	Peroneus tertius muscle
Flexor digitorum longus muscle	

Subtalar and talocalcaneonavicular joints

Supination (inversion and flexion)

Gastrocnemius muscle	Peronei muscles
Soleus muscle	Peroneus tertius muscle
Flexor digitorum longus muscle	Extensor digitorum longus muscle
Flexor hallucis longus muscle	

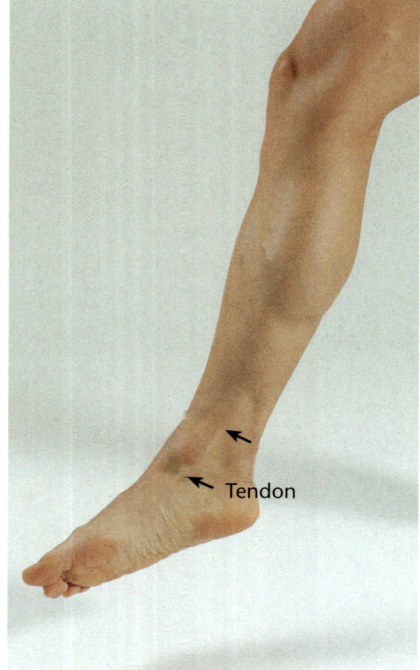

Tendon

Muscle function testing

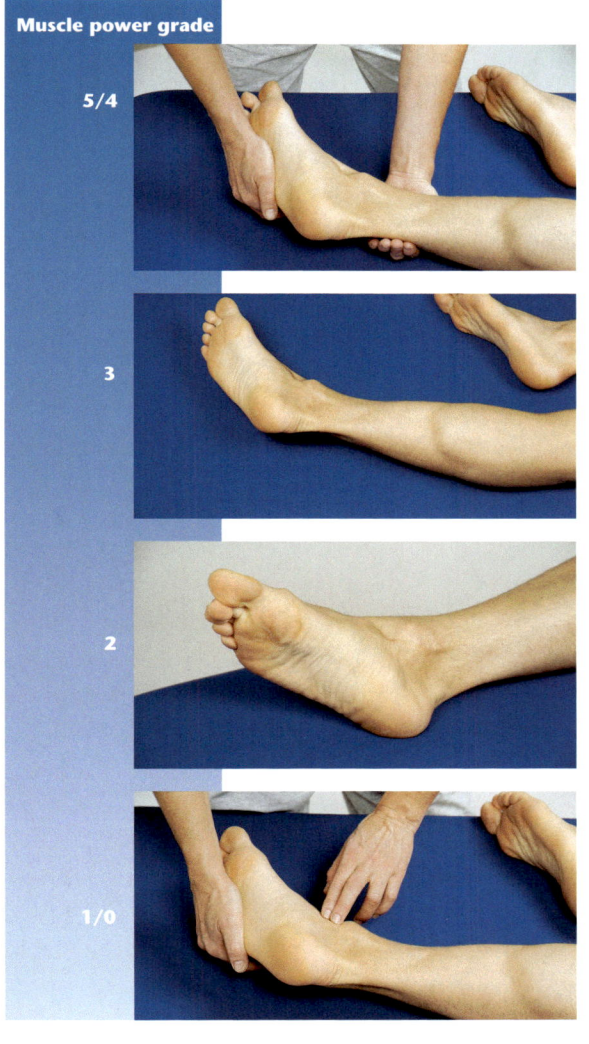

Muscle power grade

5/4

3

2

1/0

Starting position: The patient lies on the side being tested.

Test procedure: The examiner holds the distal part of the patient's lower leg in place with one hand and with the other applies pressure to the plantar surface of the little toe side of the foot in the direction of dorsal extension and pronation.

Instruction: "Push down the little toe side of your foot against my resistance and hold your final position."

Starting position: The patient lies on the side being tested.

Test procedure: The examiner looks at the movement of the foot.

Instruction: "Push down the little toe side of your foot."

Starting position: The patient lies on his back.

Test procedure: The examiner looks at the movement of the foot.

Instruction: "Push down the little toe side of your foot."

Starting position: The patient lies on the side being tested.

Test procedure: The examiner palpates the tibialis posterior muscle.

Instruction: "Try to push down the little toe side of your foot."

[!] **Problems/comments**

- The tibialis posterior muscle is aided in its function by the gastrocnemius, flexor digitorum longus and flexor hallucis longus muscles.

Tibialis anterior muscle

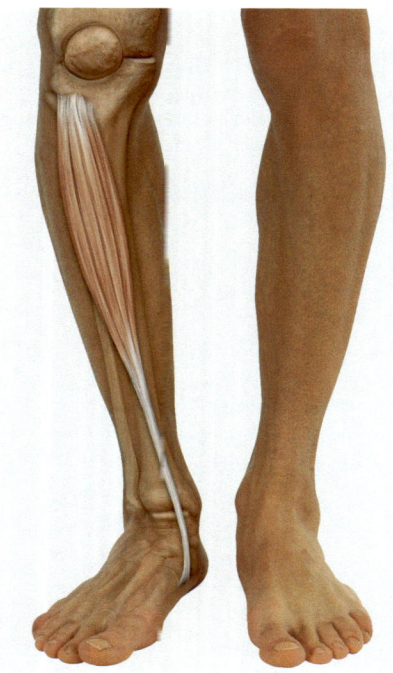

The tibialis anterior (anterior tibial) muscle produces extension of the joints of the ankle and can elevate the medial edge of the foot. This muscle is active particularly when the foot is lifted off the ground during walking and when the heel is placed back on the ground. In the standing posture, the tibialis anterior, together with its antagonist the soleus muscle, balances the lower leg on the trochlea of the talus.

Origin	Lateral condyle of the tibia
	Proximal, lateral half of the tibia
	Interosseous membrane of the leg
	Fascia of the leg, lateral intermuscular septum
Insertion	Medial and plantar surface of the anterior cuneiform bone
	Base of the first metatarsal bone
Innervation	Deep peroneal nerve, L4–L5
Special features	The tibialis anterior is an indicator muscle for spinal cord segment L4

Functions

 Synergists

 Antagonists

Talocrural joint

Extension

Synergists	Antagonists
Extensor digitorum longus muscle	Gastrocnemius muscle
Extensor hallucis longus muscle	Soleus muscle
	Flexor hallucis longus muscle
	Peronei muscles
	Tibialis posterior muscle
	Flexor digitorum longus muscle

Subtalar and talocalcaneonavicular joints

Supination (inversion and flexion)

Synergists	Antagonists
Gastrocnemius muscle	Peronei muscles
Soleus muscle	Peroneus tertius muscle
Tibialis posterior muscle	Extensor digitorum longus muscle
Flexor digitorum longus muscle	
Flexor hallucis longus muscle	

Tendon

Muscle function testing

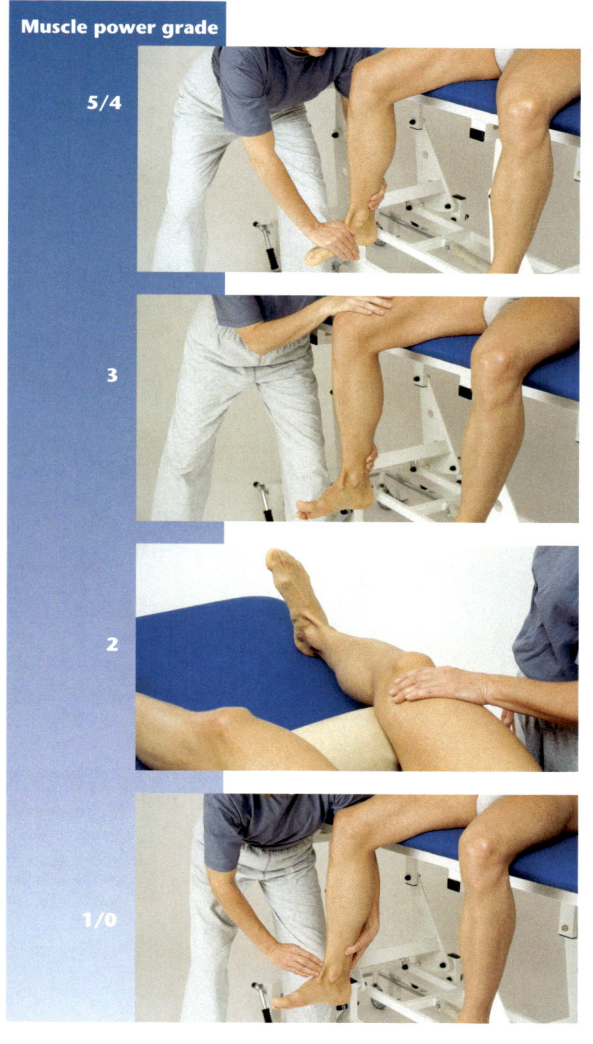

Muscle power grade

5/4

3

2

1/0

Starting position: The patient sits on the examination couch with legs hanging over the edge.

Test procedure: The examiner holds the distal part of the patient's lower leg in place with one hand and with the other applies pressure to the medial side of the dorsum of the foot in the direction of plantar flexion.

Instruction: "Pull up the big toe side of your foot, keeping your toes relaxed. Hold your final position."

Starting position: The patient sits on the examination couch with legs hanging over the edge.

Test procedure: The examiner holds the distal part of the patient's lower leg in place with one hand from the dorsal side.

Instruction: "Pull up the big toe side of your foot."

Starting position: The patient lies on his back. The leg is flexed at the knee and supported with a knee bolster.

Test procedure: The examiner holds the distal part of the patient's lower leg in place.

Instruction: "Pull up the big toe side of your foot."

Starting position: The patient sits on the examination couch, with his legs hanging over the edge.

Test procedure: The examiner palpates the tibialis anterior muscle.

Instruction: "Try to pull up the big toe side of your foot."

 Problems/comments

- Shortening of the calf muscles may mean that contraction of the tibialis anterior muscle is inhibited. Therefore, the muscle power test needs to be carried out with the leg flexed at the knee.

- The toes need to be relaxed to ensure that the extensor digitorum longus and extensor hallucis longus muscle do not contribute to the required movement.

Peroneus longus muscle

The peroneus longus (long peroneal) muscle pronates and flexes the foot. In doing so, its tendon supports the arch of the foot, together with the tibialis posterior muscle. It is also important in ensuring correct alignment of the foot when it is placed on the ground.

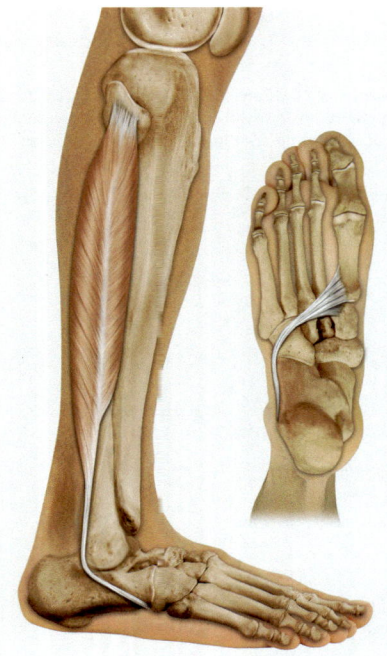

Origin	Proximal two thirds of the fibula
	Anterior and posterior crural intermuscular septum
	Fascia of the leg
Insertion	Base of the first metatarsal bone
	Medial cuneiform bone
Innervation	Superficial peroneal nerve, L5–S1
Special features	The peroneous longus is also referred to by its synonym fibularis longus.

Functions

 Synergists   Antagonists

Talocrural joint

Flexion

Gastrocnemius muscle	Tibialis anterior muscle
Soleus muscle	Extensor digitorum longus muscle
Flexor hallucis longus muscle	Extensor hallucis longus muscle
Peroneus brevis muscle	Peroneus tertius muscle
Tibialis posterior muscle	
Flexor digitorum longus muscle	

Subtalar and talocalcaneonavicular joints

Eversion

Peroneus brevis muscle	Gastrocnemius muscle
Peroneus tertius muscle	Soleus muscle
Extensor digitorum longus muscle	Tibialis posterior muscle
	Flexor digitorum longus muscle
	Flexor hallucis longus muscle
	Tibialis anterior muscle

← Tendon

Muscle function testing

5/4

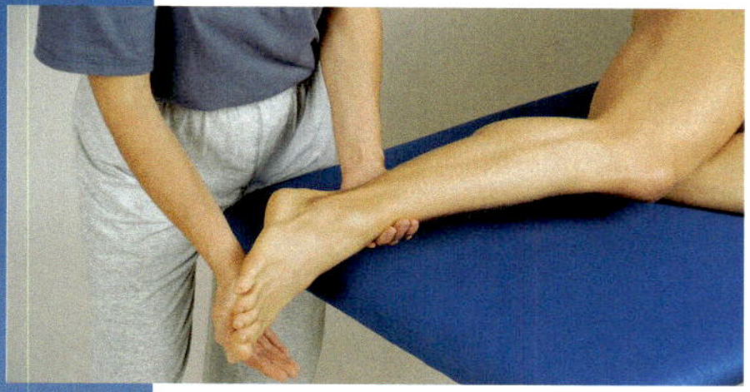

3

2

1/0

Starting position: The patient lies on the side opposite to the one being tested.

Test procedure: The examiner holds the distal part of the patient's lower leg in place with one hand and with the other applies pressure to the plantar surface of the ball of the foot under the big toe in the direction of dorsal flexion and supination.

Instruction: "Push down the big toe side of your foot against my resistance and hold your final position."

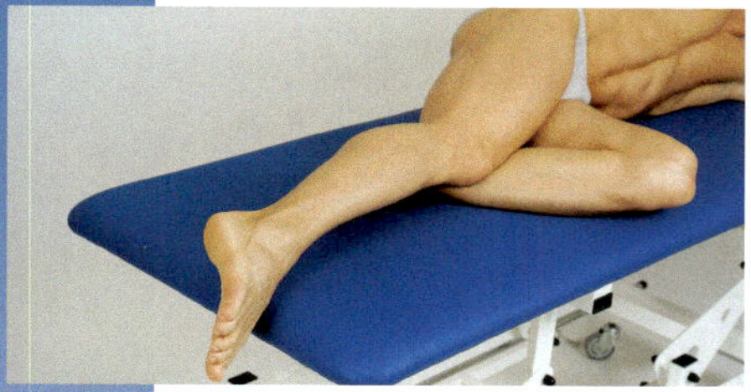

Starting position: The patient lies on the side opposite to the one being tested.

Test procedure: The examiner looks at the movement of the foot.

Instruction: "Push down the big toe side of your foot."

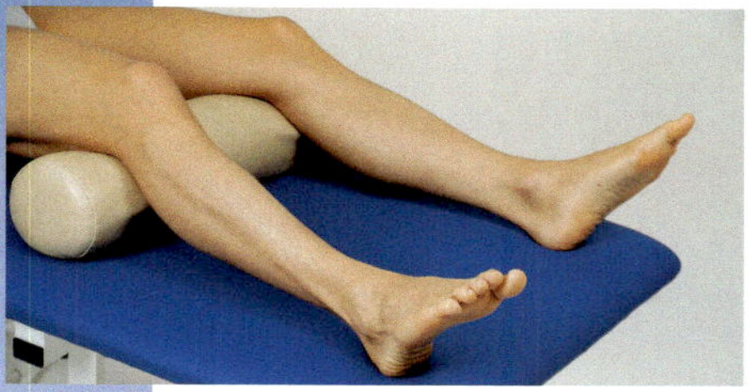

Starting position: The patient lies on his back. The leg is flexed at the knee and supported with a knee bolster.

Test procedure: The examiner looks at the movement of the foot.

Instruction: "Push down the big toe side of your foot."

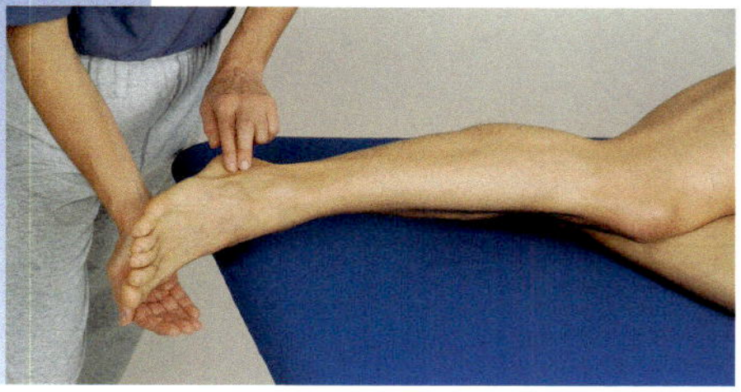

Starting position: The patient lies on the side opposite to the one being tested.

Test procedure: The examiner palpates the peroneus longus muscle, dorsally of the peroneus brevis tendon in the ankle region.

Instruction: "Try to push down the big toe side of your foot."

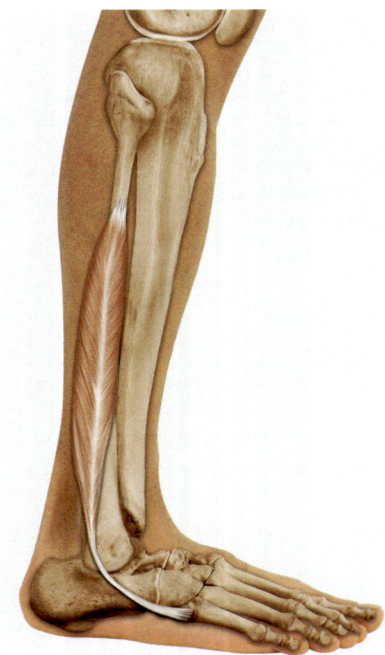

Peroneus brevis muscle

The peroneus brevis (short peroneal) muscle, like its long counterpart, is also a muscle which elevates the lateral edge of the foot and flexes the foot, as well as helping to adjust the foot to the conditions of the surface it rests on. Both peroneal muscles help to maintain balance, particularly when a person stands on one leg. The pronating component is important as it provides an equivalent antagonist to the powerful supinator muscles, thus preventing supination trauma.

Origin	Distal two thirds of the fibula Anterior and posterior crural intermuscular septum
Insertion	Tuberosity of the fifth metatarsal bone
Innervation	Superficial peroneal nerve, S1
Special features	The peroneous brevis is also known under its synonym fibularis brevis.

Functions

 Synergists

Antagonists

Talocrural joint

Flexion

Gastrocnemius muscle
Soleus muscle
Flexor hallucis longus muscle
Peroneus longus muscle
Tibialis posterior muscle
Flexor digitorum longus muscle

Tibialis anterior muscle
Extensor digitorum longus muscle
Extensor hallucis longus muscle
Peroneus tertius muscle

Subtalar and talocalcaneonavicular joints

Eversion

Peroneus longus muscle
Peroneus tertius muscle
Extensor digitorum longus muscle

Gastrocnemius muscle
Soleus muscle
Tibialis posterior muscle
Flexor digitorum longus muscle
Flexor hallucis longus muscle
Tibialis anterior muscle

Tendon

Muscle function testing

Muscle power grade

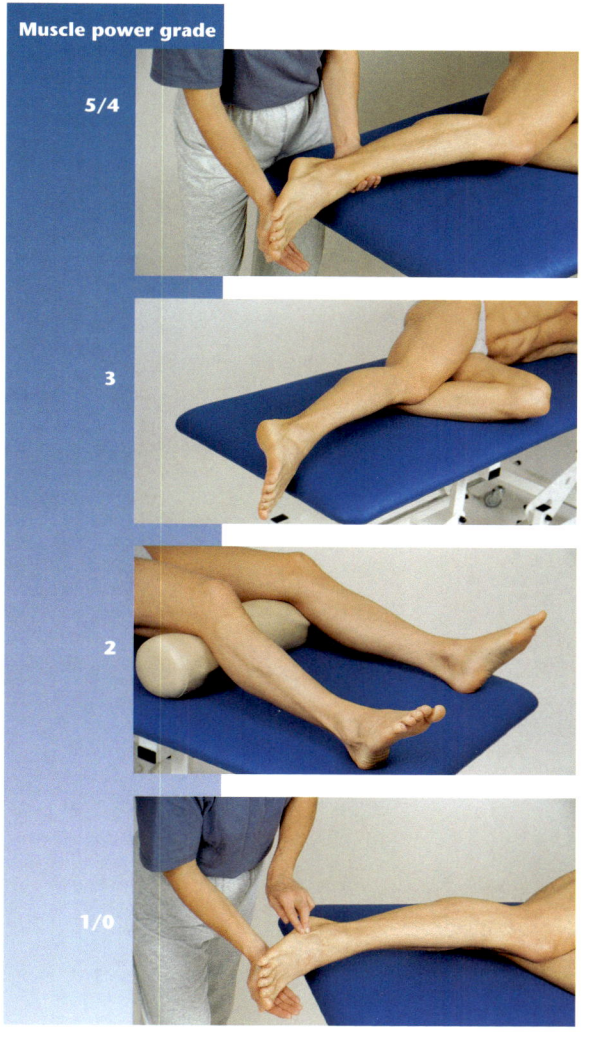

5/4

3

2

1/0

Starting position: The patient lies on the side opposite to the one being tested.

Test procedure: The examiner holds the distal part of the patient's lower leg in place with one hand and with the other applies pressure to the plantar surface of the ball of the foot under the big toe in the direction of dorsal flexion and supination.

Instruction: "Push down the big toe side of your foot against my resistance and hold your final position."

Starting position: The patient lies on the side opposite to the one being tested.

Test procedure: The examiner looks at the movement of the foot.

Instruction: "Push down the big toe side of your foot."

Starting position: The patient lies on his back. The leg is flexed at the knee and supported with a knee bolster.

Test procedure: The examiner looks at the movement of the foot.

Instruction: "Push down the big toe side of your foot."

Starting position: The patient lies on the side opposite to the one being tested.

Test procedure: The examiner palpates the peroneus brevis muscle over the fifth metatarsal bone.

Instruction: "Try to push down the big toe side of your foot."

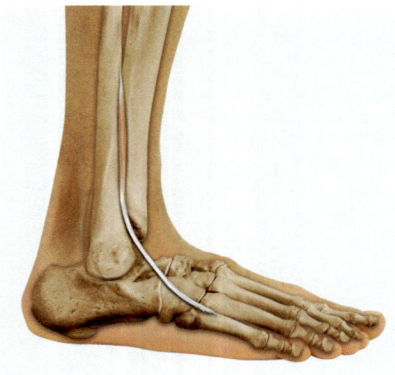

Peroneus tertius muscle

The peroneus tertius (third peroneal) muscle is a branch of the extensor digitorum longus muscle, although its tendon runs to the lateral edge of the foot. Therefore, it elevates the lateral edge of the foot, or everts it, while also producing extension of the joints of the ankle. Overall, it leads to pronation and thus acts to prevent supination trauma.

Origin	Distal third of the fibula Interosseous membrane of the leg
Insertion	Dorsal side of the base of the fifth metatarsal bone
Innervation	Deep peroneal nerve, L5–S1
Special features	The peroneous tertius is not always present; it branches off from the extensor group of muscles

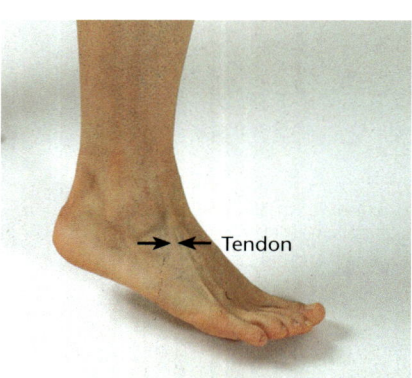

 Tendon

Functions

Synergists

Antagonists

Talocrural joint

Extension

Tibialis anterior muscle	Gastrocnemius muscle
Extensor digitorum longus muscle	Soleus muscle
Extensor hallucis longus muscle	Flexor hallucis longus muscle
	Peroneus longus muscle
	Peroneus brevis
	Tibialis posterior muscle
	Flexor digitorum longus muscle

Subtalar and talocalcaneonavicular joints

Eversion

Peroneus longus muscle	Gastrocnemius muscle
Peroneus brevis muscle	Soleus muscle
Extensor digitorum longus muscle	Tibialis posterior muscle
	Flexor digitorum longus muscle
	Flexor hallucis longus muscle
	Tibialis anterior muscle

Stretch tests

Gastrocnemius, plantaris and soleus muscles

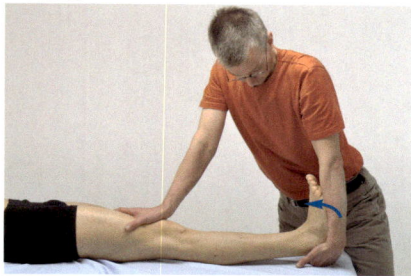

Method
The knee joint is in extension. The therapist moves the patient's foot into maximum dorsal extension at the talocrural joint.

Finding
Muscle shortening is present if the movement cannot be performed to its full extent and if the endpoint feels soft and elastic.
The patient reports a stretching sensation along the course of the muscle.
The soleus muscle is tested in isolation with the knee flexed.

Tibialis posterior muscle

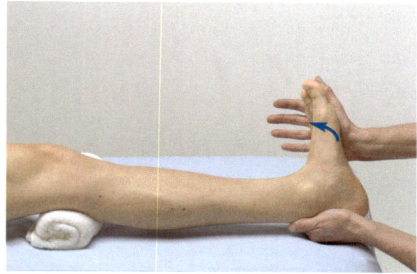

Method
The therapist moves the patient's foot into maximum dorsal extension and pronation at the talocrural, subtalar and talocalcaneonavicular joints. The knee is slightly flexed.

Finding
Muscle shortening is present if the movement cannot be performed to its full extent and if the endpoint feels soft and elastic.
The patient reports a stretching sensation along the course of the muscle.

Tibialis anterior muscle

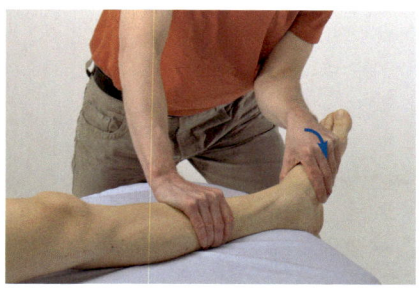

Method
The therapist moves the patient's foot into maximum plantar flexion and pronation at the talocrural, subtalar and talocalcaneonavicular joints.

Finding
Muscle shortening is present if the movement cannot be performed to its full extent and if the endpoint feels soft and elastic.
The patient reports a stretching sensation along the course of the muscle.

3 Lower extremity

Muscles of the toe joints

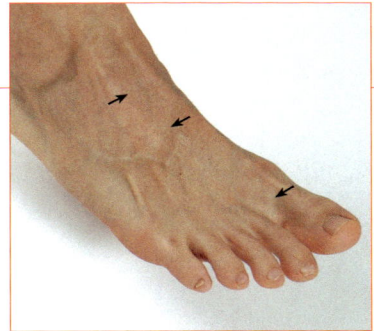

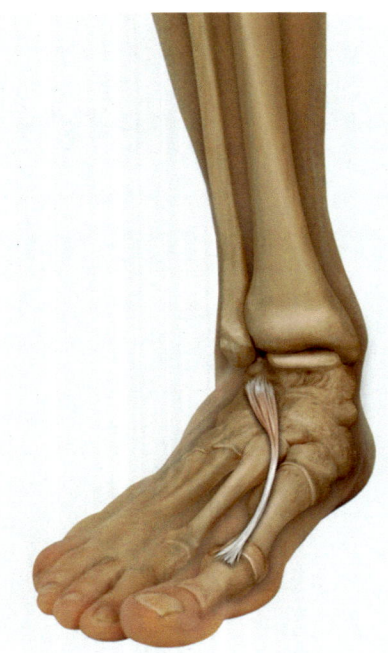

Extensor hallucis brevis muscle

The extensor hallucis brevis muscle (short extensor of the big toe) extends the big toe, as does the extensor hallucis longus muscle.

Origin	Dorsolateral surface of the calcaneal bone Tarsal sinus
Insertion	Proximal phalanx of the big toe
Innervation	Deep fibular (peroneal) nerve, L5–S1

Functions

 Synergists   Antagonists

Metatarsophalangeal joint I

Extension
Extensor hallucis longus muscle

Flexor hallucis longus muscle
Flexor hallucis brevis muscle
Abductor hallucis muscle
Adductor hallucis muscle

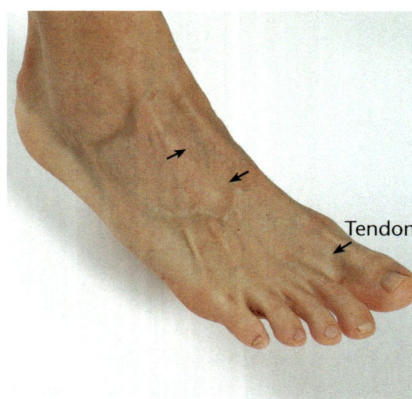

Tendon

Muscle function testing

Muscle power grade	
5/4	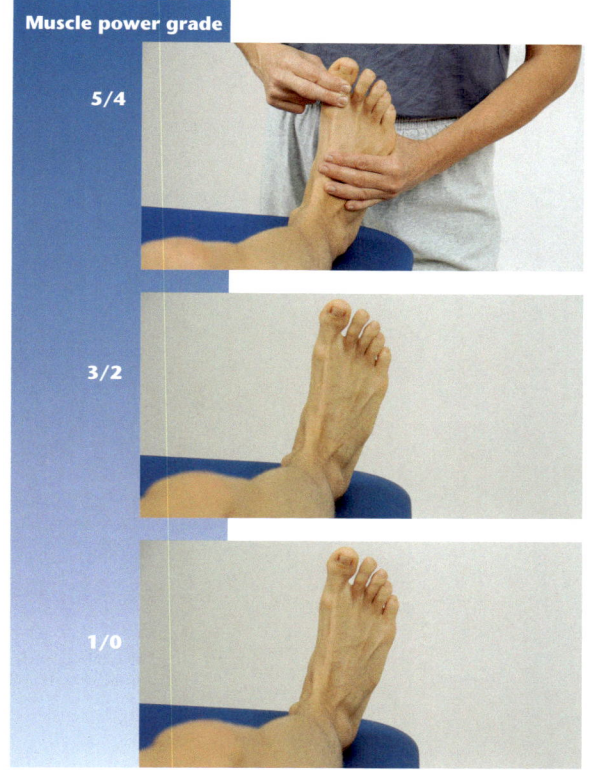
3/2	
1/0	

Starting position: The patient lies on his back. The leg is flexed at the knee and supported with a knee bolster. The foot being tested is in the neutral position.

Test procedure: The examiner holds the patient's metatarsus in place with one hand and with the other applies pressure to the proximal phalanx of the big toe in the direction of flexion of metatarsophalangeal joint I.

Instruction: "Extend your big toe against my resistance and hold your final position."

Starting position: The patient lies on his back. The leg is flexed at the knee and supported with a knee bolster. The foot being tested is in the neutral position.

Test procedure: The examiner looks at the movement of the toe.

Instruction: "Extend your big toe."

Starting position: The patient lies on his back. The leg is flexed at the knee and supported with a knee bolster. The foot being tested is in the neutral position.

Test procedure: The examiner looks at the movement of the toe.

Instruction: "Try to extend your big toe."

 Problems/comments

- The extensor hallucis brevis and extensor hallucis longus muscles work together. It is not possible to identify the tightening of the extensor hallucis brevis in isolation.

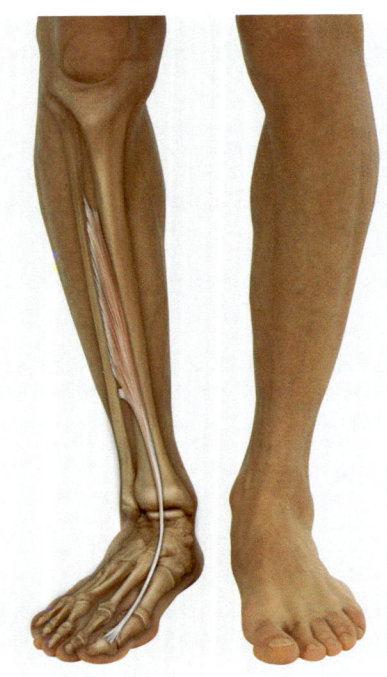

Extensor hallucis longus muscle

The extensor hallucis longus muscle (long extensor muscle of the big toe) is the only powerful extensor of the big toe; it is also, of course, an extensor of the ankle.

Origin	Middle third of the ventral surface of the fibula Interosseous membrane of the leg
Insertion	Dorsal surface of the distal phalanx of the big toe
Innervation	Deep peroneal nerve, L5–S1
Special features	The extensor hallucis longus is an indicator muscle for spinal cord segment L5

Functions

 Synergists    Antagonists

Talocrural, subtalar and talocalcaneonavicular joints

Extension

Tibialis anterior muscle	Gastrocnemius muscle
Extensor digitorum longus muscle	Soleus muscle
	Flexor hallucis longus muscle
	Peroneus longus muscle
	Peroneus brevis muscle
	Tibialis posterior muscle
	Flexor digitorum longus muscle

Metatarsophalangeal joint I

Extension

Extensor hallucis brevis muscle	Flexor hallucis brevis muscle
	Flexor hallucis longus muscle
	Abductor hallucis muscle
	Adductor hallucis muscle

Interphalangeal joint of the foot I

Extension

None	Flexor hallucis longus muscle

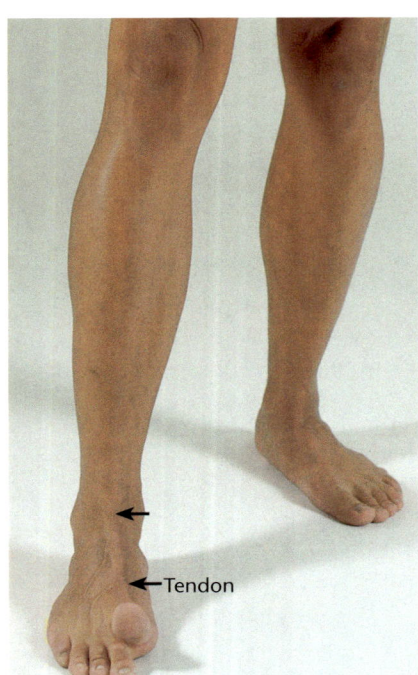

←Tendon

Muscle function testing

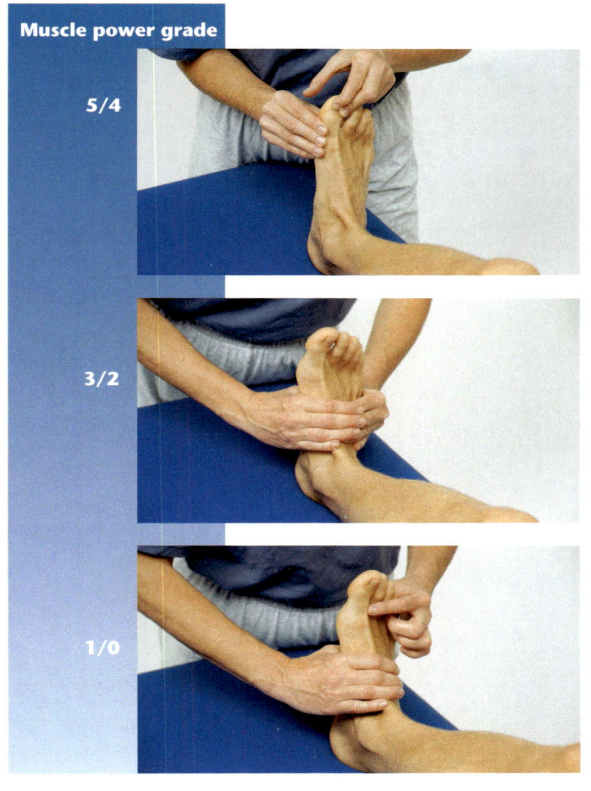

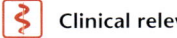

Starting position: The patient lies on his back. The leg is flexed at the knee and supported with a knee bolster. The foot of the extremity being tested is in the neutral position.

Test procedure: The examiner holds the proximal phalanx of the patient's big toe in place with one hand and with the other applies pressure to the distal phalanx of the big toe in the direction of flexion of the distal interphalangeal joint.

Instruction: "Extend your big toe against my resistance and hold your final position."

Starting position: The patient lies on his back. The leg is flexed at the knee and supported with a knee bolster. The foot of the extremity being tested is in the neutral position.

Test procedure: The examiner holds the metatarsus in place.

Instruction: "Extend your big toe."

Starting position: The patient lies on his back. The leg is flexed at the knee and supported with a knee bolster. The foot of the extremity being tested is in the neutral position.

Test procedure: The examiner palpates the extensor hallucis longus muscle.

Instruction: "Try to extend your big toe."

Clinical relevance

- The dorsalis pedis artery pulse can be palpated in the middle of the dorsum of the foot, laterally of the extensor hallucis longus tendon.

Problems/comments

- The extensor hallucis longus muscle aids the tibialis anterior muscle in its function at the ankle.

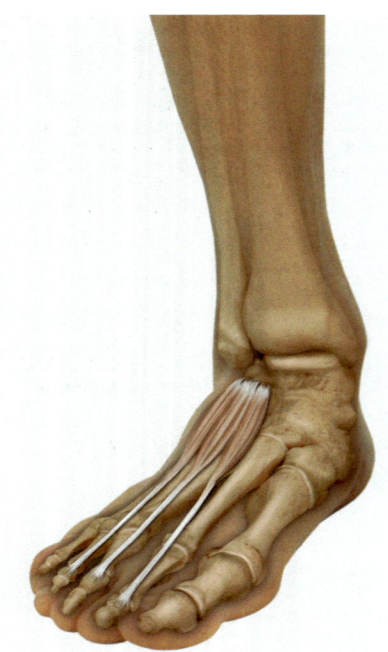

Extensor digitorum brevis muscle

The extensor digitorum brevis muscle (short extensor muscle of the toes) extends toes II to IV, aiding the extensor digitorum longus in this function.

Origin   Upper, lateral surface of the calcaneal bone

Insertion   Dorsal aponeurosis of toes II–IV

Innervation   Deep peroneal nerve, L5–S1

Functions

 Synergists

Antagonists

Metatarsophalangeal and interphal-angeal joints of the foot II to IV

Extension
Extensor digitorum longus muscle (II to IV)

Flexor digitorum longus muscle
Flexor digitorum brevis
Interossei muscles
Lumbricales muscles
Quadratus plantae muscle

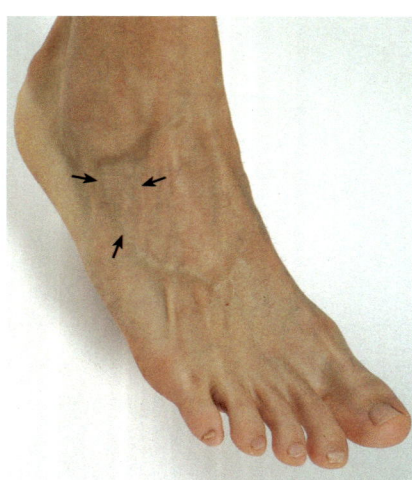

Muscle function testing

Muscle power grade

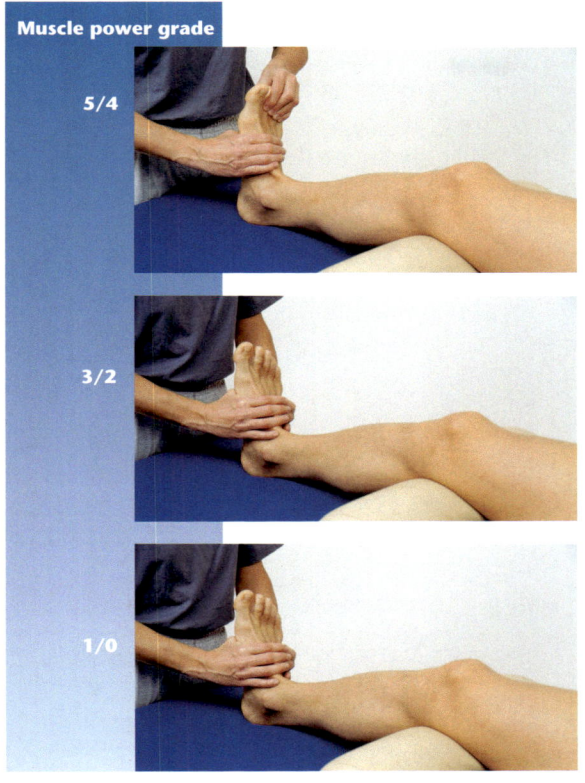

5/4

3/2

1/0

Starting position: The patient lies on his back. The leg is flexed at the knee and supported with a knee bolster. The foot of the extremity being tested is in the neutral position.

Test procedure: The examiner holds the patient's metatarsus in place with one hand and with the other applies pressure across all the toe joints of toes II to IV in the direction of flexion.

Instruction: "Extend your toes against my resistance and hold your final position."

Starting position: The patient lies on his back. The leg is flexed at the knee and supported with a knee bolster. The foot of the extremity being tested is in the neutral position.

Test procedure: The examiner holds the metatarsus in place.

Instruction: "Extend your toes."

Starting position: The patient lies on his back. The leg is flexed at the knee and supported with a knee bolster. The foot of the extremity being tested is in the neutral position.

Test procedure: The examiner looks at the movement of the toes.

Instruction: "Try to extend your toes."

 Problems/comments

• The extensor digitorum brevis is tested together with the extensor digitorum longus muscle.

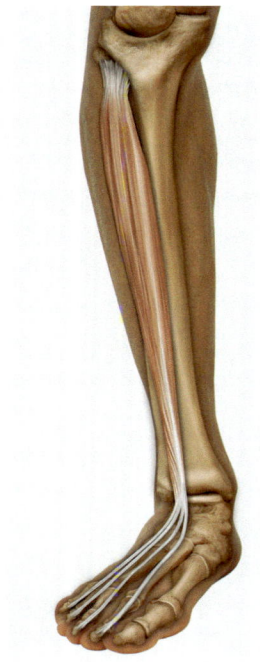

Extensor digitorum longus muscle

The extensor digitorum longus muscle (long extensor muscle of the toes) produces extension of the interphalangeal joints, the metatarsophalangeal joints and the talocrural joint.

Origin	Lateral epicondyle of the tibia
	Proximal three quarters of the ventral surface of the fibula
	Interosseous membrane of the leg
	Crural fascia, anterior intermuscular septum of the leg
Insertion	Via four tendons to the dorsal aponeuroses of toes II–V (middle and distal phalanx)
Innervation	Deep peroneal nerve, L5–S1

Functions

 Synergists

Antagonists

Talocrural joint

Extension
Tibialis anterior muscle
Extensor hallucis longus muscle

Gastrocnemius muscle
Soleus muscle
Flexor hallucis longus muscle
Peroneus longus muscle
Peroneus brevis muscle
Tibialis posterior muscle
Flexor digitorum longus muscle

Subtalar and talocalcaneonavicular joints

Pronation (eversion and extension)
Peroneus longus muscle
Peroneus brevis muscle
Peroneus tertius muscle

Gastrocnemius muscle
Soleus muscle
Tibialis posterior muscle
Flexor digitorum longus muscle
Flexor hallucis longus muscle
Tibialis anterior muscle

Metatarsophalangeal joints II to V

Extension
Extensor digitorum brevis muscle (II–IV)

Flexor digitorum longus muscle (II–IV)
Flexor digitorum brevis muscle
Dorsal interossei muscles of the foot 1–4 (II–IV)
Plantar interossei muscle of the foot 1–3 (III–V)
Lumbricales muscles of the foot 1–4
Flexor digiti minimi brevis muscle (V)
Opponens digiti minimi (V)
Abductor digiti minimi (V)

Proximal and distal interphalangeal joints of the foot II to V

Extension
Extensor digitorum brevis (II–IV) (not at the distal interphalangeal joints)

Flexor digitorum brevis (not at the distal interphalangeal joints)
Flexor digitorum longus muscle

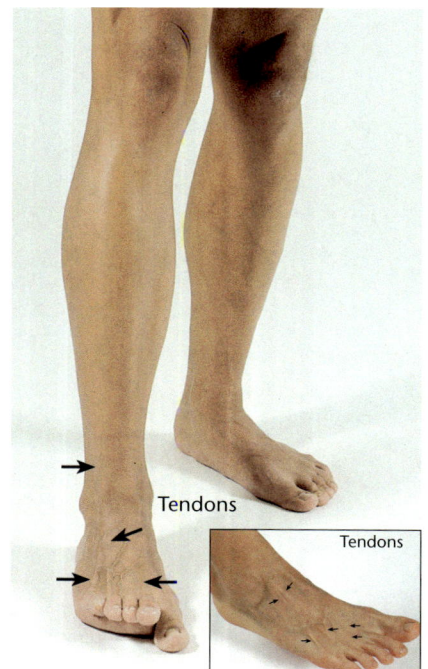

Tendons

Tendons

Muscle function testing

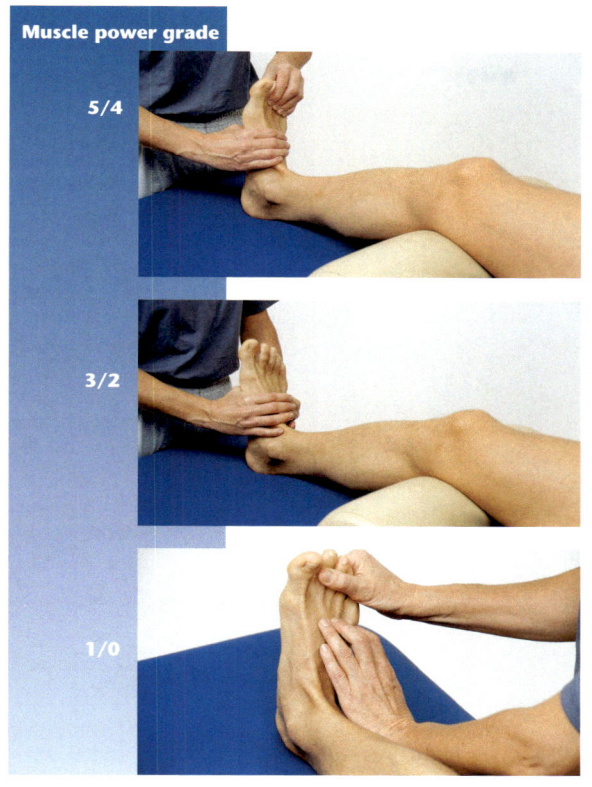

5/4

3/2

1/0

Starting position: The patient lies on his back. The leg is flexed at the knee and supported with a knee bolster. The foot of the extremity being tested is in the neutral position.

Test procedure: The examiner holds the patient's metatarsus in place with one hand and, with the other, applies pressure across all the toe joints of toes II to V in the direction of flexion.

Instruction: "Extend your toes against my resistance and hold your final position."

Starting position: The patient lies on his back. The leg is flexed at the knee and supported with a knee bolster. The foot of the extremity being tested is in the neutral position.

Test procedure: The examiner holds the metatarsus in place.

Instruction: "Extend your toes."

Starting position: The patient lies on his back. The leg is flexed at the knee and supported with a knee bolster. The foot of the extremity being tested is in the neutral position.

Test procedure: The examiner palpates the tendons of the extensor digitorum longus muscle.

Instruction: "Try to extend your toes."

 Problems/comments

- The extensor digitorum longus is tested together with the extensor digitorum brevis muscle.
- The extensor digitorum longus muscle aids the tibialis anterior muscle in its function at the ankle

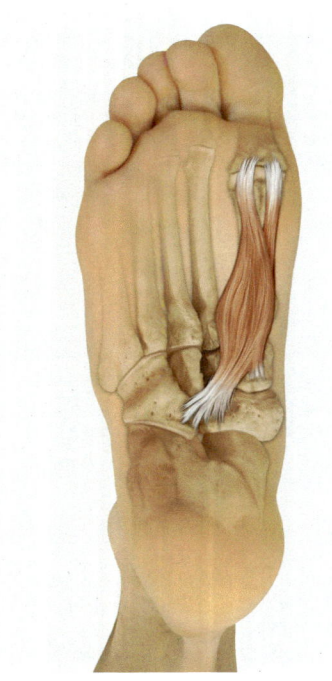

Flexor hallucis brevis muscle

The flexor hallucis brevis muscle (short flexor muscle of the big toe) flexes the big toe powerfully during the final phase of the foot's lift-off from the ground. It contributes towards propulsion and helps to ensure that the foot does not tilt medially during this phase. Furthermore, it tones the medial arch of the foot.

Origin	Medial head and lateral head: plantar side of the cuneiform bones, plantar calcaneocuboid ligament Tibialis posterior tendon
Insertion	Medial head: base of the proximal phalanx of the big toe, medial sesamoid bone Lateral head: base of the proximal phalanx of the big toe, lateral sesamoid bone
Innervation	Medial head: medial plantar nerve, S1–S3 Lateral head: lateral plantar nerve, S1–S3

Functions

 Synergists

 Antagonists

Metatarsophalangeal joint I

Flexion
Flexor hallucis longus muscle
Abductor hallucis muscle
Adductor hallucis muscle

Extensor hallucis longus muscle
Extensor hallucis brevis muscle

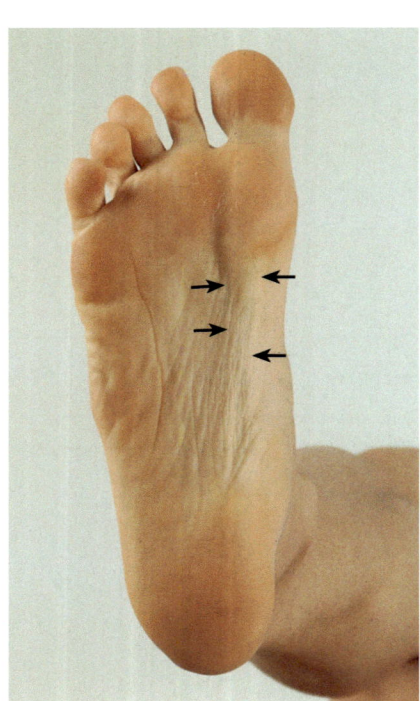

Muscle function testing

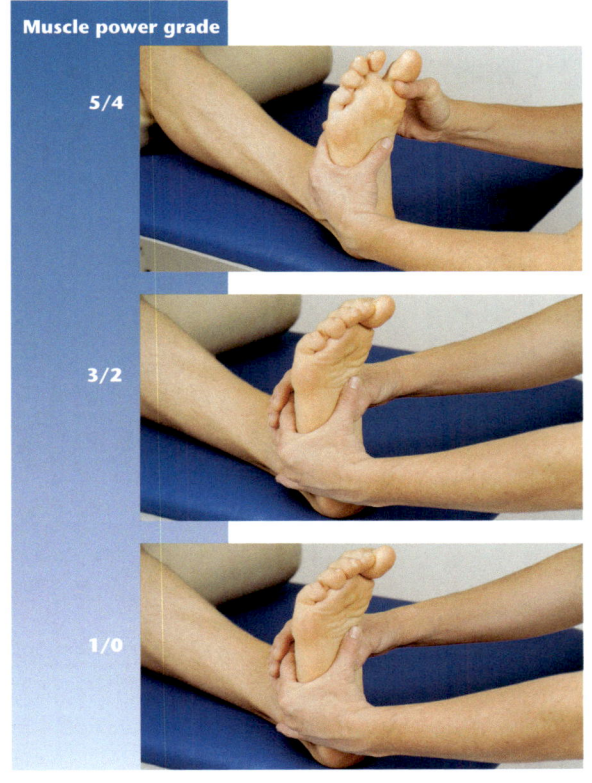

Muscle power grade

5/4

3/2

1/0

Starting position: The patient lies on his back. The leg is flexed at the knee and supported with a knee bolster. The foot being tested is in the neutral position.

Test procedure: The examiner holds the patient's metatarsus in place with one hand and with the other applies pressure to the proximal phalanx of the big toe in the direction of extension of metatarsophalangeal joint I.

Instruction: "Flex your big toe against my resistance and hold your final position."

Starting position: The patient lies on his back. The leg is flexed at the knee and supported with a knee bolster. The foot being tested is in the neutral position.

Test procedure: The examiner holds the patient's metatarsus in place.

Instruction: "Flex your big toe."

Starting position: The patient lies on his back. The leg is flexed at the knee and supported with a knee bolster. The foot of the extremity being tested is in the neutral position.

Test procedure: The examiner looks at the movement of the big toe.

Instruction: "Try to extend your big toe."

 Problems/comments

- The function of the flexor hallucis brevis cannot be distinguished from that of the flexor hallucis longus muscle.

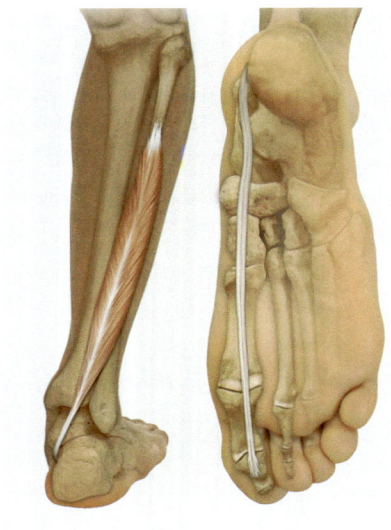

Flexor hallucis longus muscle

The flexor hallucis longus muscle (long flexor muscle of the big toe) flexes the big toe and the ankle joints. It also has a supinating effect on the subtalar and talocalcaneonavicular joints.

Origin	Distal two thirds of the dorsal surface of the fibula
	Interosseous membrane of the leg
Insertion	Distal phalanx of the big toe
Innervation	Tibial nerve, L5–S2

Functions

 Synergists　　　　　　 Antagonists

Talocrural joint

Flexion

Gastrocnemius muscle	Tibialis anterior muscle
Soleus muscle	Extensor digitorum longus muscle
Peronei muscles	Extensor hallucis longus muscle
Tibialis posterior muscle	
Flexor digitorum longus muscle	

Subtalar and talocalcaneonavicular joints

Supination

Gastrocnemius muscle	Peronei muscles
Soleus muscle	Extensor digitorum longus muscle
Tibialis posterior muscle	
Flexor digitorum longus muscle	
Tibialis anterior muscle	

Metatarsophalangeal joint I

Flexion

Flexor hallucis brevis muscle	Extensor hallucis longus muscle
Abductor hallucis muscle	Extensor hallucis brevis muscle
Adductor hallucis muscle	

Interphalangeal joint of the foot I

Flexion

None	Extensor hallucis longus muscle

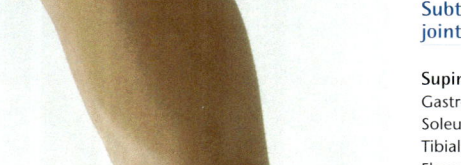

← Tendon

Muscle function testing

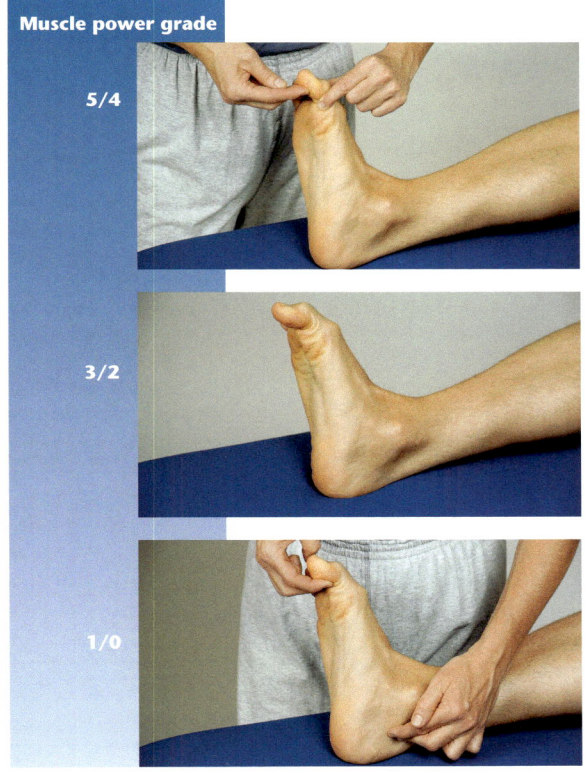

Muscle power grade

5/4

3/2

1/0

Starting position: The patient lies on his back. The leg is flexed at the knee and supported with a knee bolster. The foot being tested is in the neutral position.

Test procedure: The examiner holds the proximal phalanx of the patient's big toe in place with one hand and with the other applies pressure to the distal phalanx of the big toe in the direction of extension.

Instruction: "Flex your big toe against my resistance and hold your final position."

Starting position: The patient lies on his back. The leg is flexed at the knee and supported with a knee bolster. The foot being tested is in the neutral position.

Test procedure: The examiner looks at the movement of the toes.

Instruction: "Flex your big toe."

Starting position: The patient lies on his back. The leg is flexed at the knee and supported with a knee bolster. The foot being tested is in the neutral position.

Test procedure: The examiner palpates the tendons of the flexor hallucis longus muscle.

Instruction: "Try to flex your big toe."

 Problems/comments

- The flexor hallucis longus is the only muscle that can flex the distal interphalangeal joint of the big toe.

- In the region of the medial malleolus, the tendons run in the following order, from malleolus to heel ("Tom, Dick and Harry"):
  - tibialis posterior,
  - flexor digitorum longus and
  - flexor hallucis longus

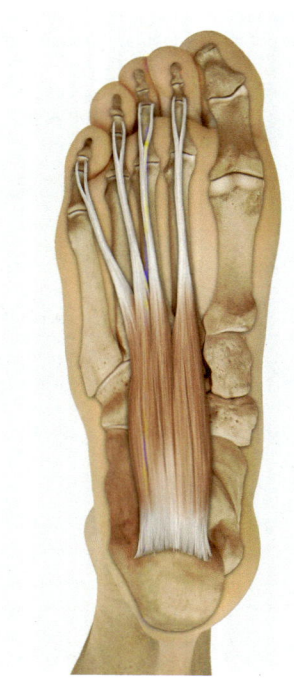

Flexor digitorum brevis muscle

The flexor digitorum brevis muscle (short flexor muscle of the toes) complements the flexor digitorum longus in flexing the toes. As the short flexor, it can focus the action of the extensor digitorum longus onto the joints of the ankle.

Origin	Plantar side of the calcaneal tuberosity Plantar aponeurosis
Insertion	Middle phalanx of toes II–V
Innervation	Medial plantar nerve, S1–S2
Special features	The tendons of the flexor digitorum longus run to the distal phalanges between the bisected tendons of the flexor digitorum brevis

Functions

 Synergists

 Antagonists

Metatarsophalangeal joints II to IV

Flexion

Flexor digitorum longus muscle
Dorsal interossei muscles of the foot 1–4 (II–IV)
Plantar interossei muscles 1–3 (III–V)
Lumbricales muscles of the foot 1–4
Flexor digiti minimi muscle (V)
Opponens digiti minimi muscle (V)
Abductor digiti minimi muscle (V)

Extensor digitorum longus muscle
Extensor digitorum brevis muscle

Interphalangeal joints of the foot II to V

Flexion

Flexor digitorum longus muscle

Extensor digitorum longus muscle
Extensor digitorum brevis muscle (not at the distal interphalangeal joints)

Muscle function testing

Muscle power grade

5/4

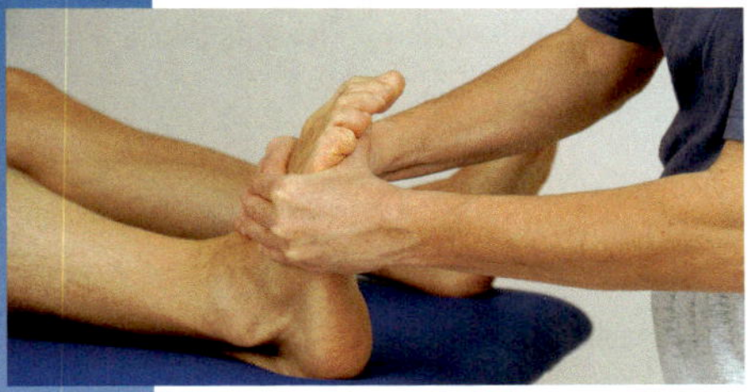

Starting position: The patient lies on his back. The leg is flexed at the knee and supported with a knee bolster. The foot being tested is in the neutral position.

Test procedure: The examiner holds the patient's metatarsus in place with one hand and with the other applies pressure to the middle phalanges of toes II to V in the direction of extension of the metatarsophalangeal and proximal interphalangeal joints.

Instruction: "Flex your toes against my resistance and hold your final position."

3/2

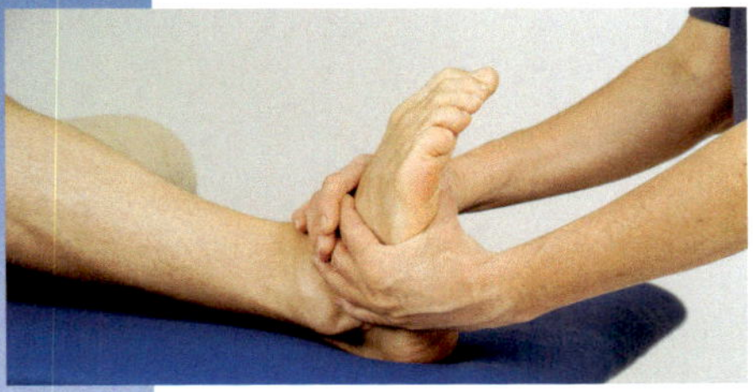

Starting position: The patient lies on his back. The leg is flexed at the knee and supported with a knee bolster. The foot being tested is in the neutral position.

Test procedure: The examiner holds the patient's metatarsus in place.

Instruction: "Flex your toes."

1/0

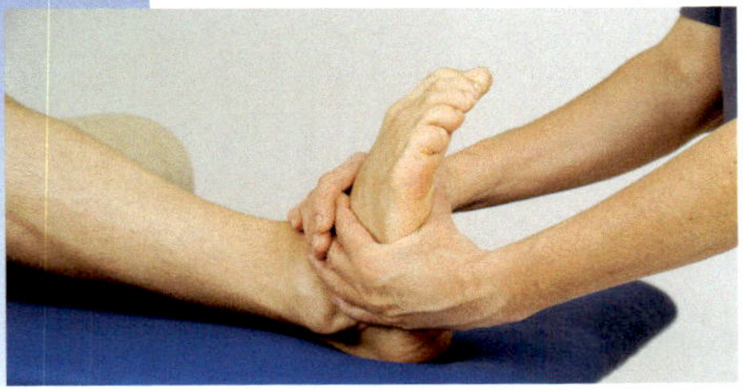

Starting position: The patient lies on his back. The leg is flexed at the knee and supported with a knee bolster. The foot being tested is in the neutral position.

Test procedure: The examiner looks at the movement of the toes.

Instruction: "Try to flex your toes."

 Problems/comments

- Flexion of the metatarsophalangeal and proximal interphalangeal joints is performed jointly by the flexor digitorum brevis, flexor digitorum longus and flexor digiti minimi muscles.

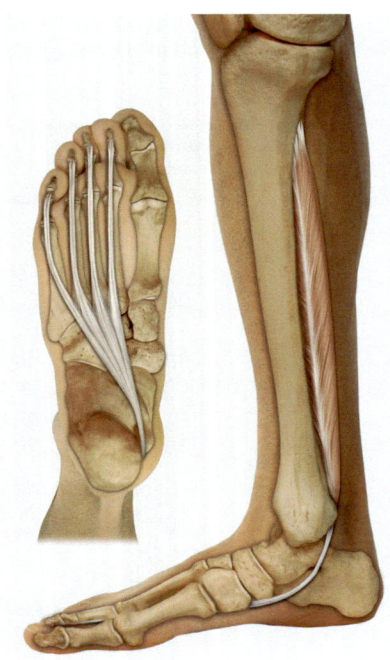

Tendon→

Flexor digitorum longus muscle

The flexor digitorum longus muscle (long flexor muscle of the toes) flexes the toes and the foot. In doing so, it is important in pushing the foot off the ground and in maintaining balance in the standing posture.

Origin   Dorsal surface of the tibia

Insertion   Distal phalanges of toes II–V

Innervation   Tibial nerve, L5–S2

Functions

Synergists

Antagonists

Talocrural joint

Flexion
Gastrocnemius muscle
Soleus muscle
Flexor hallucis longus muscle
Peroneus longus muscle
Peroneus brevis muscle
Tibialis posterior muscle

Tibialis anterior muscle
Extensor digitorum longus muscle
Extensor hallucis longus muscle

Subtalar and talocalcaneonavicular joints

Supination (inversion and flexion)
Gastrocnemius muscle
Soleus muscle
Tibialis posterior muscle
Flexor hallucis longus muscle
Tibialis anterior muscle

Peroneus longus muscle
Peroneus brevis muscle
Extensor digitorum longus muscle
Peroneus tertius muscle

Metatarsophalangeal joints II to V

Flexion
Flexor digitorum brevis muscle
Dorsal interossei muscles of the foot 1-4 (II–IV)
Plantar interossei muscles 1–3 (III–V)
Lumbricales muscles of the foot 1–4
Quadratus plantae muscle

Extensor digitorum longus muscle
Extensor digitorum brevis muscle (II–IV)

Interphalangeal joints of the foot II to V

Flexion
Flexor digitorum brevis
Quadratus plantae muscle
Flexor digiti minimi brevis muscle (V)
Abductor digiti minimi muscle (V)
Opponens digiti minimi muscle

Extensor digitorum longus muscle
Extensor digitorum brevis muscle (II–IV)

Muscle function testing

Muscle power grade	
5/4	
3/2	
1/0	

Starting position: The patient lies on his back. The leg is flexed at the knee and supported with a knee bolster. The foot is in the neutral position.

Test procedure: The examiner holds the patient's metatarsus in place with one hand and, with the other, applies pressure to the distal phalanges of toes II to V in the direction of extension.

Instruction: "Flex your toes against my resistance and hold your final position."

Starting position: The patient lies on his back. The leg is flexed at the knee and supported with a knee bolster. The foot is in the neutral position.

Test procedure: The examiner looks at the movement of the toes.

Instruction: "Flex your toes."

Starting position: The patient lies on his back. The leg is flexed at the knee and supported with a knee bolster.

Test procedure: The examiner palpates the flexor digitorum longus muscle.

Instruction: "Try to flex your toes."

 Problems/comments

- In the region of the medial malleolus, the tendons run in the following order, from malleolus to heel ("Tom, Dick and Harry"):
  - tibialis posterior,
  - flexor digitorum longus and
  - flexor hallucis longus
- The flexor digitorum longus is tested together with the quadratus plantae muscle.

219

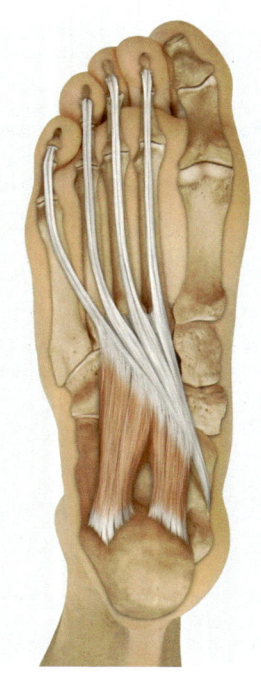

Quadratus plantae muscle

The quadratus plantae muscle (quadrate muscle of the sole) exerts a dorsolateral pull on the tendons of the flexor digitorum longus, which run into the sole of the foot from the medial side. This allows the toes to be bent towards the heel. It also adds its contraction to that of the flexor digitorum longus, allowing powerful flexion of the toes even if the foot is already flexed at the ankle.

Origin	Plantar side of the calcaneus, long plantar ligament
Insertion	Lateral margin of the flexor digitorum longus muscle, before it branches off into its end tendons
Innervation	Lateral plantar nerve, S2–S3
Special features	This muscle is also known as the flexor accessories muscle

Functions

 Synergists   Antagonists

Metatarsophalangeal joints II to V

Flexion

Flexor digitorum brevis muscle
Flexor digitorum longus muscle

Extensor digitorum longus muscle
Extensor digitorum brevis muscle (II–IV)

Interphalangeal joints of the foot II to V

Flexion

Flexor digitorum longus muscle
Flexor digitorum brevis muscle (not at the distal interphalangeal joints)
Dorsal interossei muscles of the foot 1-4 (II–IV)
Plantar interossei muscles 1–3 (III–V)
Lumbricales muscles of the foot 1–4
Flexor digiti minimi brevis muscle (V)
Opponens digiti minimi muscle
Abductor digiti minimi muscle (V)

Extensor digitorum longus muscle
Extensor digitorum brevis muscle (II–IV) (not at the distal interphalangeal joints)

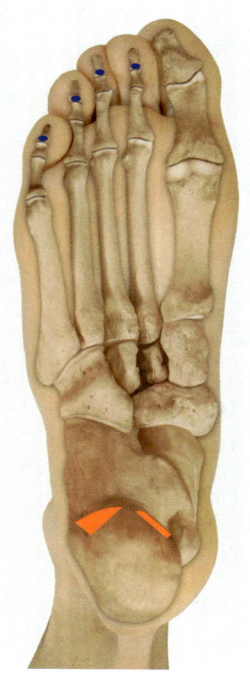

Muscle function testing

Muscle power grade
5/4
3/2
1/0

Starting position: The patient lies on his back. The leg is flexed at the knee and supported with a knee bolster. The foot is in the neutral position.

Test procedure: The examiner holds the patient's metatarsus in place with one hand and with the other applies pressure to the distal phalanges of toes II to V in the direction of extension.

Instruction: "Flex all the joints of your toes against my resistance and hold your final position."

Starting position: The patient lies on his back. The leg is flexed at the knee and supported with a knee bolster. The foot is in the neutral position.

Test procedure: The examiner looks at the movement of the toes.

Instruction: "Flex your toes."

Starting position: The patient lies on his back. The leg is flexed at the knee and supported with a knee bolster.

Test procedure: The examiner looks at the movement of the toes.

Instruction: "Try to flex your toes."

 Problems/comments

- It is not possible to palpate the quadratus plantae muscle.
- The quadratus plantae is tested together with the flexor digitorum longus muscle.

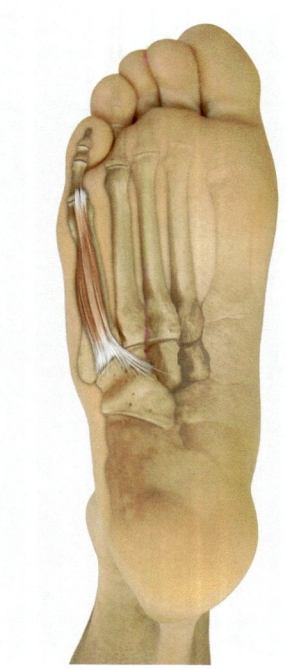

Flexor digiti minimi brevis muscle

The flexor digiti minimi brevis muscle (short flexor muscle of the little toe) flexes the little toe and supports the arch of the foot.

Origin	Base of the fifth metatarsal bone
	Long plantar ligament
	Tendon sheath of the peroneus (fibularis) longus muscle
Insertion	Proximal phalanx of the little toe
Innervation	Lateral plantar nerve, S2–S3

Functions

 Synergists Antagonists

Metatarsophalangeal joint V

Flexion

Flexor digitorum longus muscle	Extensor digitorum longus muscle
Flexor digitorum brevis muscle	
Opponens digiti minimi muscle	
Abductor digiti minimi muscle	
Plantar interosseous muscle 3	
Lumbrical muscle of the foot 4	

Abduction

Abductor digiti minimi muscle	Plantar interosseous muscle 3
Opponens digiti minimi muscle	

Muscle function testing

Muscle power grade	
5/4	
3/2	
1/0	

Starting position: The patient lies on his back. The leg is flexed at the knee and supported with a knee bolster. The foot being tested is in the neutral position.

Test procedure: The examiner holds the patient's metatarsus in place with one hand and with the other applies pressure to the proximal phalanx of the little toe in the direction of extension.

Instruction: "Flex your toes against my resistance and hold your final position."

Starting position: The patient lies on his back. The leg is flexed at the knee and supported with a knee bolster. The foot being tested is in the neutral position.

Test procedure: The examiner holds the patient's metatarsus in place.

Instruction: "Flex your toes."

Starting position: The patient lies on his back. The leg is flexed at the knee and supported with a knee bolster. The foot being tested is in the neutral position.

Test procedure: The examiner looks at the movement of the little toe.

Instruction: "Try to flex your toes."

 Problems/comments

- Flexion at the metatarsophalangeal and proximal interphalangeal joints is carried out jointly by the flexor digitorum brevis, the flexor digitorum longus, the quadratus plantae and the flexor digiti minimi muscles.

Dorsal interossei muscles of the foot

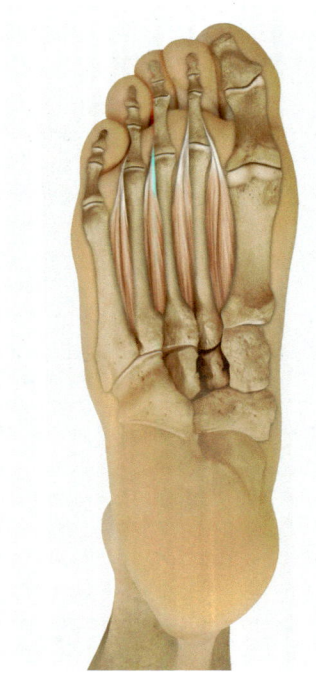

The dorsal interossei (dorsal interosseous) muscles of the foot abduct the toes and cause flexion at the metatarsophalangeal joints. In doing so, acting in conjunction with the lumbricales and plantar interossei muscles, they help to conduct the force of the long extensors onto the interphalangeal joints and the joints of the ankle. They themselves have virtually no effect on the interphalangeal joints

Origin	Two heads each, originating from two adjoining metatarsal bones in each case
Insertion	Lateral surface of the proximal phalanges of toes II–IV (fibular side of toes II–IV, tibial side of toe II)
Innervation	Lateral plantar nerve, S2–S3

Functions

 Synergists   Antagonists

Metatarsophalangeal joints II to IV

Flexion

Flexor digitorum longus muscle	Extensor digitorum longus muscle
Flexor digitorum brevis muscle	Extensor digitorum brevis muscle
Plantar interossei muscles 1–2 (III–IV)	
Lumbricales muscles of the foot 1-3 (II–IV)	

Abduction (III and IV only)

None	Plantar interossei muscles 1–2 (III–IV)

Muscle function testing

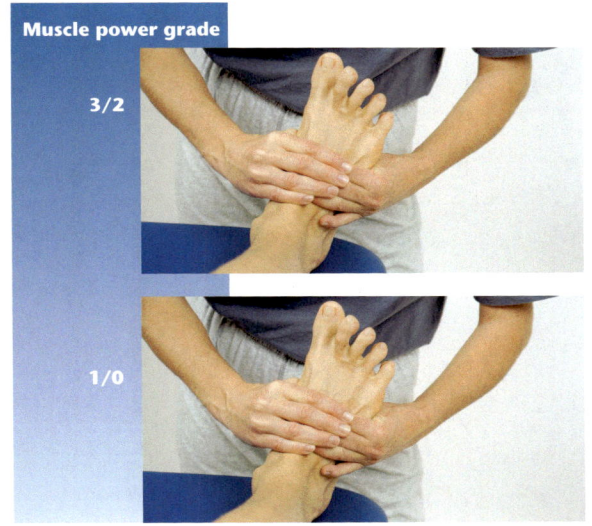

Muscle power grade

3/2

1/0

Starting position: The patient lies on his back. The leg is flexed at the knee and supported with a knee bolster.

Test procedure: The examiner holds the foot in the neutral position.

Instruction: "Spread out your toes."

Starting position: The patient lies on his back. The leg is flexed at the knee and supported with a knee bolster.

Test procedure: The examiner holds the foot in the neutral position and looks at the movement of the toes.

Instruction: "Try to spread out your toes."

 Problems/comments

- The absence of resistance in this test is intentional. It is common for people not to be able to move these muscles voluntarily from lack of use in daily life.

- The four dorsal interossei muscles are inserted at the proximal phalanges of toes II to IV. The big toe and little toe have their own abductors.

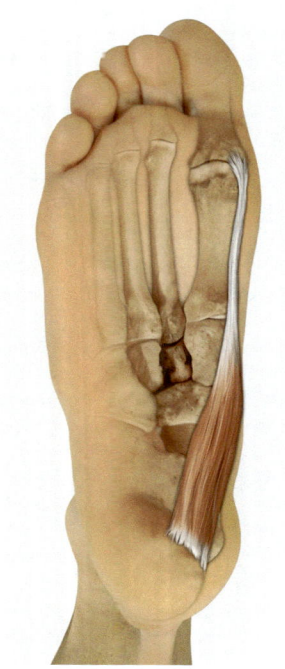

Abductor hallucis muscle

The abductor hallucis muscle (abductor muscle of the big toe) abducts the big toe and flexes it slightly at its metatarsophalangeal joint. Above all, it stabilizes the medial arch of the foot, particularly when the ball of the foot is lifted off the ground and the weight of the body tends to flatten the arch.

Origin	Medial part of the calcaneal tuberosity
	Plantar aponeurosis
Insertion	Proximal phalanx of the big toe
Innervation	Medial plantar nerve, S1–S2

Functions

![Synergists icon] Synergists

![Antagonists icon] Antagonists

Metatarsophalangeal joint I

Abduction

None

Adductor hallucis muscle
Extensor hallucis brevis muscle
Flexor hallucis longus muscle (with the big toe adducted)
Extensor hallucis longus muscle (with the big toe adducted)

Flexion

Flexor hallucis longus muscle
Flexor hallucis brevis muscle
Adductor hallucis muscle

Extensor hallucis longus muscle
Extensor hallucis brevis muscle

← Tendon

Muscle function testing

Muscle power grade

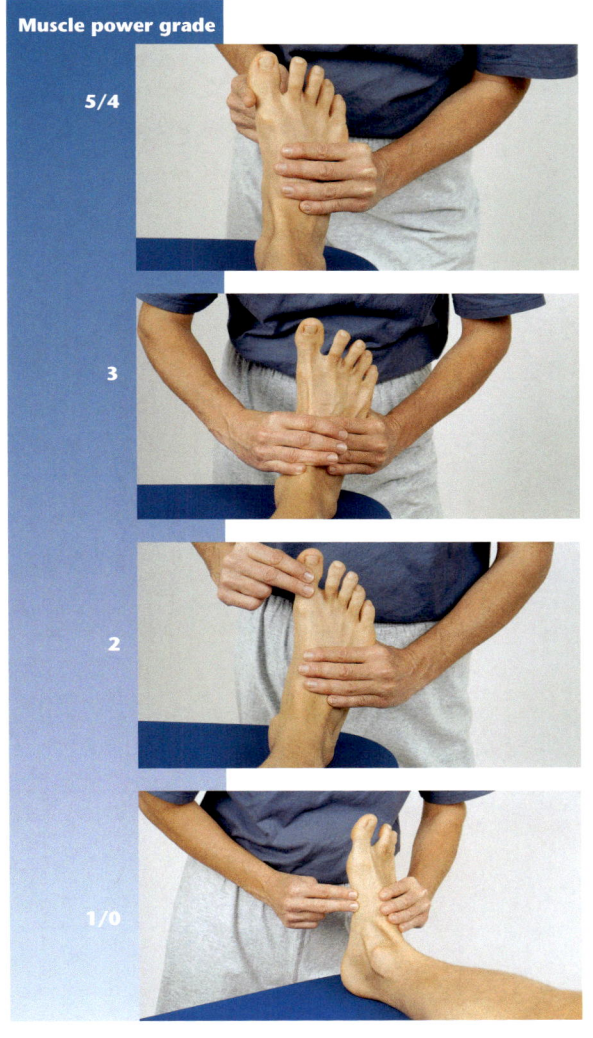

5/4

3

2

1/0

Starting position: The patient lies on his back. The leg is flexed at the knee and supported with a knee bolster. The foot being tested is in the neutral position.

Test procedure: The examiner holds the patient's metatarsus in place with one hand and with the other applies pressure to the outer side of the proximal phalanx of the big toe in the direction of adduction of metatarsophalangeal joint I.

Instruction: "Splay out your big toe against my resistance and hold your final position."

Starting position: The patient lies on his back. The leg is flexed at the knee and supported with a knee bolster. The foot being tested is in the neutral position.

Test procedure: The examiner holds the patient's metatarsus in place.

Instruction: "Splay out your big toe."

Starting position: The patient lies on his back. The leg is flexed at the knee and supported with a knee bolster.

Test procedure: The examiner holds the patient's metatarsus in place and supports the big toe.

Instruction: "Splay out your big toe."

Starting position: The patient lies on his back. The leg is flexed at the knee and supported with a knee bolster. The foot being tested is in the neutral position.

Test procedure: The examiner palpates the abductor hallucis muscle.

Instruction: "Try to splay out your big toe."

⚠️ Problems/comments

- The big toe often exhibits a valgus deformity, and so the abductor hallucis muscle becomes atrophied. This makes it much more difficult to tighten this muscle. If this is the case, it is helpful to bring the big toe passively into the neutral position and only then to allow the abductor hallucis to tighten (see test for muscle power grade 2).

- In hallux valgus, the extensors and flexors change position and become adductors of the big toe. Tightening of these muscles also acts against contraction of the abductor hallucis.

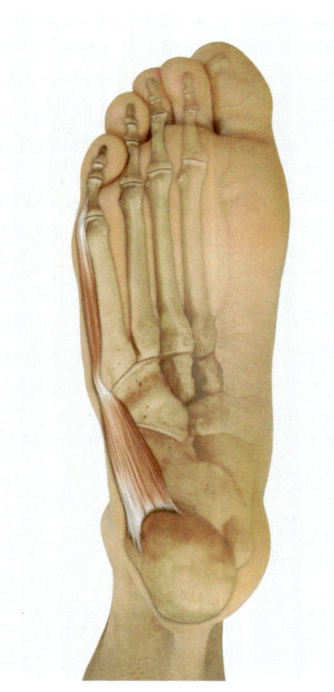

Abductor digiti minimi muscle

The abductor digiti minimi muscle (abductor muscle of the little toe) abducts and flexes the little toe. This muscle also tones the lateral arch of the foot.

Origin   Lateral process of the calcaneal tuberosity, adjoining fascias

Insertion   Lateral surface of the proximal phalanx of the little toe

Innervation   Lateral plantar nerve, S2–S3

Functions

 Synergists   Antagonists

Metatarsophalangeal joint V

Flexion
Flexor digitorum longus muscle   Extensor digitorum longus muscle
Flexor digitorum brevis muscle
Flexor digiti minimi muscle
Opponens digiti minimi muscle
Plantar interosseous muscle 3
Lumbrical muscle of the foot 4

Abduction
Opponens digiti minimi muscle   Plantar interosseous muscle 3
                                              Lumbrical muscle of the foot 4

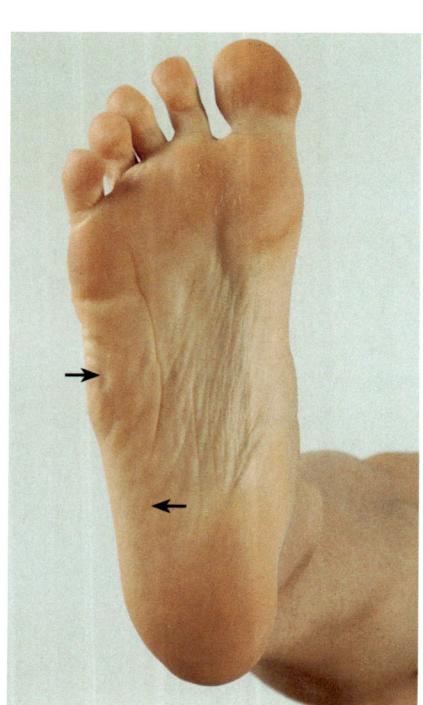

Muscle function testing

5/4

Starting position: The patient lies on his back. The leg is flexed at the knee and supported with a knee bolster. The foot being tested is in the neutral position.

Test procedure: The examiner holds the patient's metatarsus in place with one hand and with the other applies pressure to the outer side of the proximal phalanx of the little toe in the direction of adduction of metatarsophalangeal joint V.

Instruction: "Splay out your little toe against my resistance and hold your final position."

3

Starting position: The patient lies on his back. The leg is flexed at the knee and supported with a knee bolster. The foot being tested is in the neutral position.

Test procedure: The examiner holds the patient's metatarsus in place.

Instruction: "Splay out your little toe."

2

Starting position: The patient lies on his back. The leg is flexed at the knee and supported with a knee bolster. The foot being tested is in the neutral position.

Test procedure: The examiner holds the patient's metatarsus in place and supports the little toe.

Instruction: "Splay out your little toe."

1/0

Starting position: The patient lies on his back. The leg is flexed at the knee and supported with a knee bolster. The foot being tested is in the neutral position.

Test procedure: The examiner palpates the abductor digiti minimi muscle.

Instruction: "Try to splay out your big toe."

⚠ Problems/comments

- The little toe often exhibits a varus deformity, and so the abductor digiti minimi muscle becomes atrophied. This makes it much more difficult to tighten this muscle. If this is the case, it is helpful to bring the little toe passively into the neutral position and only then to allow the abductor digiti minimi to tighten (see the test for muscle power grade 2).

- In varus deformity of the little toe, the extensors and flexors change position and become adductors of the little toe. Tightening of these muscles also acts against contraction of the abductor digiti minimi.

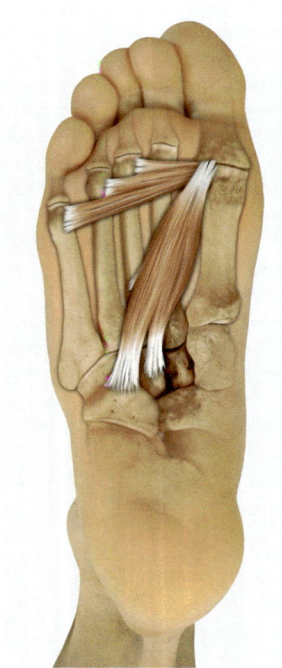

Adductor hallucis muscle

The abductor hallucis muscle (adductor muscle of the big toe) adducts the big toe and flexes it slightly. Importantly, it runs across the sole, toning the arch of the foot.

Origin	Oblique head: cuboid bone, lateral cuneiform bone, plantar calcaneocuboid ligament, long plantar ligament Transverse head: capsules of metatarsophalangeal joints III to V, deep transverse metatarsal ligament
Insertion	Proximal phalanx and fibular sesamoid bone of the big toe
Innervation	Lateral plantar nerve, S2–S3

Functions

 Synergists

 Antagonists

Metatarsophalangeal joint I

Adduction

Flexor hallucis longus muscle (with the big toe already adducted)
Extensor hallucis longus muscle (with the big toe already adducted)

Abductor hallucis muscle

Flexion

Flexor hallucis longus muscle
Flexor hallucis brevis muscle
Abductor hallucis muscle

Extensor hallucis longus muscle
Extensor hallucis brevis muscle

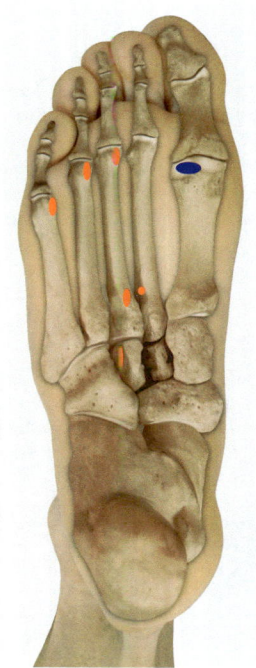

Muscle function testing

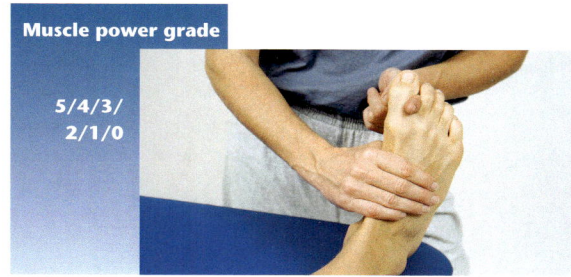

Muscle power grade

5/4/3/
2/1/0

Starting position: The patient lies on his back. The leg is flexed at the knee and supported with a knee bolster. The foot being tested is in the neutral position.

Test procedure: The examiner holds the patient's metatarsus in place with one hand. He places a finger of the other hand between the first and second toes.

Instruction: "Try to squeeze my finger with your toes."

 Problems/comments

- In many patients, the big toe is already in adduction (hallux valgus). As a result, there is no visible movement when the patient is asked to bring the big toe back towards the other toes.

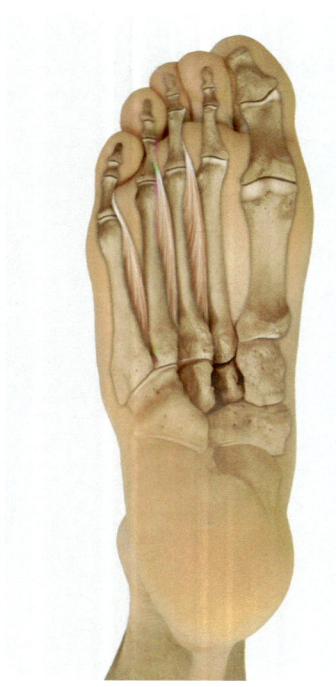

Plantar interossei muscles

The plantar interossei (plantar interosseous) muscles adduct toes III, IV and V towards toe II. As a result of their flexor effect on the relevant metatarsophalangeal joints, they focus the effect of the long extensors on the interphalangeal joints or the joints of the ankle.

Origin	Bases and medial parts of the third to fifth metatarsal bones
Insertion	Medial parts of the proximal phalanges of toes III–V
Innervation	Lateral plantar nerve, S2–S3

Functions

 Synergists

 Antagonists

Metatarsophalangeal joints III to V

Flexion
Flexor digitorum longus muscle
Flexor digitorum brevis muscle
Dorsal interossei muscles of the foot 3–4 (III–IV)
Lumbricales muscles of the foot 2–4
Quadratus plantae muscle
Flexor digiti minimi brevis muscle (V)
Opponens digiti minimi muscle (V)
Abductor digiti minimi muscle (V)

Extensor digitorum longus muscle
Extensor digitorum brevis muscle (III–IV)

Adduction
None

Dorsal interossei muscles of the foot 3–4 (III–IV)

Muscle function testing

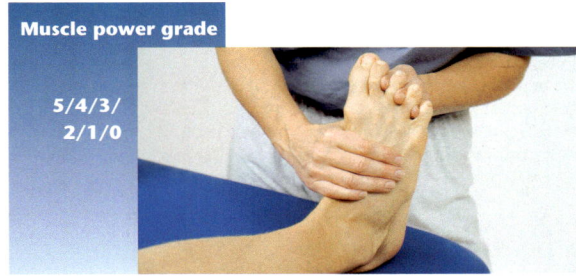

Muscle power grade

5/4/3/
2/1/0

Starting position: The patient lies on his back. The leg is flexed at the knee and supported with a knee bolster. The foot being tested is in the neutral position.

Test procedure: The examiner holds the patient's metatarsus in place with one hand. He places one finger each of the other hand between toes II and III, III and IV and IV and V, respectively.

Instruction: "Try to squeeze my fingers with your toes."

 Problems/comments

- Many people find it difficult to perform this movement voluntarily.
- The plantar interossei muscles are supported in their function by the lumbricales muscle of the foot.

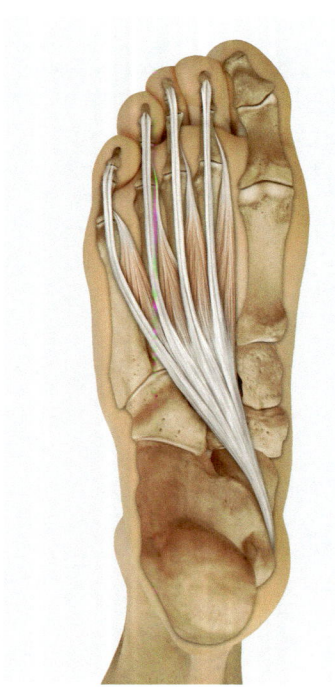

Lumbricales muscles of the foot

The lumbricales (lumbrical) muscles of the foot flex the metatarsophalangeal joints. This function is important, as it prevents hyperextension of these joints, thus concentrating the force of the extensorsonto the interphalangeal joints. Thus, contraction of the lumbricales muscles can determine the joints onto which the force of the extensors is to be exerted. In contrast to their counterparts in the hand, the lumbricales and interossei muscles of the foot have little or no effect on extension of the proximal and distal interphalangeal joints.

Origin	Tendons of the flexor digitorum longus muscle
Insertion	Proximal phalanges of toes II–V, dorsal aponeuroses of these toes
Innervation	Lumbrical muscle 1: medial plantar nerve, S1–S2
	Lumbrical muscles 2–4: lateral plantar nerve, S2–S3

Functions

 Synergists

 Antagonists

Metatarsophalangea

Flexion

Flexor digitorum longus muscle	Extensor digitorum longus muscle
Flexor digitorum brevis muscle	Extensor digitorum brevis muscle (II–IV)
Dorsal interossei muscles of the foot 1–4 (II–IV)	
Plantar interossei muscles 1–3 (III–V)	
Quadratus plantae muscle	
Flexor digiti minimi brevis muscle (V)	
Abductor digiti minimi muscle (V)	

Stretch tests

Flexor digitorum longus and flexor hallucis longus muscles

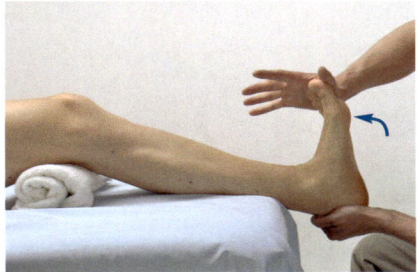

Method

The patient's toes are extended at the metatarsophalangeal, proximal interphalangeal and distal interphalangeal joints. The therapist moves the foot into maximum dorsal extension and pronation at the talocrural, subtalar and talocalcaneonavicular joints.

Finding

Muscle shortening is present if the movement cannot be performed to its full extent, or if the toes cannot be kept extended, and if the endpoint feels soft and elastic.
The patient reports a stretching sensation along the course of the muscle.

Extensor digitorum longus and extensor hallucis longus muscles

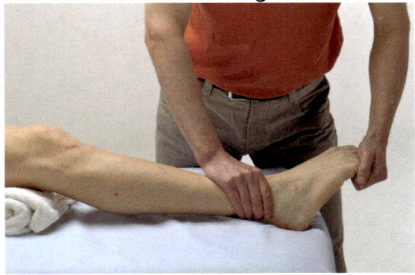

Method

The patient's toes are flexed at the metatarsophalangeal, proximal interphalangeal and distal interphalangeal joints. The therapist moves the patient's foot into maximum plantar flexion and pronation to test the extensor hallucis longus, and into maximum plantar flexion with supination to test the extensor digitorum longus.

Finding

Muscle shortening is present if the movement cannot be performed to its full extent, or if the toes cannot be kept flexed, and if the endpoint feels soft and elastic.
The patient reports a stretching sensation along the course of the muscle.

Adductor hallucis muscle

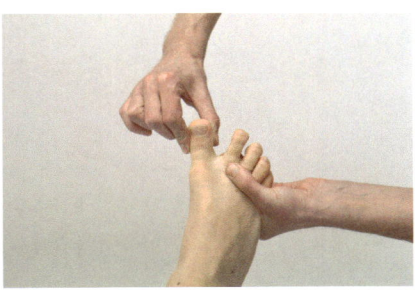

Method

The therapist moves the patient's big toe into maximum abduction at the metatarsophalangeal joint.

Finding

Muscle shortening is present if the movement cannot be performed to its full extent and if the endpoint feels soft and elastic. The patient reports a stretching sensation along the course of the muscle

Innervation of the lower extremity									
Nerve	**Innervated muscles**	**Segments of origin**							
		L1	L2	L3	L4	L5	S1	S2	S3
Femoral nerve		X	X	X	X				
	Psoas muscle		X	X	X				
	Iliacus muscle		X	X	X				
	Pectineus muscle		X	X	X				
	Sartorius muscle		X	X	X				
	Quadriceps femoris muscle		X	X	X				
Obturator nerve			X	X	X				
	Pectineus muscle		X	X	X				
	Adductor longus muscle		X	X	X				
	Adductor brevis muscle		X	X	X				
	Adductor magnus muscle, deep part Superficial part: tibial nerve L4–L5		X	X	X				
	Gracilis muscle		X	X	X				
	Obturatorius externus muscle		X	X	X				
Superior gluteal nerve					X	X	X		
	Tensor fasciae latae muscle				X	X	X		
	Gluteus medius muscle				X	X	X		
	Gluteus minimus muscle				X	X	X		
Inferior gluteal nerve						X	X	X	
	Gluteus maximus muscle					X	X	X	
	Quadratus femoris muscle (partially)					X	X		
Sciatic nerve					X	X	X	X	X
	Quadratus femoris muscle (partially)				X	X	X		
Direct branches from the sacral plexus					X	X	X	X	X
	Piriformis muscle						X	X	
	Obturatorius internus muscle					X	X	X	
	Gemelli muscles					X	X	X	
Tibial nerve					X	X	X	X	
	Biceps femoris muscle, long head					X	X	X	
	Semimembranosus muscle				X	X	X		
	Semitendinosus muscle				X	X	X		
Common fibular (peroneal) nerve					X	X	X		
	Biceps femoris muscle, short head					X	X	X	

Nerves of the lower extremity 2						
Nerve	**Innervated muscles**	**Segments of origin**				
		L4	L5	S1	S2	S3
Femoral nerve		■	■	■	■	■
	Biceps femoris muscle, long head		■	■		
	Semimembranosus muscle		■	■		
	Semitendinosus muscle		■	■		
	Triceps surae muscle			■	■	
	Plantaris muscle			■	■	
	Popliteus muscle		■	■		
	Tibialis posterior muscle	■	■	■		
	Flexor digitorum longus muscle		■	■	■	
	Flexor hallucis longus muscle		■	■	■	
Common fibular (peroneal) nerve		■	■	■	■	
	Biceps femoris muscle, short head		■	■	■	
Deep fibular (peroneal) nerve		■	■	■	■	
	Tibialis anterior muscle	■	■			
	Extensor digitorum longus muscle	■	■	■		
	Extensor hallucis longus muscle		■	■		
	Extensor digitorum brevis muscle		■	■		
	Peroneus tertius muscle	■	■	■		
Superficial fibular (peroneal) nerve		■	■	■	■	
	Peroneus (fibularis) longus muscle		■	■		
	Peroneus (fibularis) brevis muscle		■	■		
Lateral plantar nerve		■	■	■	■	■
	Abductor digit minimi muscle			■	■	
	Flexor digiti minimi brevis muscle			■	■	
	Opponens digiti minimi muscle			■	■	
	Plantar interossei muscles			■	■	
	Lumbricales muscles of the foot III–IV			■	■	
	Dorsal interossei muscles of the foo			■	■	
	Flexor hallucis brevis muscle, lateral head			■	■	
	Adductor hallucis muscle			■	■	
	Quadratus plantae muscle			■	■	
Medial plantar nerve		■	■	■	■	
	Abductor hallucis muscle		■	■		
	Lumbricales muscles of the foot I and II			■	■	
	Flexor hallucis brevis muscle, medial head		■	■		
	Flexor digitorum brevis		■	■		

4 Trunk

Deep muscles of the back, lumbar

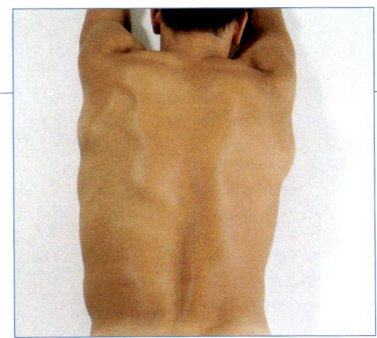

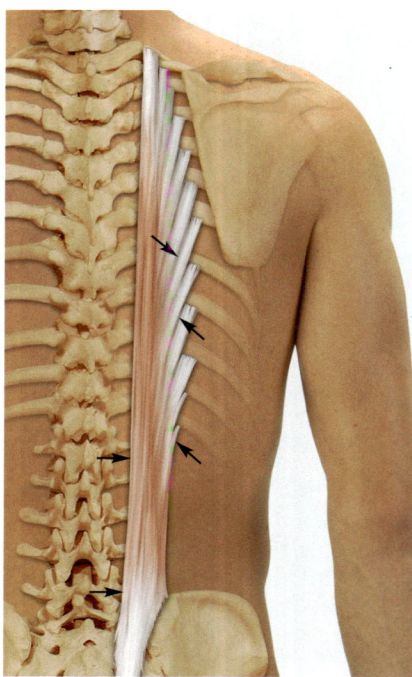

Iliocostalis lumborum muscle
Lateral tract, sacrospinal system

The iliocostalis or iliocostal muscle (also known as the sacrolumbalis in this region) can extend the whole spine powerfully if it tightens on both sides. When contracting unilaterally, the iliocostalis inclines the spine to the same side. Its rotatory effect on the trunk is negligible.

Origin Sacral bone, iliac crest, spinous processes of all the lumbar vertebrae, thoracolumbar fascia

Insertion Costal angles of ribs 7–12

Innervation Dorsal branches of spinal nerves T5–L5

Special features The iliocostalis thoracis muscle is not considered part of the iliocostalis lumborum muscle in the text below, but is listed separately

Functions

 Synergists

 Antagonists

Intervertebral joints and disks (lumbar spine)

Extension (bilateral contraction)
All the other deep back muscles in the region

Rectus abdominis muscle
Obliquus externus abdominis muscle
Obliquus internus abdominis muscle

Inclination to the same side
All the other deep back muscles in the region
Quadratus lumborum muscle (without the spinal and interspinal muscles)

Rectus abdominis muscle
Obliquus externus abdominis muscle
Obliquus internus abdominis muscle
All the muscles which act as synergists on the same side become antagonists when contracting on the opposite side

Rotation to the same side
Longissimus lumborum muscle
All the muscles which act as antagonists on the same side become synergists when contracting on the opposite side

Obliquus externus abdominis muscle
Multifidus lumborum muscle
Rotatores lumborum muscles
All the muscles which act as synergists on the same side become antagonists when contracting on the opposite side

Costovertebral and sternocostal joints

Depression of the ribs
Rectus abdominis muscle
Obliquus externus abdominis muscle
Obliquus internus abdominis muscle
Transversus abdominis muscle
Internal intercostal muscles
Transversus thoracis muscle
Serratus posterior inferior muscle
Quadratus lumborum muscle
Longissimus thoracis muscle

External intercostal muscles
Internal intercostal muscles (intercartilaginous part)

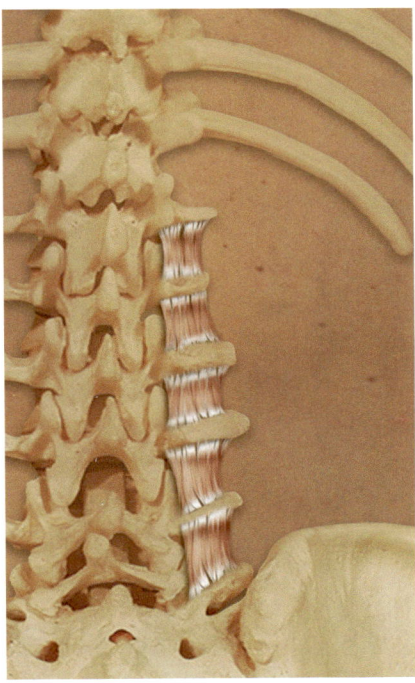

Lateral lumbar intertransversarii muscles
Lateral tract, intertransverse system

The lateral lumbar intertransversarii (intertransverse) muscles extend the spine when they tighten on both sides and incline it to the same side when they contract unilaterally. Like the medial intertransversarii muscles, they stabilize the lumbar spine and stop the vertebrae from slipping sideways.

Origin	Costal processes of all the lumbar vertebrae
	Transverse process of thoracic vertebra 12
Insertion	Costal processes of lumbar vertebrae 5 to 1
	Transverse process of thoracic vertebra 11
	Iliac tuberosity
Innervation	Ventral branches of spinal nerves T12–L5
Special features	These muscles are of ventral origin and are, therefore, innervated by the ventral branches of the spinal nerves.

Functions

 Synergists

 Antagonists

Intervertebral joints and disks (lumbar spine)

Extension (bilateral contraction)
Medial lumbar intertransversarii muscles
All the other deep back muscles in the region

Rectus abdominis muscle
Obliquus externus abdominis muscle
Obliquus internus abdominis muscle

Inclination to the same side
Medial lumbar intertransversarii muscles
All the other deep back muscles in the region
(without the spinal and interspinal muscles)
Rectus abdominis muscle
Obliquus externus abdominis muscle
Obliquus internus abdominis muscle
Quadratus lumborum muscle

All the muscles which act as synergists on the same side become antagonists when contracting on the opposite side

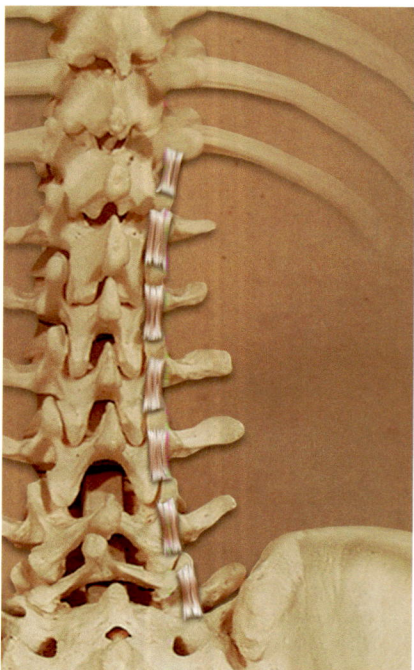

Medial lumbar intertransversarii muscles
Lateral tract, intertransverse system

The medial lumbar intertransversarii (intertransverse) muscles extend the spine when they tighten on both sides and incline it to the same side when they contract unilaterally. They owe their strong development to the important task of stabilizing the lumbar spine and stopping the vertebrae from slipping sideways. For this reason, they are at their strongest at the iliosacral junction, where this risk is particularly high.

Origin	Iliac tuberosity, accessory processes of lumbar vertebrae 4 to 1
Insertion	Mamillary tubercles of lumbar vertebrae 4 to 2
Innervation	Dorsal branches of spinal nerves L1–L5

Functions

Synergists   Antagonists

Intervertebral joints and disks (lumbar spine)

Extension (bilateral contraction)
Lateral lumbar intertransversarii muscles
All the other deep back muscles in the region

Rectus abdominis muscle
Obliquus externus abdominis muscle
Obliquus internus abdominis muscle

Inclination to the same side
Lateral lumbar intertransversarii muscles
All the other deep back muscles in the region
(without the spinal and interspinal muscles)
Rectus abdominis muscle
Obliquus externus abdominis muscle
Obliquus internus abdominis muscle
Quadratus lumborum muscle

All the muscles which act as synergists on the same side become antagonists when contracting on the opposite side

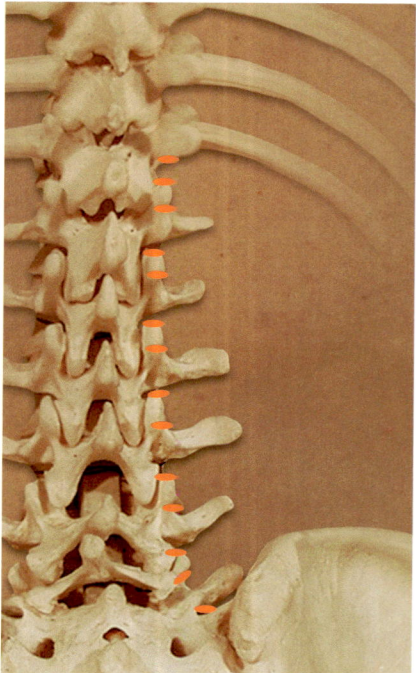

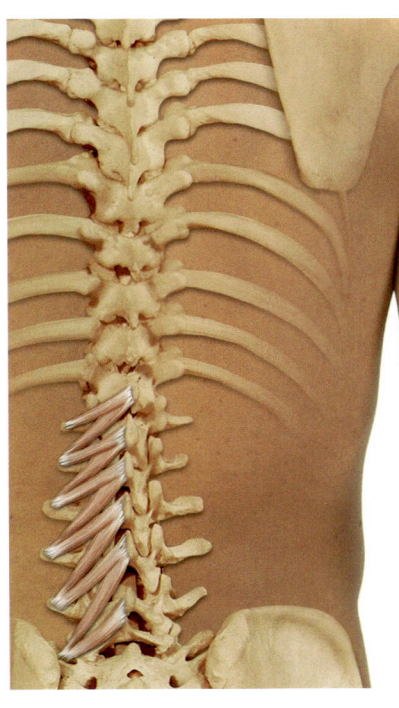

Rotatores lumborum breves and longi muscles
Medial tract, transversospinal system

The rotatores lumborum breves and longi muscles (short and long lumbar rotators) are weak, and are replaced by the mighty multifidus muscle in this region. When they contract bilaterally, their effect consists of extension of the spine; on unilateral contraction, they have an ipsilateral inclination component which increases, and a contralateral rotation component which decreases, with increasing muscle length. The rotatores breves muscles run to the next vertebra up, while the rotatores longi skip 2–3 vertebrae. If muscles along the same tract skip more than 3–4 vertebrae, they are grouped with the multifidus muscle (see p. 244); if they skip 5–6 vertebrae, they are referred to as the semispinalis muscle.

Origin	Basal mamillary tubercles of the lumbar vertebrae
Insertion	Basal spinous processes and vertebral arches of the lumbar vertebrae
Innervation	Dorsal branches of spinal nerves L1–L5

Functions

 Synergists

 Antagonists

Intervertebral joints and disks (lumbar spine)

Extension (bilateral contraction)

All the other deep back muscles in the region

Rectus abdominis muscle
Obliquus externus abdominis muscle
Obliquus internus abdominis muscle

Inclination to the same side

Lateral lumbar intertransversarii muscles
All the other deep back muscles in the region (without the spinal and interspinal muscles)
Rectus abdominis muscle
Obliquus externus abdominis muscle
Obliquus internus abdominis muscle
Quadratus lumborum muscle

All the muscles which act as synergists on the same side become antagonists when contracting on the opposite side

Rotation to the opposite side

Obliquus internus abdominis muscle
Multifidus lumborum muscle
All the muscles which act as antagonists on the same side become synergists when contracting on the opposite side

Obliquus externus abdominis muscle
Iliocostalis lumborum muscle
Longissimus lumborum muscle
All the muscles which act as synergists on the same side become antagonists when contracting on the opposite side

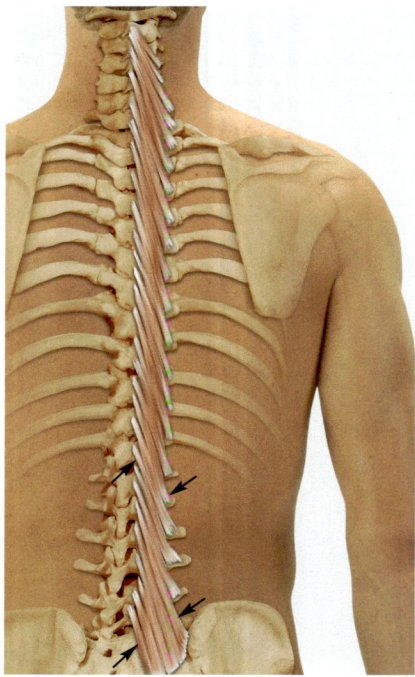

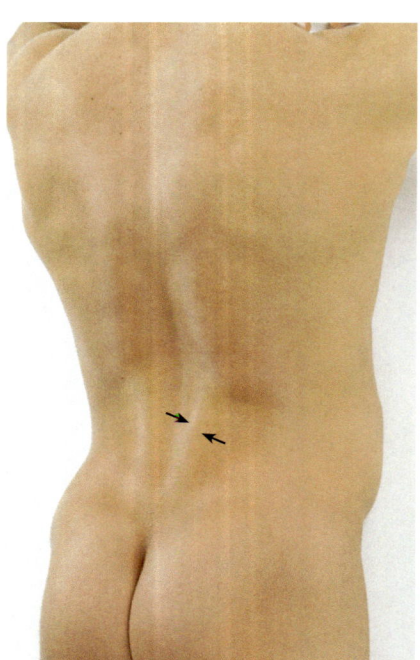

Multifidus lumborum muscle
Medial tract, transversospinal system

The multifidus lumborum (lumbar multifidus) is a mighty muscle which fills most of the lordosis of the lumbar spine, straightening it out visually. Its main effect is to extend the spine through bilateral contraction. On unilateral contraction, it has an ipsilateral inclination component which increases, and a rotatory component which decreases, with increasing muscle length. Both of these, components, however, are of secondary importance with regard to the lumbar spine.

Origin	Mamillary tubercles of the lumbar vertebrae, sacral bone (dorsal surface up to S4) Posterior sacroiliac ligament Iliac crest
Insertion	Spinous processes of the upper lumbar and lower thoracic vertebrae
Innervation	Dorsal branches of spinal nerves L1–S1

Functions

 Synergists

Antagonists

Intervertebral joints and disks (lumbar spine)

Extension (bilateral contraction)
All the other deep back muscles in the region

Rectus abdominis muscle
Obliquus externus abdominis muscle
Obliquus internus abdominis muscle

Inclination to the same side
Lateral lumbar intertransversarii muscles
All the other deep back muscles in the region (without the spinal and interspinal muscles)
Rectus abdominis muscle
Obliquus externus abdominis muscle
Obliquus internus abdominis muscle
Quadratus lumborum muscle

All the muscles which act as synergists on the same side become antagonists when contracting on the opposite side

Rotation to the opposite side
Obliquus internus abdominis muscle
Rotatores lumborum muscles
All the muscles which act as antagonists on the same side become synergists when contracting on the opposite side

Obliquus externus abdominis muscle
Iliocostalis lumborum muscle
Longissimus lumborum muscle
All the muscles which act as synergists on the same side become antagonists when contracting on the opposite side

The flat tendons of the iliocostalis lumborum and longissimus thoracis muscles lie on top of the multifidus lumborum, but the bulging contour in this region is determined by the multifidus muscle itself.

Muscle function testing

Muscle power grade

5/4

Starting position: The patient lies on his front. The arms are placed alongside the body, with the palms facing down.

Test procedure: The examiner holds the patient's pelvis in place with one hand and, with the other, applies downward pressure (towards the couch) to the thoracic spine.

Instruction: "Raise your upper body from the couch against my resistance and hold your final position."

3

Starting position: The patient lies on his front. The arms are placed alongside the body, with the palms facing down.

Test procedure: The examiner looks at the movement of the trunk.

Instruction: "Raise your upper body from the couch."

2

Starting position: The patient lies on his side with the spine flexed.

Test procedure: The examiner looks at the movement of the trunk.

Instruction: "Stretch out your spine. While doing so, grasp the base of the back of your head with both hands and raise your sternum."

1/0

Starting position: The patient lies on his front. The arms are placed alongside the body, with the palms facing down.

Test procedure: The examiner palpates the lumbar spine extensors.

Instruction: "Try to raise your upper body from the couch."

⚕ Clinical relevance

- A poor spinal posture often manifests in the form of knots or tension in the deep back muscles of the affected region.
- Lower back pain may indicate pathological processes in the renal bed.

⚠ Problems/comments

- The patient needs to be able to extend the thoracic and lumbar spine.
- A patient with powerful back extensors and weak hip extensors can hyperextend the lumbar spine, but may not be capable of raising the trunk from the couch.
- The interspinales lumborum muscles are not included in this muscle function test.

Stretch test

Lumbar back extensors

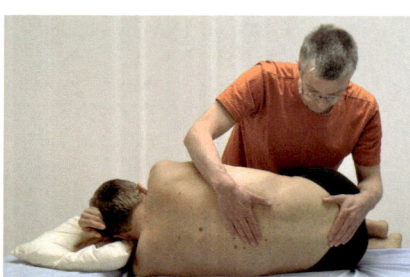

Method

The therapist moves the patient's trunk into maximum flexion of the lumbar spine, by flexing the hips and moving the pelvis into the upright position.

Finding

Muscle shortening is present if the movement cannot be performed to its full extent and if the endpoint feels soft and elastic.

The patient reports a stretching sensation along the course of the muscle.

4 Trunk

Deep muscles of the back, thoracic

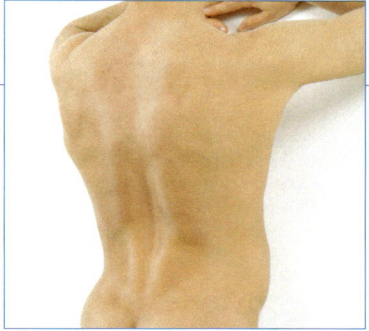

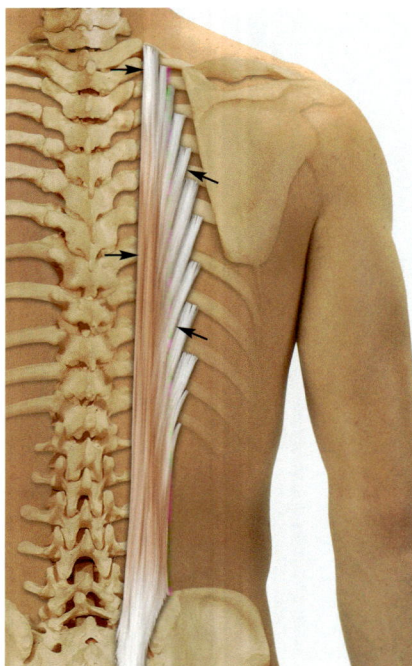

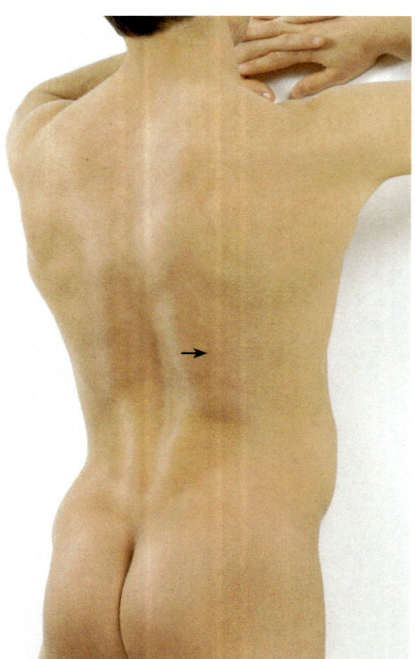

Iliocostalis thoracis muscle
Lateral tract, sacrospinal system

The iliocostalis thoracis muscle (iliocostal muscle of the back) can extend the whole spine powerfully when it tightens on both sides. Since the muscle only acts indirectly on the spine via the hips, it also causes depression of the ribs. If it tightens on one side, it has an ipsilateral rotatory component.

Origin	Medially of the costal angle of ribs 7–12
Insertion	Costal angle of ribs 1–7
Innervation	Dorsal branches of spinal nerves T1–L1

Functions

 Synergists   Antagonists

Intervertebral joints and disks (thoracic spine)

Extension (bilateral contraction)
All the other autochthonous back muscles in the region

The antagonistic abdominal muscles only have a very indirect effect via their origins at the lower thorax

Inclination to the same side
All the other autochthonous back muscles in the region (without the spinal and interspinal muscles)
The antagonistic abdominal muscles only have a very indirect effect via their origins at the lower thorax

All the muscles which act as synergists on the same side become antagonists when contracting on the opposite side

Rotation to the same side
The antagonistic abdominal muscles only have a very indirect effect via their origins at the lower thorax
All the muscles which act as antagonists on the same side become synergists when contracting on the opposite side

Rotatores thoracis muscles
Multifidus lumborum muscle
The antagonistic abdominal muscles only have a very indirect effect via their origins at the lower thorax
All the muscles which act as synergists on the same side become antagonists when contracting on the opposite side

Costovertebral and sternocostal joints

Depression of the ribs
Rectus abdominis muscle
Obliquus externus abdominis muscle
Obliquus internus abdominis muscle
Transversus abdominis muscle
Internal intercostal muscles
Transversus thoracis muscle
Serratus posterior inferior muscle
Quadratus lumborum muscle
Longissimus thoracis muscle
Iliocostalis lumborum muscle

External intercostal muscles
Internal intercostal muscles (intercartilaginous part)
Scalene muscles
Serratus posterior superior muscle

Longissimus thoracis muscle
Lateral tract, sacrospinal system

Since it also originates from the sacrum and the ilium, the longissimus thoracis muscle (long muscle of the back), acting in conjunction with the iliocostalis muscle, can tilt the pelvis ventrally on the heads of the femurs. When it contracts unilaterally, the longissimus inclines the spine to the same side. When it contracts bilaterally, the spine is extended powerfully. In contrast, its rotatory component on the trunk is negligible.

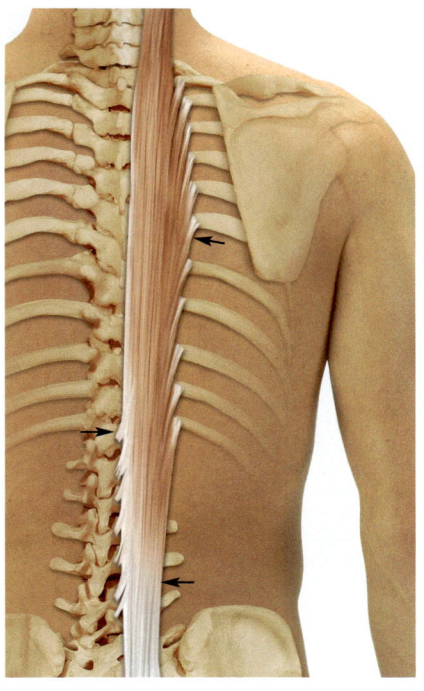

Origin	Dorsal surface of the transverse processes of the lumbar vertebrae Deep lamina of the thoracolumbar fascia, dorsal surface of the sacral bone
Insertion	Transverse processes of the thoracic vertebrae At the lower 9 or 10 ribs between the tubercle and the costal angle
Innervation	Dorsal branches of spinal nerves T1–L5

Functions

 Synergists

 Antagonists

Intervertebral joints and disks (thoracic spine)

Extension (bilateral contraction)
All the other deep back muscles in the region

The antagonistic abdominal muscles only have a very indirect effect via their origins at the lower thorax

Inclination to the same side
All the other deep back muscles in the region (without the spinal and interspinal muscles)
The antagonistic abdominal muscles only have a very indirect effect via their origins at the lower thorax

All the muscles which act as synergists on the same side become antagonists when contracting on the opposite side

Costovertebral and sternocostal joints

Depression of the ribs
Rectus abdominis muscle
Obliquus externus abdominis muscle
Obliquus internus abdominis muscle
Transversus abdominis muscle
Internal intercostal muscles
Transversus thoracis muscle
Serratus posterior inferior muscle
Quadratus lumborum muscle
Iliocostalis thoracis muscle

External intercostal muscles
Internal intercostal muscles (intercartilaginous part)
Scalene muscles
Serratus posterior superior muscle

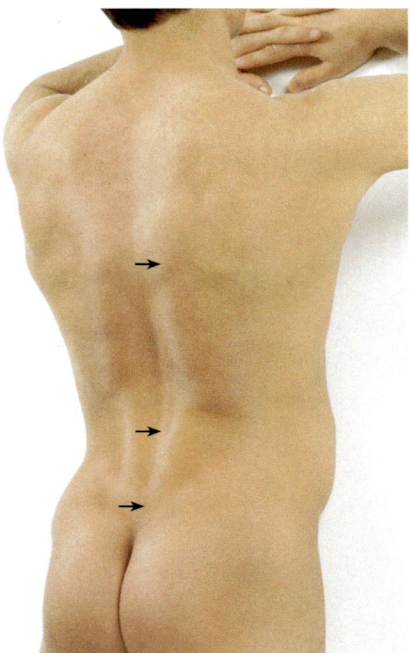

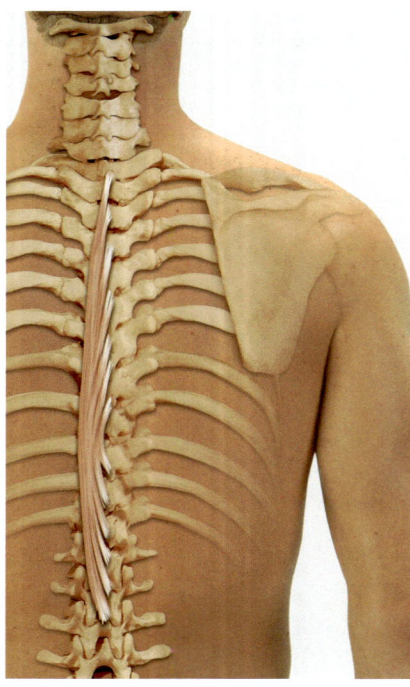

Spinalis thoracis muscle
Medial tract, spinal system

The spinalis thoracis muscle (spinal muscle of the back) is strongly developed. Its effect consists of extension and stabilization of the thoracic spine.

Origin	Spinous processes of lumbar vertebrae 2 and 1 and thoracic vertebrae 12–10
Insertion	Spinous processes of thoracic vertebrae 9–2
Innervation	Dorsal branches of spinal nerves T3–L1

Functions

 Synergists

Intervertebral joints and disks (thoracic spine)

Extension (bilateral contraction)
All the other deep back muscles in the region

Antagonists

The antagonistic abdominal muscles only have a very indirect effect via their origins at the lower thorax

Rotatores thoracis muscles
Medial tract, transversospinal system

The rotatores thoracis breves and longi muscles (long and short thoracic rotators) are strongly developed. When they contract bilaterally, their effect consists of extension of the spine. On unilateral contraction, they have an ipsilateral inclination component which increases, and a contralateral rotatory component which decreases, with increasing muscle length. The rotatores breves muscles run to the next vertebra up, while the rotatores longi skip two vertebrae. If muscles along the same tract skip more than two vertebrae, they are grouped with the multifidus muscle.

Origin	Basal transverse processes of thoracic vertebrae 12–2
Insertion	Basal spinous processes and vertebral arches of thoracic vertebrae 11–1 and cervical vertebra 7
Innervation	Dorsal branches of spinal nerves C7–T12

Functions

 Synergists

Antagonists

Intervertebral joints and disks (thoracic spine)

Extension (bilateral contraction)

All the other deep back muscles in the region

The antagonistic abdominal muscles only have a very indirect effect via their origins at the lower thorax

Inclination to the same side

All the other ipsilateral deep back muscles in the region (without the spinal and interspinal muscles)

The antagonistic abdominal muscles only have a very indirect effect via their origins at the lower thorax
All the muscles which act as synergists on the same side become antagonists when contracting on the opposite side

Rotation to the opposite side

Semispinalis thoracis muscle
Obliquus externus abdominis muscle
All the muscles which act as antagonists on the same side become synergists when contracting on the opposite side

Obliquus internus abdominis muscle
Iliocostalis thoracis muscle
Longissimus thoracis muscle
The antagonistic abdominal muscles only have a very indirect effect via their origins at the lower thorax
All the muscles which act as synergists on the same side become antagonists when contracting on the opposite side

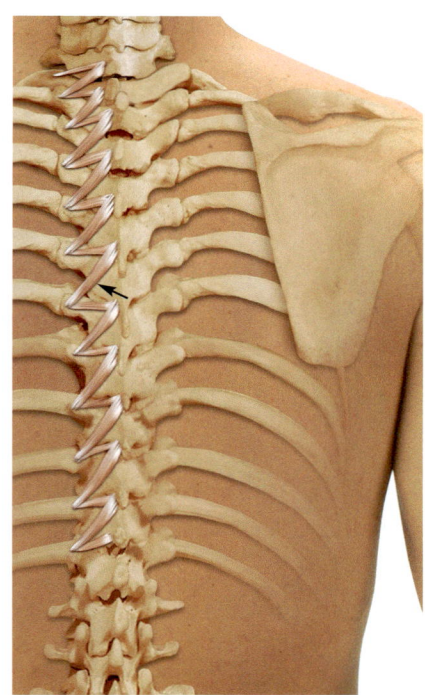

Rotatores thoracis breves muscles

Rotatores thoracis longi muscles

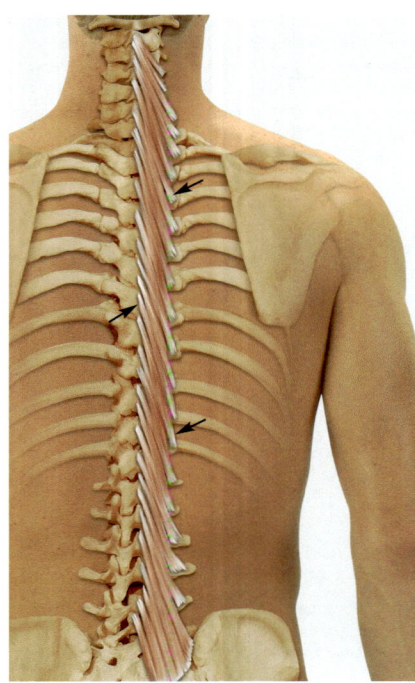

Multifidus thoracis muscle
Medial tract, transversospinal system

The multifidus thoracis (thoracic multifidus) is a very powerful muscle. It also skips 3–4 vertebrae. On bilateral contraction, its effect consists of extension of the spine. On unilateral contraction, it has an ipsilateral inclination component which increases, and a contralateral rotatory component which decreases, with increasing muscle length.

Origin	Transverse processes of the thoracic vertebrae
Insertion	Spinous processes of the upper thoracic vertebrae and lower cervical vertebrae
Innervation	Dorsal branches of spinal nerves C3–T5

Functions

 Synergists

 Antagonists

Intervertebral joints and disks (thoracic spine)

Extension (bilateral contraction)
All the other deep back muscles in the region

The antagonistic abdominal muscles only have a very indirect effect via their origins at the lower thorax

Inclination to the same side
All the other deep back muscles in the region (without the spinal and interspinal muscles)

All the muscles which act as synergists on the same side become antagonists when contracting on the opposite side

Rotation to the opposite side
Rotatores thoracis muscles
Obliquus externus abdominis muscle
All the muscles which act as antagonists on the same side become synergists when contracting on the opposite side

Obliquus internus abdominis muscle
Iliocostalis thoracis muscle
Longissimus thoracis muscle
All the muscles which act as synergists on the same side become antagonists when contracting on the opposite side

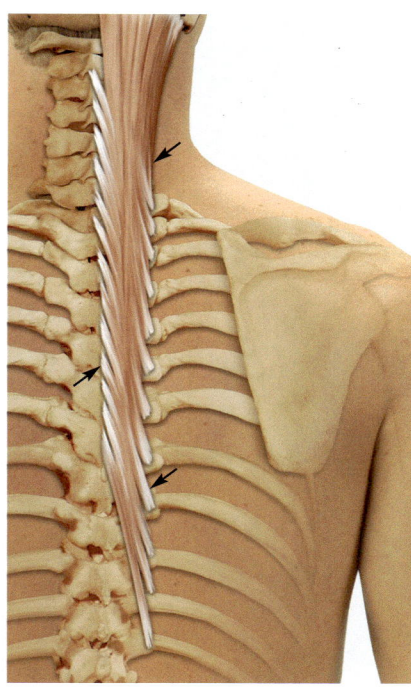

Semispinalis thoracis muscle
Medial tract, transversospinal system

The semispinalis thoracis muscle (semispinal muscle of the thorax) skips 5–6 vertebrae and can extend the spine if it tightens on both sides, while inclining it to the same side if contracting unilaterally. Its rotatory component is very weak in the thoracic region and will not be considered here.

Origin	Transverse processes of thoracic vertebra 12 to cervical verterbra 7
Insertion	Spinous processes of thoracic vertebra 3 to cervical vertebra 6
Innervation	Dorsal branches of spinal nerves C6–T12

Functions

 Synergists

 Antagonists

Intervertebral joints and disks (thoracic spine)

Extension (bilateral contraction)

All the other deep back muscles in the region

The antagonistic abdominal muscles only have a very indirect effect via their origins at the lower thorax

Inclination to the same side

All the other deep back muscles in the region (without the spinal and interspinal muscles)

The antagonistic abdominal muscles only have a very indirect effect via their origins at the lower thorax
All the muscles which act as synergists on the same side become antagonists when contracting on the opposite side

Muscles tested together:

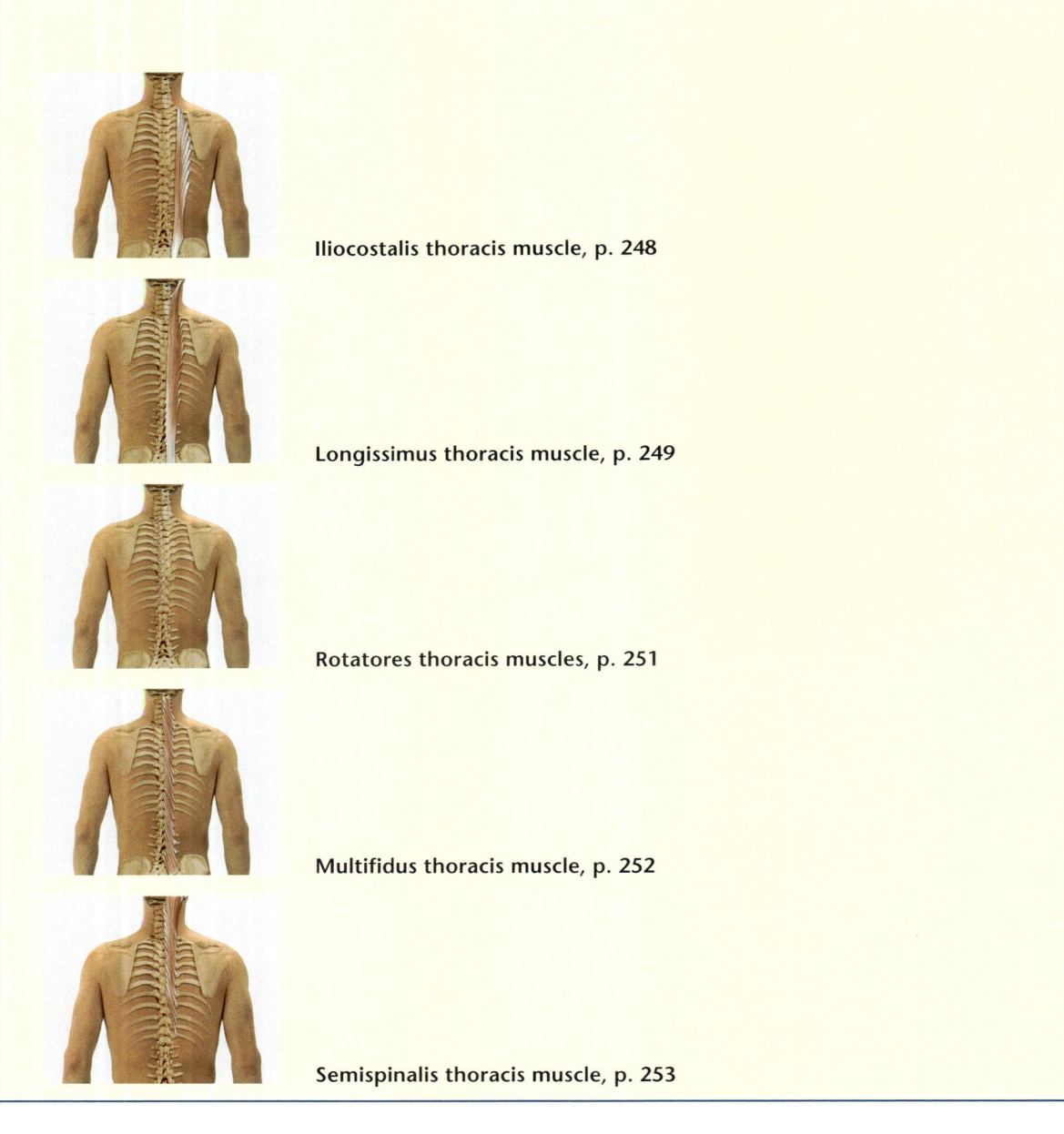

Iliocostalis thoracis muscle, p. 248

Longissimus thoracis muscle, p. 249

Rotatores thoracis muscles, p. 251

Multifidus thoracis muscle, p. 252

Semispinalis thoracis muscle, p. 253

Muscle function testing

Muscle power grade

5/4

3

2

1/0

Starting position: The patient lies on his front. The arms are placed alongside the body, with the palms facing down.

Test procedure: The examiner holds the patient's pelvis in place with one hand and with the other applies downward pressure (towards the couch) to the thoracic spine.

Instruction: "Raise your upper body from the couch against my resistance and hold your final position."

Starting position: The patient lies on his front. The arms are placed alongside the body, with the palms facing down.

Test procedure: The examiner looks at the movement of the trunk.

Instruction: "Raise your upper body from the couch."

Starting position: The patient lies on his side with the spine flexed.

Test procedure: The examiner looks at the movement of the trunk.

Instruction: "Stretch out your spine. While doing so, grasp the base of the back of your head with both hands and raise your sternum."

Starting position: The patient lies on his front. The arms are placed alongside the body, with the palms facing down.

Test procedure: The examiner palpates the thoracic spine extensors.

Instruction: "Try to raise your upper body from the couch."

⚡ Clinical relevance

- A poor spinal posture often manifests in the form of knots or tension in the deep back muscles of the affected region.

❗ Problems/comments

- The patient needs to be able to extend the thoracic spine.
- A patient with powerful back extensors and weak hip extensors can hyperextend the lumbar spine but may not be capable of raising the trunk from the couch.

4 Trunk

Deep muscles of the back, cervical

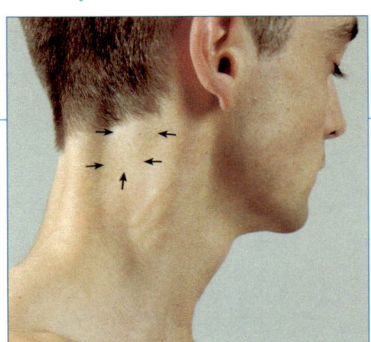

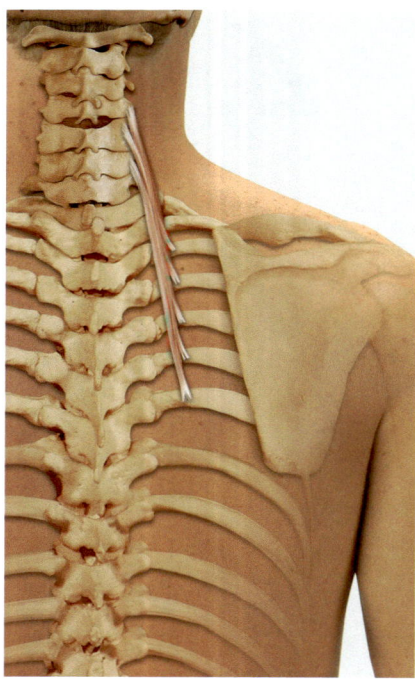

Iliocostalis cervicis muscle
Lateral tract, sacrospinal system

The iliocostalis cervicis muscle (iliocostal muscle of the neck) can extend the cervical spine powerfully when it contracts on both sides.

Origin	Medially of the costal angle of ribs 3–7
Insertion	Transverse processes and posterior tubercles of cervical vertebrae 3–6
Innervation	Dorsal branches of spinal nerves C3–T7

Functions

 Synergists

 Antagonists

Intervertebral joints and disks (cervical spine)

Extension (bilateral contraction)

Sternocleidomastoideus muscle (with the head extended)	Sternocleidomastoideus muscle (with the head flexed)
Trapezius muscle (descending part)	Longus capitis muscle
Levator scapulae muscle	Longus colli muscle
All the other deep back muscles in the region	Scalenus anterior muscle
	Infra- and suprahyoid muscles

Inclination to the same side

Sternocleidomastoideus muscle	All the muscles which act as synergists on the same side become antagonists when contracting on the opposite side
Scalene muscles	
Trapezius muscle (descending part)	
Levator scapulae muscle	
All the other deep back muscles in the region (without the spinal and interspinal muscles)	

Rotation to the same side

Splenius capitis muscle	Sternocleidomastoideus muscle
Splenius cervicis muscle	Semispinalis cervicis muscle
Longissimus cervicis muscle	Semispinalis capitis muscle
Longissimus capitis muscle	Multifidus cervicis muscle
Rectus capitis posterior major muscle	Rotatores cervicis muscles
Obliquus capitis inferior muscle	Obliquus capitis superior muscle
All the muscles which act as antagonists on the same side act become synergists when contracting on the opposite side	All the muscles which act as synergists on the same side become antagonists when contracting on the opposite side

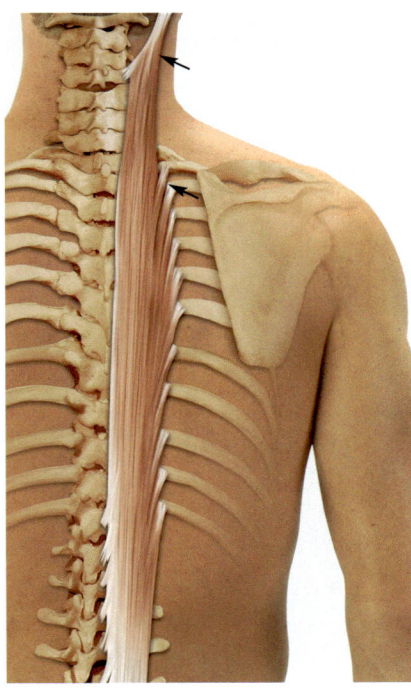

Longissimus capitis muscle
Lateral tract, sacrospinal system

The longissimus capitis muscle (longissimus muscle of the head) can extend the cervical spine and head powerfully when it tightens on both sides. When it contracts unilaterally, it inclines the head and cervical spine to the same side and can also rotate the head to the same side.

Origin	Transverse processes of cervical vertebra 3 to thoracic vertebra 3
Insertion	Mastoid process
Innervation	Dorsal branches of spinal nerves C3–T3

Functions

 Synergists

 Antagonists

Atlanto-occipital joint and intervertebral joints and disks (cervical spine)

Extension (bilateral contraction)

Sternocleidomastoideus muscle (with the head extended)	Sternocleidomastoideus muscle (with the head flexed)
Trapezius muscle (descending part)	Longus capitis muscle
Levator scapulae muscle	Longus colli muscle
All the other deep back muscles in the region	Rectus capitis anterior
	Scalenus anterior muscle
	Infra- and suprahyoid muscles

Inclination to the same side

Sternocleidomastoideus muscle	
Scalene muscles	All the muscles which act as synergists on the same side become antagonists when contracting on the opposite side
Trapezius muscle (descending part)	
Levator scapulae muscle	
All the other deep back muscles in the region (without the spinal and interspinal muscles)	

Rotation to the same side

Splenius capitis muscle	Sternocleidomastoideus muscle
Splenius cervicis muscle	Semispinalis cervicis muscle
Iliocostalis cervicis muscle	Multifidus cervicis muscle
Longissimus cervicis muscle	Rotatores cervicis muscles
Rectus capitis posterior major muscle	Obliquus capitis superior muscle
Obliquus capitis inferior muscle	All the muscles which act as synergists on the same side become antagonists when contracting on the opposite side
All the muscles which act as antagonists on the same side act become synergists when contracting on the opposite side	

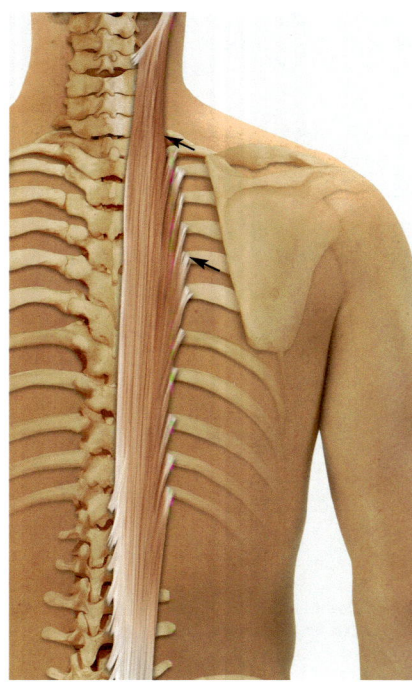

Longissimus cervicis muscle
Lateral tract, sacrospinal system

The longissimus cervicis muscle (longissimus muscle of the neck) can extend the cervical spine powerfully when it tightens on both sides and incline it to the same side when acting unilaterally.

Origin	Transverse processes of thoracic vertebrae 1-6 and cervical vertebrae 3–7
Insertion	Posterior tubercles of the transverse processes of cervical vertebrae 2–5
Innervation	Dorsal branches of spinal nerves C3–T6

Functions

 Synergists  Antagonists

Intervertebral joints and disks (cervical spine)

Extension (bilateral contraction)

Sternocleidomastoideus muscle (with the head extended)	Sternocleidomastoideus muscle (with the head flexed)
Trapezius muscle (descending part)	Longus capitis muscle
Levator scapulae muscle	Longus colli muscle
All the other deep back muscles in the region	Scalenus anterior muscle
	Infra- and suprahyoid muscles

Inclination to the same side

Sternocleidomastoideus muscle	All the muscles which act as synergists on the
Scalene muscles	same side become antagonists when contract-
Trapezius muscle (descending part)	ing on the opposite side
Levator scapulae muscle	
All the other deep back muscles in the region (without the spinal and interspinal muscles)	

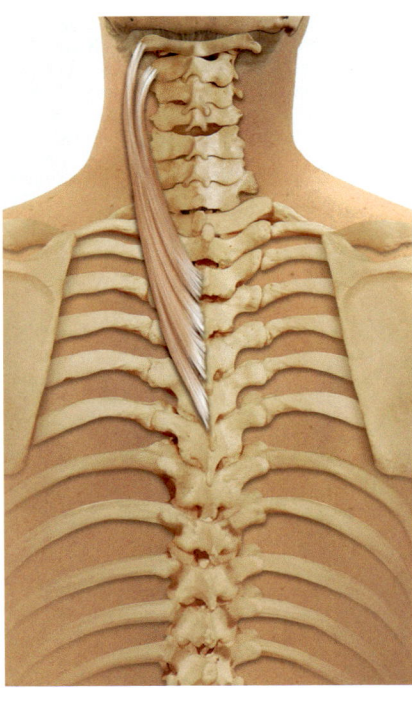

Splenius cervicis muscle
Spinotransversal system

The splenius cervicis muscle (splenius muscle of the neck) can extend the cervical spine when it tightens on both sides, rotate it to the same side when contracting unilaterally and, if its rotatory component is balanced out by its antagonists, can also incline the neck to the same side. This muscle runs dorsally of the other deep muscles of this region, fixing them to the spine like a retinaculum.

Origin	Spinous processes of thoracic vertebrae 3–6
Insertion	Posterior tubercles and transverse processes of cervical vertebrae 1 and 2
Innervation	Dorsal branches of spinal nerves C5–C7

Functions

 Synergists

Antagonists

Intervertebral joints and disks (cervical spine)

Extension (bilateral contraction)

Synergists	Antagonists
Sternocleidomastoideus muscle (with the head extended)	Sternocleidomastoideus muscle (with the head flexed)
Trapezius muscle (descending part)	Longus capitis muscle
Levator scapulae muscle	Longus colli muscle
All the other deep back muscles in the region	Scalenus anterior muscle
	Infra- and suprahyoid muscles

Rotation to the same side

Synergists	Antagonists
Splenius capitis muscle	Sternocleidomastoideus muscle
Iliocostalis muscle	Rotatores muscles
Longissimus muscle	Semispinalis muscle
Rectus capitis posterior major muscle	Obliquus capitis superior muscle
Obliquus capitis inferior muscle	All the muscles which act as antagonists on the
All the muscles which act as synergists on the same side act become antagonists when contracting on the opposite side	same side become synergists when contracting on the opposite side

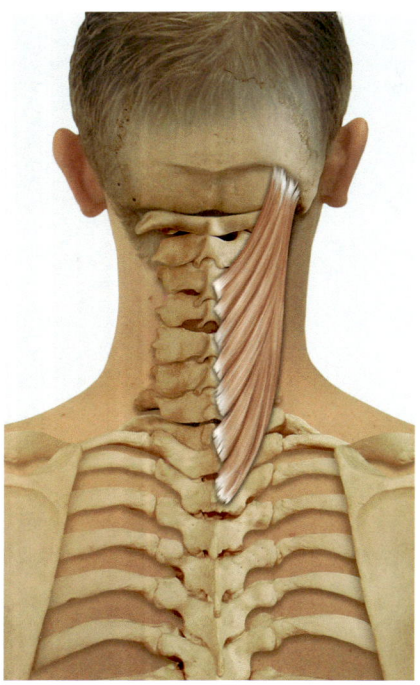

Splenius capitis muscle
Spinotransversal system

Like the splenius cervicis, the splenius capitis muscle (splenius muscle of the head) can extend the cervical spine when it tightens on both sides, rotate it to the same side when contracting unilaterally and, if its rotatory component is balanced out by its antagonists, can also incline the neck to the same side. In contrast to the splenius cervicis, however, the head part of the spinotransversal system acts on the joints of the head and can thus help to rotate it to the same side, incline it or, if acting bilaterally, tip it backwards against the neck.

Origin	Caudal half of the nuchal ligament
	Spinous processes of cervical vertebra 7 to thoracic vertebra 3
Insertion	Mastoid process
	Superior arcuate line of the occipital bone
Innervation	Dorsal branches of spinal nerves C3–C5

Functions

 Synergists

Antagonists

Atlanto-occipital joint and intervertebral joints and disks (cervical spine)

Extension (bilateral contraction)

Sternocleidomastoideus muscle (with the head extended)
Trapezius muscle (descending part)
Levator scapulae muscle
All the other deep back muscles in the region

Sternocleidomastoideus muscle (with the head flexed)
Longus capitis muscle
Longus colli muscle (cervical spine only)
Rectus capitis anterior muscle
Scalenus anterior muscle (cervical spine only)
Infra- and suprahyoid muscles

Intervertebral joints (cervical spine)

Rotation to the same side

Splenius cervicis muscle
Iliocostalis muscle
Longissimus muscle
Rectus capitis posterior major muscle
Obliquus capitis inferior muscle
All the muscles which act as synergists on the same side become antagonists when contracting on the opposite side

Sternocleidomastoideus muscle
Rotatores muscles
Semispinalis muscle
Obliquus capitis superior muscle
All the muscles which act as antagonists on the same side act become synergists when contracting on the opposite side

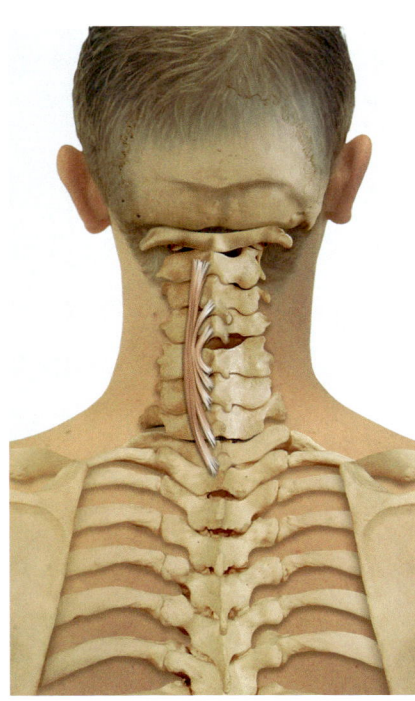

Spinalis cervicis muscle
Medical tract, spinal system

The weak spinalis cervicis muscle (spinal muscle of the neck) extends and stabilizes the cervical spine.

Origin	Spinous processes of thoracic vertebra 2 to cervical vertebra 6
Insertion	Spinous processes of cervical vertebrae 4–2
Innervation	Dorsal branches of spinal nerves C2–T6

Functions

Synergists

Antagonists

Intervertebral joints and discs (cervical spine)

Extension (bilateral contraction)

Sternocleidomastoideus muscle (with the head extended)	Sternocleidomastoideus muscle (with the head flexed)
Trapezius muscle (descending part)	Longus capitis muscle
Levator scapulae muscle	Longus colli muscle
All the other deep back muscles in the region	Scalenus anterior muscle
	Infra- and suprahyoid muscles

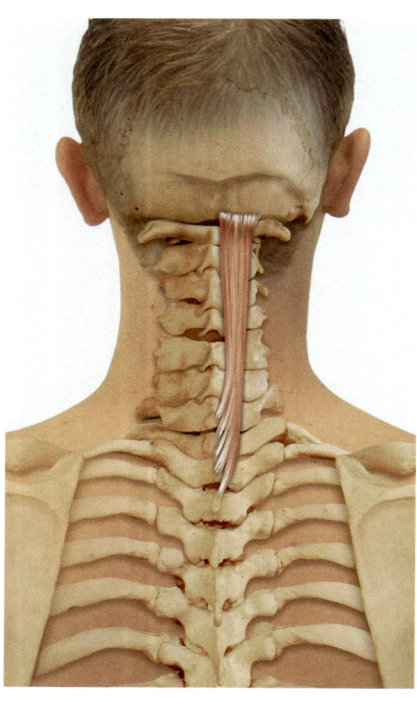

Spinalis capitis muscle
Medial tract, spinal system

A few of the fibers of the spinalis capitis muscle (spinal muscle of the head) reach the head. Its action consists of extending the head and the cervical spine and stabilizing them.

Origin	Spinous processes of thoracic vertebrae 3–1 and cervical vertebrae 7 and 6
Insertion	Occipital squama
Innervation	Dorsal branches of spinal nerves C6–T3

Functions

 Synergists

Antagonists

Atlanto-occipital joint and interverte-bral joints and disks (cervical spine)

Extension (bilateral contraction)
Sternocleidomastoideus muscle (with the head extended)
Trapezius muscle (descending part)
Levator scapulae muscle
All the other deep back muscles in the region

Sternocleidomastoideus muscle (with the head flexed)
Longus capitis muscle
Longus colli muscle (cervical spine only)
Rectus capitis anterior muscle
Scalenus anterior muscle (cervical spine only)
Infra- and suprahyoid muscles

Rotatores cervicis muscles
Medial tract, transversospinal system

The rotatores breves and longi muscles (long and short rotators) are weakly developed in the cervical spine region. When they contract bilaterally, their effect consists of extension of the spine. On unilateral contraction, they have an ipsilateral inclination component which increases, and a contralateral rotatory component which decreases, with increasing muscle length. The rotatores breves muscles run to the next vertebra up. The rotatores longi skip two vertebrae. If muscles along the same tract skip more than two vertebrae, they are grouped with the multifidus muscle.

Origin	Inferior articular processes of the cervical vertebrae
Insertion	Basal spinous processes and vertebral arches of the cervical vertebrae
Innervation	Dorsal branches of spinal nerves C1–C8

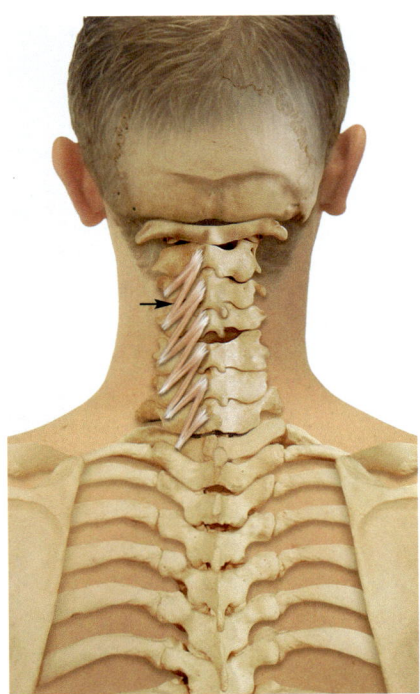

Rotatores cervicis breves muscles

Rotatores cervicis longi muscles

Functions

 Synergists   Antagonists

Intervertebral joints and disks (cervical spine)

Extension (bilateral contraction)

Synergists	Antagonists
Sternocleidomastoideus muscle (with the head extended)	Sternocleidomastoideus muscle (with the head flexed)
Trapezius muscle (descending part)	Longus capitis muscle
Levator scapulae muscle	Longus colli muscle
All the other deep back muscles in the region	Rectus capitis anterior muscle
	Scalenus anterior muscle
	Infra- and suprahyoid muscles

Inclination to the same side

Synergists	Antagonists
Sternocleidomastoideus muscle	All the muscles which act as synergists on the same side become antagonists when contracting on the opposite side
Scalene muscles	
Trapezius muscle (descending part)	
Levator scapulae muscle	
All the other deep back muscles in the region (without the spinal and interspinal muscles)	

Rotation to the opposite side

Synergists	Antagonists
Sternocleidomastoideus muscle	Splenius capitis muscle
Semispinalis cervicis muscle	Splenius cervicis muscle
Multifidus cervicis muscle	Iliocostalis cervicis muscle
Obliquus capitis superior muscle	Longissimus cervicis muscle
All the muscles which act as antagonists on the same side act become synergists when contracting on the opposite side	Rectus capitis posterior major muscle
	Obliquus capitis inferior muscle
	All the muscles which act as synergists on the same side become antagonists when contracting on the opposite side

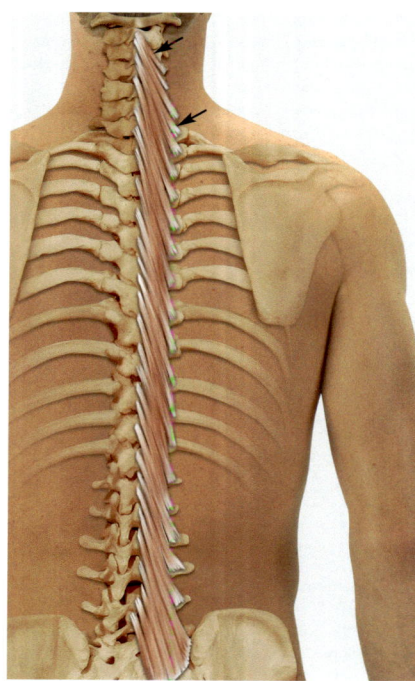

Multifidus cervicis muscle
Medial tract, transversospinal system

The multifidus cervicis (cervical multifidus muscle) is a powerful muscle which skips three or four vertebrae. On bilateral contraction, its effect consists of extension of the cervical spine. On unilateral contraction, it has an ipsilateral inclination component which increases, and a contralateral rotatory component which decreases, with increasing muscle length.

Origin	Inferior articular processes of cervical vertebrae 7–4
Insertion	Spinous processes of cervical vertebrae 7–2
Innervation	Dorsal branches of spinal nerves C3–C8

Functions

 Synergists   Antagonists

Interverte disks (cervical spine)

Extension (bilateral contraction)

Sternocleidomastoideus muscle (with the head extended)
Trapezius muscle (descending part)
Levator scapulae muscle
All the other deep back muscles in the region

Sternocleidomastoideus muscle (with the head flexed)
Longus capitis muscle
Longus colli muscle
Rectus capitis anterior muscle
Scalenus anterior muscle
Infra- and suprahyoid muscles

Inclination to the same side

Sternocleidomastoideus muscle
Scalene muscles
Trapezius muscle (descending part)
Levator scapulae muscle
All the other deep back muscles in the region
(without the spinal and interspinal muscles)

All the muscles which act as synergists on the same side become antagonists when contracting on the opposite side

Rotation to the opposite side

Sternocleidomastoideus muscle
Semispinalis cervicis muscle
Multifidus cervicis muscle
Obliquus capitis superior muscle
Rotatores cervicis muscles
All the muscles which act as antagonists on the same side act become synergists when contracting on the opposite side

Splenius capitis muscle
Splenius cervicis muscle
Iliocostalis cervicis muscle
Longissimus cervicis muscle
Rectus capitis posterior major muscle
Obliquus capitis inferior muscle
All the muscles which act as synergists on the same side become antagonists when contracting on the opposite side

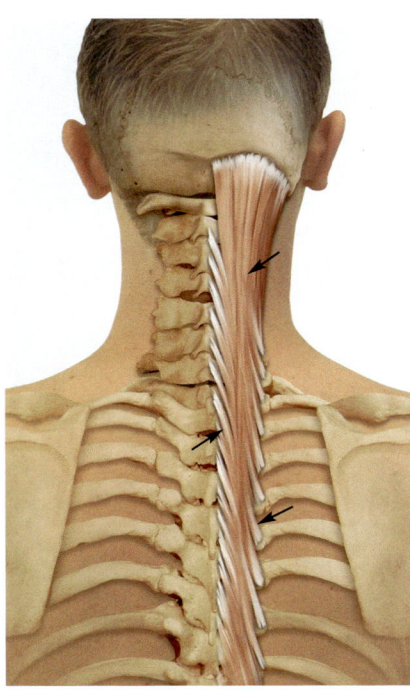

Semispinalis cervicis muscle
Medial tract, transversospinal system

The semispinalis cervicis muscle (semispinal muscle of the neck) can extend the cervical spine if it tightens on both sides, and incline it to the same side if contracting unilaterally. Its rotatory component will not be considered here as it is of little significance.

Origin	Transverse processes of thoracic vertebra 6 to cervical verterbra 7
Insertion	Spinous processes of cervical vertebrae 6 to 2
Innervation	Dorsal branches of spinal nerves C1–T6

Functions

 Synergists

Antagonists

Intervertebral joints and disks (cervical spine)

Extension (bilateral contraction)
Sternocleidomastoideus muscle (with the head extended)
Trapezius muscle (descending part)
Levator scapulae muscle
All the other deep back muscles in the region

Sternocleidomastoideus muscle (with the head flexed)
Longus capitis muscle
Longus colli muscle
Scalenus anterior muscle
Infra- and suprahyoid muscles

Inclination to the same side
Sternocleidomastoideus muscle
Semispinalis capitis muscle
Scalene muscles
Trapezius muscle (descending part)
Levator scapulae muscle
All the other deep back muscles in the region
(without the spinal and interspinal muscles)

All the muscles which act as synergists on the same side become antagonists when contracting on the opposite side.

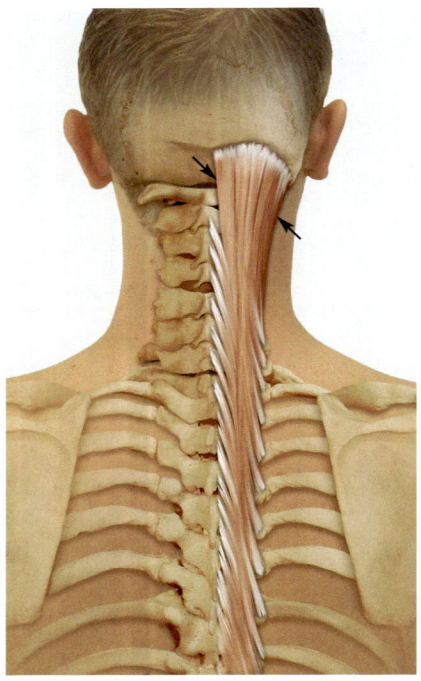

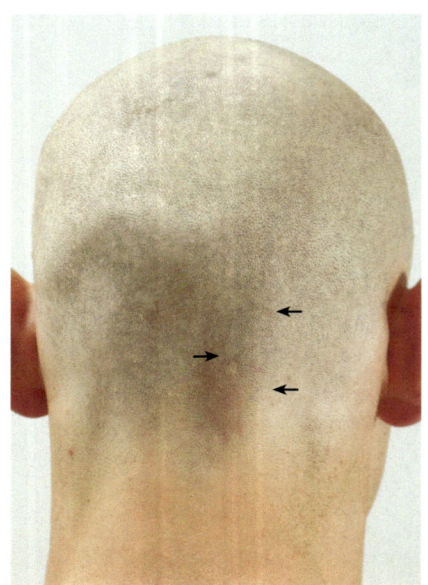

Semispinalis capitis muscle
Medial tract, transversospinal system

The semispinalis capitis muscle (semispinal muscle of the head) can extend the head and cervical spine if it tightens on both sides, and incline them to the same side if contracting unilaterally. Its contralateral rotatory component, which tends to be weak, is at its strongest in the neck region.

Origin	Transverse processes of thoracic vertebra 7 to cervical vertebrae 3
Insertion	Occipital squama
Innervation	Dorsal branches of spinal nerves C4–C8

Functions

Synergists   Antagonists

Atlanto–occipital joint and intervertebral joints and disks (cervical spine)

Extension (bilateral contraction)

Synergists	Antagonists
Sternocleidomastoideus muscle (with the head extended)	Sternocleidomastoideus muscle (with the head flexed)
Trapezius muscle (descending part)	Longus capitis muscle
Levator scapulae muscle	Longus colli muscle (cervical spine only)
All the other deep back muscles in the region	Rectus capitis anterior muscle
	Scalenus anterior muscle (cervical spine only)
	Infra- and suprahyoid muscles

Inclination to the same side

Synergists	Antagonists
Sternocleidomastoideus muscle	All the muscles which act as synergists on the same side become antagonists when contracting on the opposite side
Scalene muscles	
Trapezius muscle (descending part)	
Levator scapulae muscle	
All the other deep back muscles in the region (without the spinal and interspinal muscles)	

Rotation to the opposite side

Synergists	Antagonists
Sternocleidomastoideus muscle	Splenius capitis muscle
Multifidus cervicis muscle	Splenius cervicis muscle
Rotatores cervicis muscles	Iliocostalis cervicis muscle
Obliquus capitis superior muscle	Longissimus cervicis muscle
All the muscles which act as antagonists on the same side act become synergists when contracting on the opposite side	Rectus capitis posterior major muscle
	Obliquus capitis inferior muscle
	All the muscles which act as synergists on the same side become antagonists when contracting on the opposite side

Rectus capitis posterior major muscle
Lateral tract

The rectus capitis posterior major muscle tips the head backwards against the neck and rotates it to the same side if contracting unilaterally.

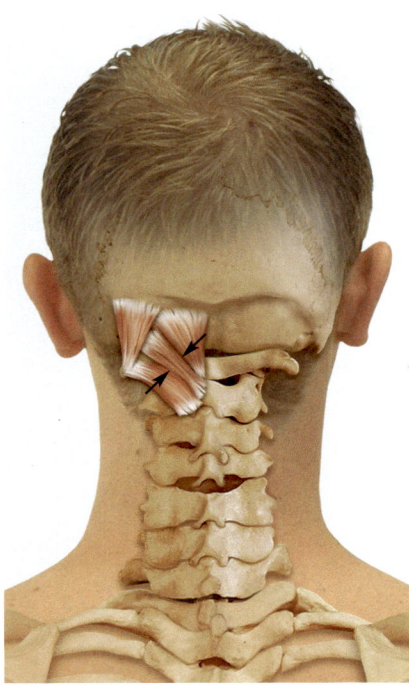

Origin	Spinous process of the axis
Insertion	Middle part of the inferior arcuate line of the occipital bone
Innervation	Suboccipital nerve from the dorsal branches of spinal nerves C1–C2

Functions

 Synergists   Antagonists

Atlanto–occipital joint

Extension (bilateral contraction)

Synergists	Antagonists
Sternocleidomastoideus muscle (with the head extended)	Sternocleidomastoideus muscle (with the head flexed)
Trapezius muscle (descending part)	Longus capitis muscle
Levator scapulae muscle	Rectus capitis anterior muscle
All the other deep back muscles which run to the head	Infra- and suprahyoid muscles

Atlanto–axial joint

Rotation to the same side

Synergists	Antagonists
Splenius capitis muscle	Sternocleidomastoideus muscle
Splenius cervicis muscle	Semispinalis capitis muscle
Longissimus capitis muscle	All the muscles which act as synergists on the same side become antagonists when contracting on the opposite side
Trapezius muscle (descending part)	
Levator scapulae muscle	
All the muscles which act as synergists on the same side become antagonists when contracting on the opposite side	

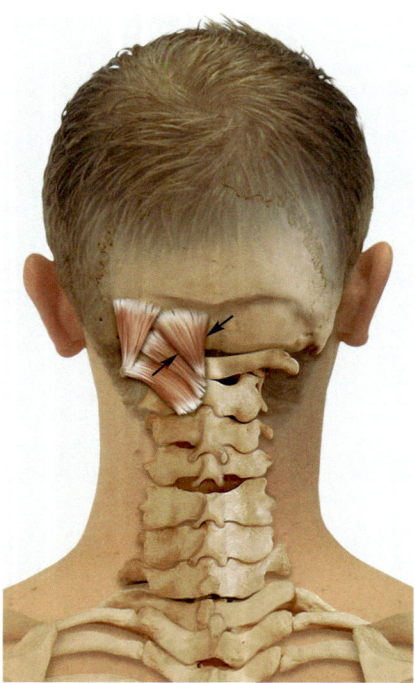

Rectus capitis posterior minor muscle
Lateral tract

| The rectus capitis posterior minor muscle causes extension at the atlanto-occipital joint.

Origin	Posterior tubercle of the atlas
Insertion	Middle part of the inferior arcuate line of the occipital bone Occipital bone
Innervation	Suboccipital nerve from the dorsal branches of spinal nerve C1
Special features	This muscle forms one of the sides of the suboccipital triangle (access point for the vertebral artery)

Functions

 Synergists

 Antagonists

Atlanto–occipital joint

Extension (bilateral contraction)

Sternocleidomastoideus muscle (with the head extended)
Trapezius muscle (descending part)
Levator scapulae muscle
All the other deep back muscles which run to the head

Sternocleidomastoideus muscle (with the head flexed)
Longus capitis muscle
Rectus capitis anterior muscle
Infra- and suprahyoid muscles

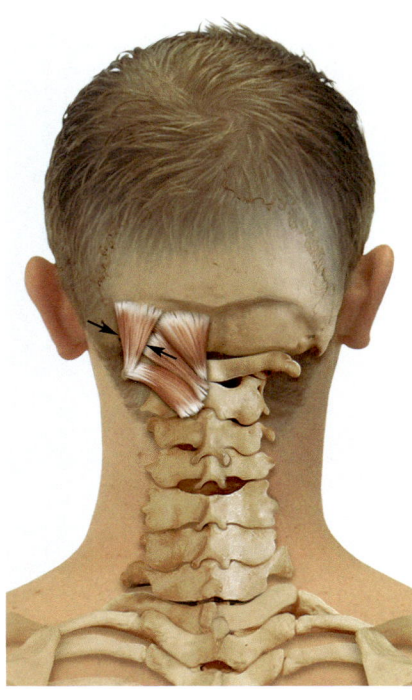

Obliquus capitis superior muscle
Lateral tract

The obliquus capitis superior muscle (superior oblique muscle of the head) runs from the transverse process of the atlas to the occiput. Although this might be expected to give it a powerful rotatory component, the range of movement of the atlanto-occipital joint only leaves the options of extension and lateral inclination. Any rotation that is possible will be minimal, and will not, therefore, be considered here.

Origin	Posterior tubercle of the atlas
Insertion	Occipital bone, above and laterally of the inferior arcuate line
Innervation	Suboccipital nerve from the dorsal branches of spinal nerve C1
Special features	This muscle forms one of the sides of the suboccipital triangle (access point for the vertebral artery)

Functions

Synergists

Antagonists

Atlanto-occipital joint

Extension (bilateral contraction)

Sternocleidomastoideus muscle (with the head extended)	Sternocleidomastoideus muscle (with the head flexed)
Trapezius muscle (descending part)	Longus capitis muscle
Levator scapulae muscle	Rectus capitis anterior muscle
All the other deep back muscles which run to the head	Infra- and suprahyoid muscles

Inclination to the same side

Sternocleidomastoideus muscle	All the muscles which act as synergists on the same side become antagonists when contracting on the opposite side
Trapezius muscle (descending part)	
Levator scapulae muscle	
All the other deep back muscles in the region (without the spinal and interspinal muscles) which run to the head	

271

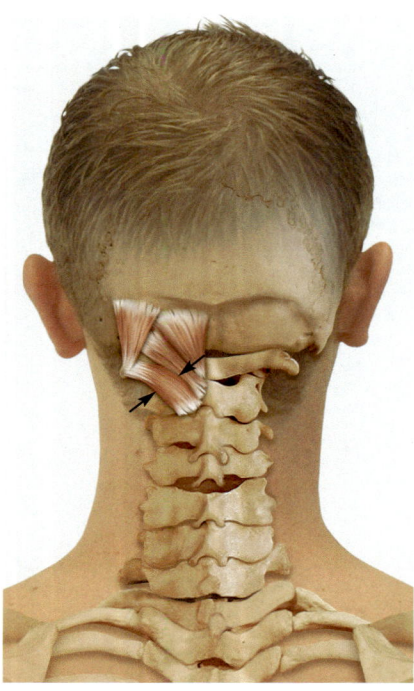

Obliquus capitis inferior muscle
Lateral tract

The obliquus capitis inferior muscle (inferior oblique muscle of the head) runs from the spinous process of the axis to the transverse process of the atlas. Although it has no insertion at the head, it rotates the head to the same side via its effect on the atlanto-axial joint. Furthermore, like all the short nuchal muscles, it stabilizes the sensitive joints of the head when it contracts bilaterally.

Origin	Spinous process of the axis
Insertion	Dorsal transverse process of the atlas
Innervation	Suboccipital nerve, C2
Special features	This muscle forms one of the sides of the suboccipital triangle (access point for the vertebral artery)

Functions

Synergists

Antagonists

Atlanto–axial joint

Rotation to the same side
Splenius capitis muscle
Splenius cervicis muscle
Longissimus capitis muscle
Rectus capitis posterior major muscle
All the muscles which act as synergists on the same side become antagonists when contracting on the opposite side

Sternocleidomastoideus muscle
All the muscles which act as synergists on the same side become antagonists when contracting on the opposite side

Muscle function testing

Muscle power grade

5/4

3

2

1/0

Starting position: The patient lies on his front, with his trunk resting on the treatment couch up to the shoulders and his head hanging over **the edge.**

Test procedure: The examiner holds the patient's chest in place and applies pressure to the back of the head in the direction of cervical spine flexion.

Instruction: "Raise your head against my resistance until it reaches the back of your neck and hold your final position."

Starting position: The patient lies on his front, his trunk resting on the treatment couch up to the shoulders and his head hanging over the edge.

Test procedure: The examiner looks at the movement of the head.

Instruction: "Raise your head until it reaches the back of your neck."

Starting position: The patient lies on his side with the cervical spine flexed.

Test procedure: The examiner holds the thorax in place.

Instruction: "Extend your neck so as to pull your head towards the back of your neck."

Starting position: The patient lies on his front, his trunk resting on the treatment couch up to the shoulders and his head hanging over the edge.

Test procedure: The examiner palpates the cervical spine extensors.

Instruction: "Try to raise your head."

Clinical relevance

- Simultaneous contraction of the anterior and posterior neck muscle groups stabilizes the cervical spine into a neutral position, as when an object is balanced on the head.
- The obliquus capitis inferior is needed to preserve the integrity of the atlanto-occipital joint, both at rest and during movement.

Problems/comments

- It is not normally possible to distinguish functionally between the muscles listed here.
- The junction between the occipital bone and the atlas is known as C0 in the physiotherapy literature.

Stretch tests

Semispinales, splenii and longissimi muscles (both sides)

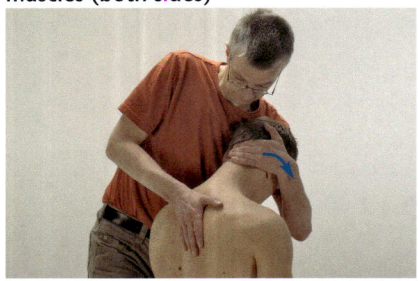

Method
Applying gentle traction, the therapist moves the patient's head and cervical spine into maximum flexion.

Finding
Muscle shortening is present if the movement cannot be performed to its full extent and if the endpoint feels soft and elastic.
The patient reports a stretching sensation along the course of the muscle.

Short cervical spine extensors

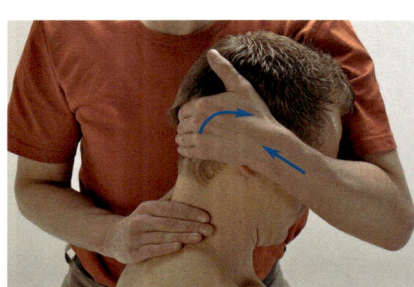

Method
The therapist holds the middle of the cervical spine in place with his thumb and index finger. Using the other hand, he grasps the patient's occiput and stabilizes the head against his shoulder. The therapist moves the head into maximum flexion at the upper joints of the head by pushing the occiput in the dorso-cranial direction.

Finding
Muscle shortening is present if the movement cannot be performed to its full extent and if the endpoint feels soft and elastic.
The patient reports a stretching sensation along the course of the muscle.

4 Trunk

Ventral musculature, abdominal

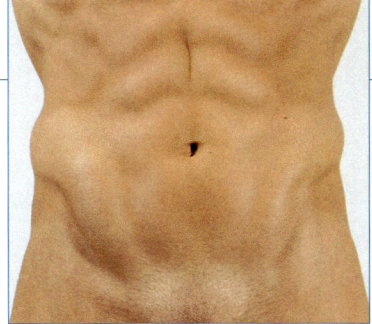

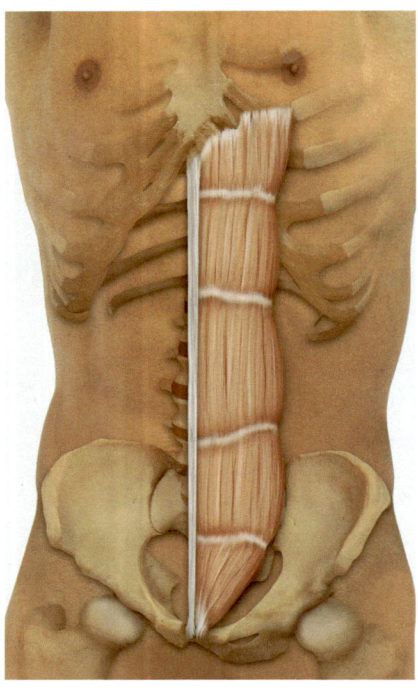

Rectus abdominis muscle

The rectus abdominis (straight abdominal) muscle is a powerful flexor of the spine of the thoracic and lumbar spine, for example when the body is lifted from a supine position. The rectus abdominis helps with abdominal straining. Its cranial part is of particular significance in the fine regulation of expiration during speech.

Origin	Outer surface of the costal cartilage of ribs 5 to 7, xiphoid process
Insertion	Pubic crest Pubic symphysis
Innervation	Intercostal nerves, T5–T11 Subcostal nerve, T12 Iliohypogastric nerve, T12–L1 Ilioinguinal nerve, L1
Special features	The anterior surface of this muscle is crossed by tendinous intersections which run into the anterior lamina of the rectus sheath. On the posterior surface, the muscle fibers run continuously with no intersections. A small portion of this muscle, which runs from the pubic bone to the linea alba (white midline of the abdomen) below the navel, is also known as the pyramidalis muscle

Functions

 Synergists

 Antagonists

Intervertebral joints and disks (primarily thoracic spine)

Flexion (bilateral contraction)
Obliquus externus abdominis muscle
Obliquus internus abdominis muscle

All the back muscles in the region

Straining (bilateral contraction)
Transversus abdominis muscle
Obliquus externus abdominis muscle
Obliquus internus abdominis muscle
Diaphragm

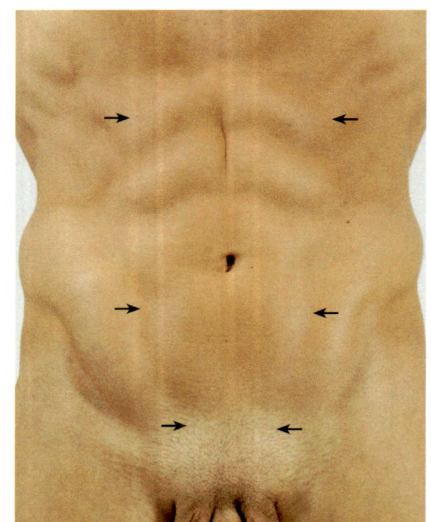

Muscle function testing

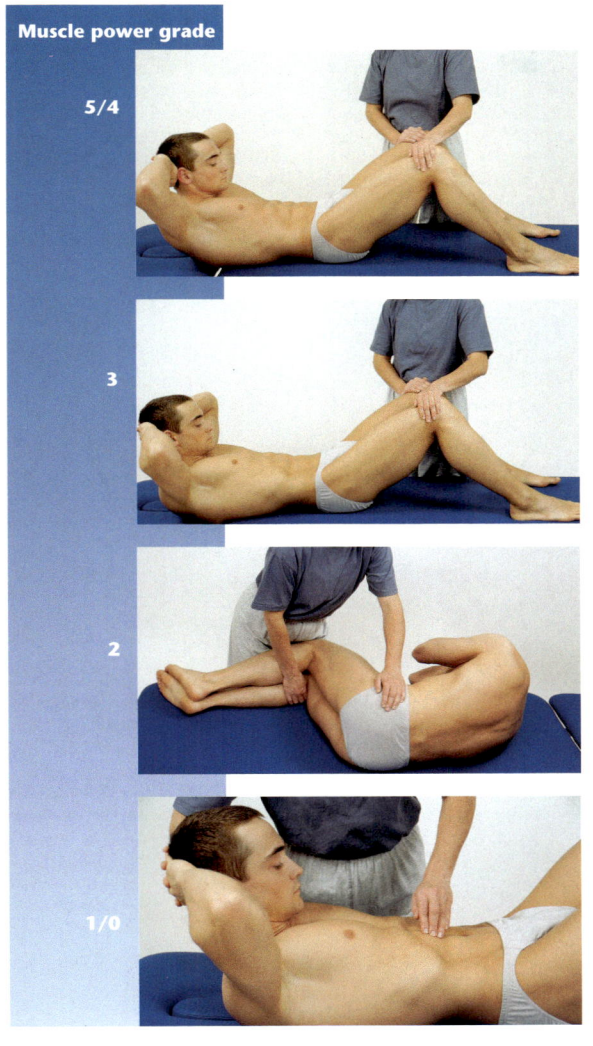

Muscle power grade

5/4

Starting position: The patient lies on his back, with the knees up and the hands grasping the back of the head.

Test procedure: The examiner holds the legs in place.

Instruction: "Roll yourself up by lifting your head, shoulders and chest well clear of the couch."

3

Starting position: The patient lies on his back, with the knees up and the hands grasping the back of the head.

Test procedure: The examiner holds the legs in place.

Instruction: "Roll yourself up by lifting your head, shoulders and chest off the couch as high as possible."

2

Starting position: The patient lies on his side with the legs flexed at the hips and knees.

Test procedure: The examiner holds the legs and pelvis in place and looks at the movement of the trunk.

Instruction: "Roll yourself up, aiming your nose towards your navel."

1/0

Starting position: The patient lies on his back, with the knees up and the hands grasping the back of the head.

Test procedure: The examiner palpates the rectus abdominis muscle.

Instruction: "Try to lift your head and shoulders off the couch."

 Clinical relevance

- Rectus diastasis is a term used to describe separation of the rectus abdominis muscles into two halves.

Problems/comments

- A few parts of the oblique abdominal muscles help with the movement shown here.

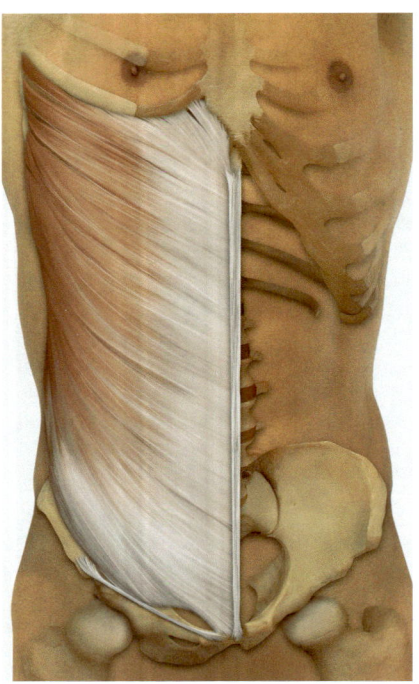

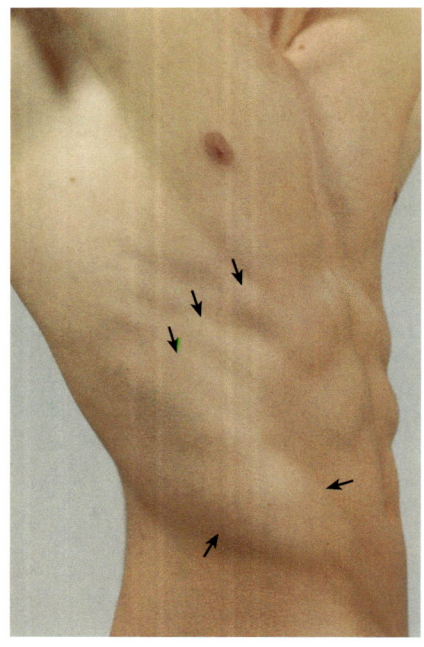

Obliquus externus abdominis muscle

The obliquus externus abdominis muscle (external oblique muscle of the abdomen) rotates the thorax to the opposite side relative to the pelvis and has a flexor effect on the spine when it contracts bilaterally. Together with the other flat abdominal muscles, it also causes abdominal straining, as would occur during childbirth or in urination and defecation. When contracting bilaterally, its cranial part can constrict the inferior thoracic aperture, producing an expiratory effect which will not be considered here.

Origin	Ribs 5–12, caudal margins and outer surfaces
Insertion	Pubic tubercle, pubic crest, external lip of the iliac crest, inguinal ligament, linea alba
Innervation	Intercostal nerves, T5–T11 Subcostal nerve, T12 Iliohypogastric nerve, T12–L1 Ilioinguinal nerve, L1
Special features	The obliquus externus abdominis muscle forms part of the anterior lamina of the rectus sheath

Functions

Synergists

Antagonists

Intervertebral joints and disks (primarily thoracic spine)

Rotation of the trunk to the opposite side

Synergists	Antagonists
Obliquus internus abdominis muscle on the opposite side	Transversus abdominis muscle
Multifidus lumborum muscle	Obliquus internus abdominis muscle
Rotatores lumborum muscles	Iliocostalis lumborum muscle
All the muscles which act as antagonists on the same side become synergists when contracting on the opposite side	Longissimus lumborum muscle
	All the muscles which act as synergists on the same side become antagonists when contracting on the opposite side

Flexion (bilateral contraction)

Synergists	Antagonists
Rectus abdominis muscle	All the deep back muscles in the region
Obliquus internus abdominis muscle	

Intervertebral joints and disks (lumbar and thoracic spine)

Inclination to the same side

Synergists	Antagonists
Obliquus internus abdominis muscle	Obliquus externus abdominis muscle on the opposite side
Quadratus lumborum muscle	Obliquus internus abdominis muscle on the opposite side
Rotatores lumborum muscles	Quadratus lumborum muscle on the opposite side
Levatores costarum muscles	All the contralateral deep back muscles in the region (without the spinal and interspinal muscles)
All the deep back muscles in the region (without the spinal and interspinal muscles)	

Straining (bilateral contraction)

Synergists	Antagonists
Transversus abdominis muscle	None
Obliquus internus abdominis muscle	
Rectus abdominis muscle	
Diaphragm	

Abdominal hollowing (bilateral contraction)

Synergists	Antagonists
Transversus abdominis muscle	Rectus abdominis muscle
Obliquus internus abdominis muscle	Diaphragm

Muscle function testing

Muscle power grade

5/4

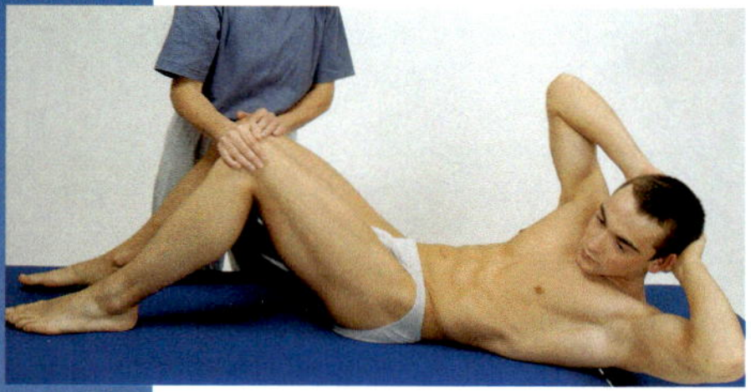

Starting position: The patient lies on his back, with the knees up and the hands grasping the back of the head.

Test procedure: The examiner looks at the movement of the trunk.

Instruction: "Lift your head and shoulders off the couch and move your right shoulder towards the left side of the pelvis. You may lift your chest as you do so."

3

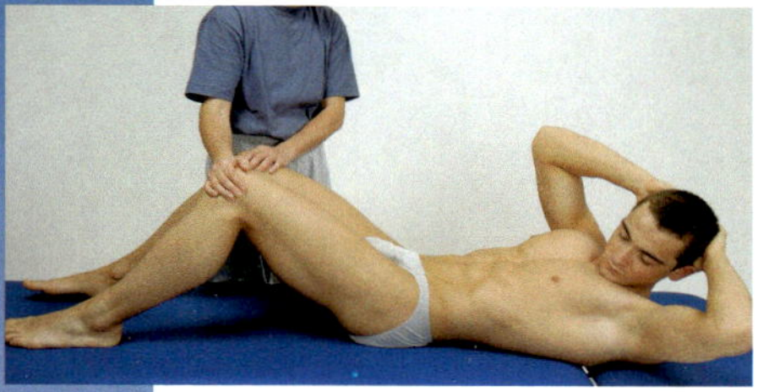

Starting position: The patient lies on his back, with the knees up and the hands grasping the back of the head.

Test procedure: The examiner looks at the movement of the trunk.

Instruction: "Lift your head and shoulders off the couch as far as possible and move your right shoulder towards the left side of the pelvis."

2

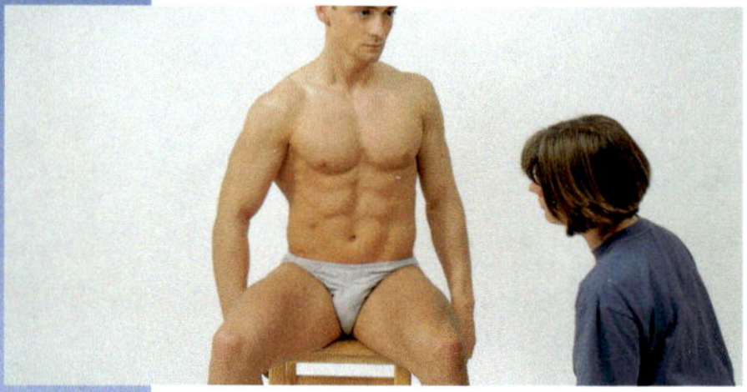

Starting position: The patient sits upright.

Test procedure: The investigator looks at the patient's trunk.

Instruction: "Rotate your body to the left"

1/0

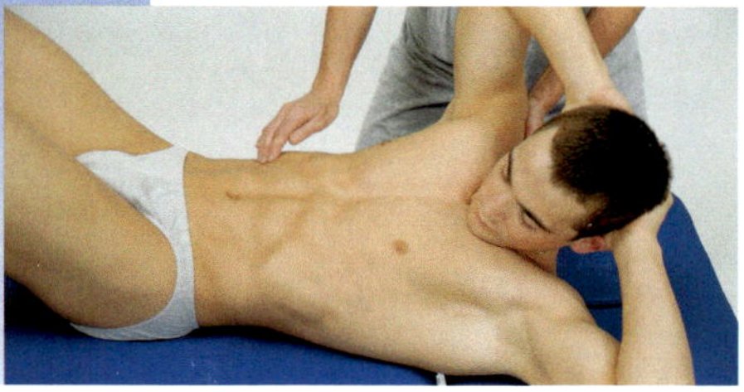

Starting position: The patient lies on his back, with the knees up and the hands grasping the back of the head.

Test procedure: The investigator palpates the oblique externus abdominus muscle.

Instruction: "Try to lift your right shoulder towards the left side of the pelvis."

 Problems/comments

- The extent to which these tests can be performed depends on the mobility of the spine and of the other parts of the body involved.
- Flexing the legs stops the iliopsoas muscle from helping this movement to any extent.
- The right obliquus externus abdominis and the left obliquus internus abdominis, together with the two transversus abdominis muscles, work in conjunction.
- Testing of the right obliquus externus abdominis is being shown here.

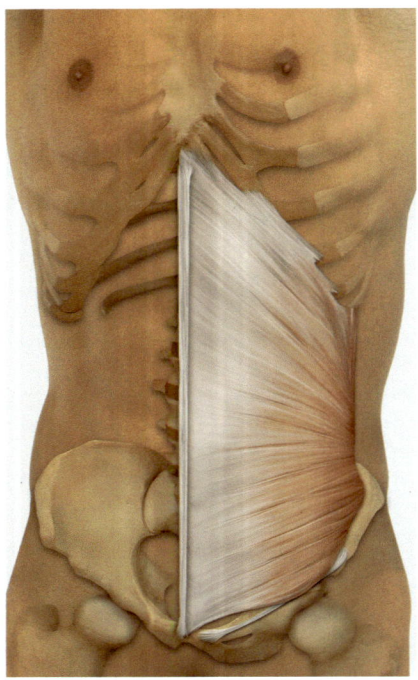

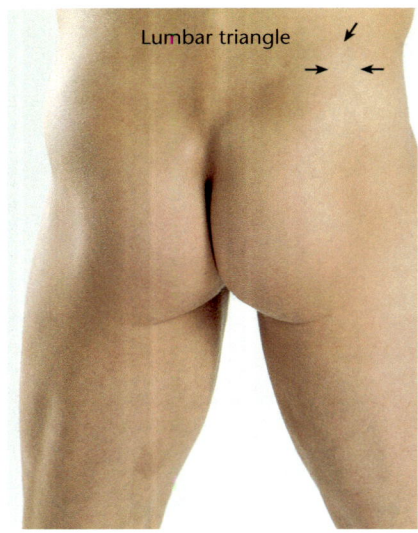

Lumbar triangle

Obliquus internus abdominis muscle

The obliquus internus abdominis muscle (internal oblique muscle of the abdomen) rotates the thorax to the opposite side relative to the pelvis and has a flexor effect on the spine when it contracts bilaterally. Together with the other flat abdominal muscles, it also causes abdominal straining, as would occur during childbirth or in urination and defecation. When contracting bilaterally, its cranial part can constrict the inferior thoracic aperture, producing an expiratory effect which will not be considered here.

Origin	Inguinal ligament, iliac crest, thoracolumbar fascia
Insertion	Pubic crest, costal cartilage of ribs 9–12 Linea alba via the rectus sheath
Innervation	Intercostal nerves, T5–T11 Subcostal nerve, T12 Iliohypogastric nerve, T12–L1 Ilioinguinal nerve, L1
Special features	The obliquus internus abdominis muscle forms part of the anterior and posterior lamina of the rectus sheath. At the lumbar triangle, this muscle may be located directly below the fascia

Functions

Synergists | Antagonists

Intervertebral joints and disks (primarily thoracic spine)

Rotation of the trunk to the same side

Synergists	Antagonists
Transversus abdominis muscle	Obliquus externus abdominis muscle
Obliquus externus abdominis muscle on the opposite side	Multifidus lumborum muscle
	Rotatores lumborum muscles
Iliocostalis lumborum muscle	All the muscles which act as synergists on the
Longissimus thoracis muscle	same side become antagonists when contract-
All the muscles which act as antagonists on the	ing on the opposite side
same side become synergists when contracting	
on the opposite side	

Flexion (bilateral contraction)

Synergists	Antagonists
Rectus abdominis muscle	All the autochthonous back muscles in the
Obliquus externus abdominis muscle	region

Intervertebral joints and disks (lumbar and thoracic spine)

Inclination to the same side

Synergists	Antagonists
Obliquus externus abdominis muscle	Obliquus externus abdominis muscle on the
Quadratus lumborum muscle	opposite side
Rotatores lumborum muscles	Obliquus internus abdominis muscle on the op-
Levatores costarum muscles	posite side
All the autochthonous back muscles in the re-	Quadratus lumborum muscle on the opposite side
gion (without the spinal and interspinal muscles)	All the contralateral autochthonous back mus-
	cles in the region (without the spinal and inter-
	spinal muscles)

Straining (bilateral contraction)

Synergists	Antagonists
Transversus abdominis muscle	None
Obliquus externus abdominis muscle	
Rectus abdominis muscle	
Diaphragm	

Abdominal hollowing (bilateral contraction)

Synergists	Antagonists
Transversus abdominis muscle	Rectus abdominis muscle
Obliquus externus abdominis muscle	Diaphragm

Muscle function testing

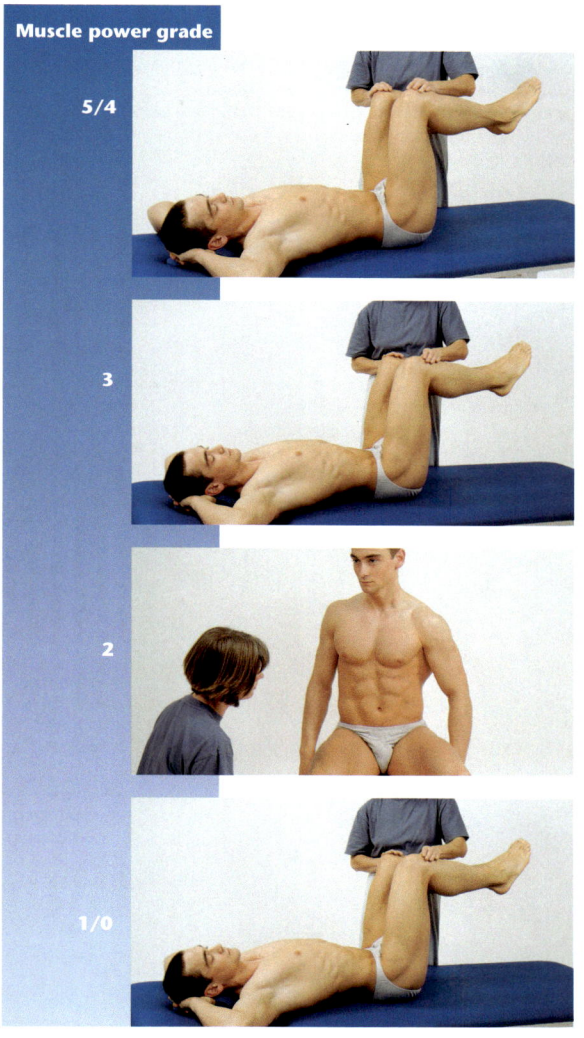

Muscle power grade

5/4

3

2

1/0

Starting position: The patient lies on his back, with the hands grasping the back of the head. The legs are flexed 90° at the hip and knee joints.

Test procedure: The examiner supports the legs.

Instruction: "Lift the right side of your pelvis well clear of the couch and move it towards the left lower edge of your ribcage. Your shoulders should stay on the couch as you do this."

Starting position: The patient lies on his back, with the hands grasping the back of the head. The legs are flexed 90° at the hip and knee joints.

Test procedure: The examiner supports the legs.

Instruction: "Lift the right side of your pelvis off the couch as high as possible and move it towards the left lower edge of your ribcage. Your shoulders should stay on the couch as you do this."

Starting position: The patient sits upright.

Test procedure: The examiner looks at the patient's trunk.

Instruction: "Rotate your body to the right."

Starting position: The patient lies on his back, with the hands grasping the back of the head. The legs are flexed 90° at the hip and knee joints.

Test procedure: The examiner supports the legs.

Instruction: "Try to lift the right side of your pelvis off the couch and move it towards the left lower edge of your ribcage.

 Problems/comments

- The right obliquus internus abdominis and the left obliquus externus abdominis, together with the two transversus abdominis muscles, work in conjunction to produce this movement.
- The best place to palpate the obliquus internus abdominis muscle is at the lumbar triangle.
- Trunk rotation, in which the main activity comes from the transverse abdominal muscles, involves not only the oblique abdominal muscles but also the oblique muscles of the back.
- Testing of the right obliquus internus abdominis is being shown here.

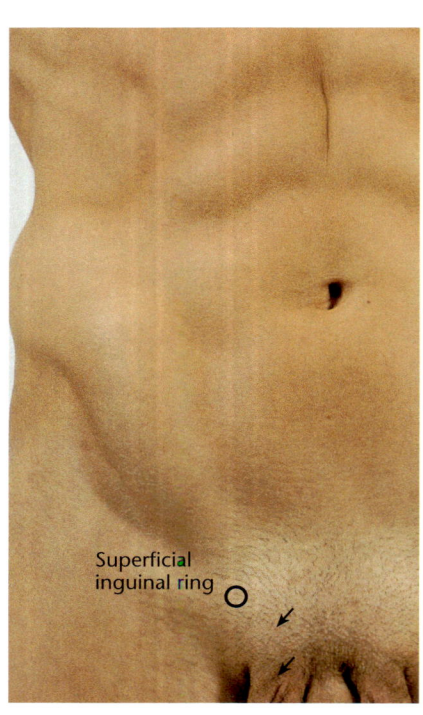

Cremaster muscle

The cremaster muscle lifts the testicles briefly within the scrotum. This movement occurs both when the lower abdominal muscles tighten voluntarily and in the cremasteric reflex, and may be regarded as an insignificant vestigial function. Continuing elevation of the testes and tightening of the scrotum for temperature regulation occurs through the dartos fascia of the scrotum and not through the cremaster muscle.

Origin	Extreme inferior fibers of the obliquus internus muscle and a few fibers of the transversus abdominis muscle
Insertion	Tunica vaginalis of the testes
Innervation	Genitofemoral nerve, genital branch, L1–L2
Special features	If there is a lesion of spinal nerve L2, the cremasteric reflex is weakened or absent. This reflex is produced by lightly stroking the skin of the upper part of the inner thigh, causing the testis on the same side to be lifted within the scrotum.

Superficial
inguinal ring

Transversus abdominis muscle

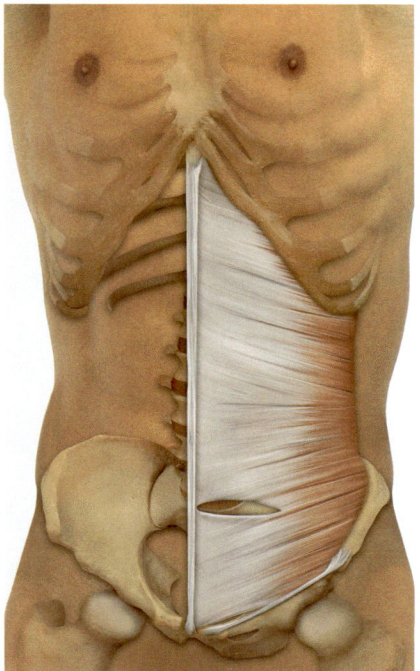

The transversus abdominis muscle (transverse muscle of the abdomen) rotates the thorax to the same side relative to the pelvis. When contracting bilaterally, its cranial part can constrict the inferior thoracic aperture, producing an expiratory effect. However, this muscle is particularly adapted to produce high intra-abdominal pressure during straining, as would occur in childbirth or in urination and defecation, and also for abdominal hollowing. Its expiratory function will not be considered in this list (see transversus thoracis muscle).

Origin	Costal cartilage of ribs 6–12 Costal processes of the lumbar vertebrae
Insertion	Linea alba
Innervation	Intercostal nerves, T5–T11 Subcostal nerve, T12 Iliohypogastric nerve, T12–L1 Ilioinguinal nerve, L1
Special features	Above the arcuate line, the transversus abdominis muscle forms part of the posterior lamina of the rectus sheath; below that line, it forms the superficial lamina of the rectus sheath

Functions

 Synergists   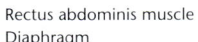 Antagonists

Intervertebral joints and disks (primarily thoracic spine)

Rotation of the trunk to the same side

Synergists	Antagonists
Obliquus internus abdominis muscle	Obliquus externus abdominis muscle
Obliquus externus abdominis muscle on the opposite side	Obliquus internus abdominis muscle on the opposite side
Iliocostalis lumborum muscle	Multifidus lumborum muscle
Longissimus thoracis muscle	Rotatores lumborum muscles
All the muscles which act as antagonists on the same side become synergists when contracting on the opposite side	All the muscles which act as synergists on the same side become antagonists when contracting on the opposite side

Straining (bilateral contraction)

Synergists	Antagonists
Obliquus internus abdominis muscle	None
Obliquus externus abdominis muscle	
Rectus abdominis muscle	
Diaphragm	

Abdominal hollowing (bilateral contraction)

Synergists	Antagonists
Obliquus internus abdominis muscle	Rectus abdominis muscle
Obliquus externus abdominis muscle	Diaphragm

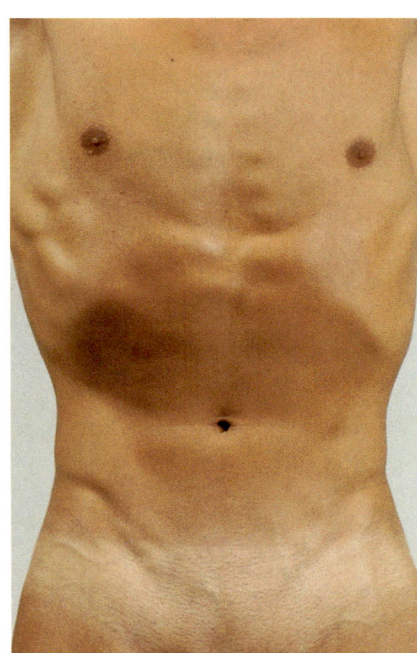

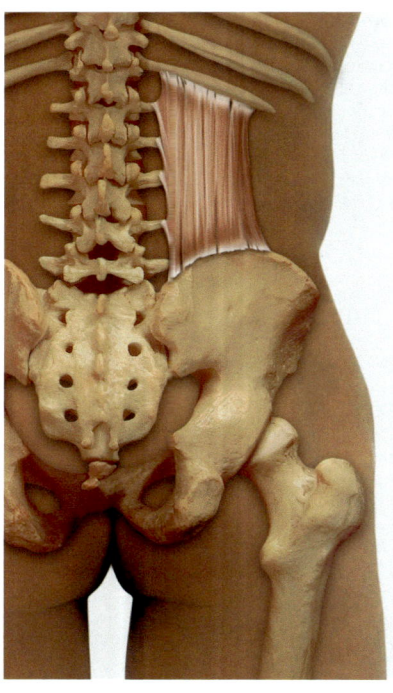

Quadratus lumborum muscle

The quadratus lumborum (quadrate lumbar) muscle contracts powerfully on the swinging leg side and helps the lesser gluteal muscles of the supporting leg to stop the pelvis from subsiding on the swinging leg side. When it contracts unilaterally, it is also capable of inclining the thoracic and lumbar spine to the same side. When contracting bilaterally, this muscle, together with the serratus posterior inferior muscle, stabilizes the inferior thoracic aperture, providing a stable origin for the diaphragm. Thus, it has an inspiratory effect, though this will not be considered in its list of functions. The quadratus lumborum is highly unlikely to have a kyphotic effect on the lordosis of the lumbar spine (like that of the long muscles of the head and neck on the lordosis of the cervical spine), because the belly of this muscle runs some way dorsally of the vertebrae.

Origin	Iliac crest
	Iliolumbar ligament
Insertion	Lower margin of rib 12
	Costal processes of lumbar vertebrae 1–4
Innervation	Intercostal nerves, T5–T11
	Subcostal nerve, T12
	Iliohypogastric nerve, T12–L1
	Ilioinguinal nerve, L1

Functions

 Synergists   Antagonists

Intervertebral joints and disks (primarily lumbar spine)

Inclination to the same side
Obliquus internus abdominis muscle
Obliquus externus abdominis muscle
Rotatores lumborum muscles
Levatores costarum muscles
All the deep back muscles (without the spinal and interspinal muscles)

Obliquus externus abdominis muscle on the opposite side
Obliquus internus abdominis muscle on the opposite side
Quadratus lumborum muscle on the opposite side
All the contralateral deep back muscles (without the spinal and interspinal muscles)

Muscle function testing

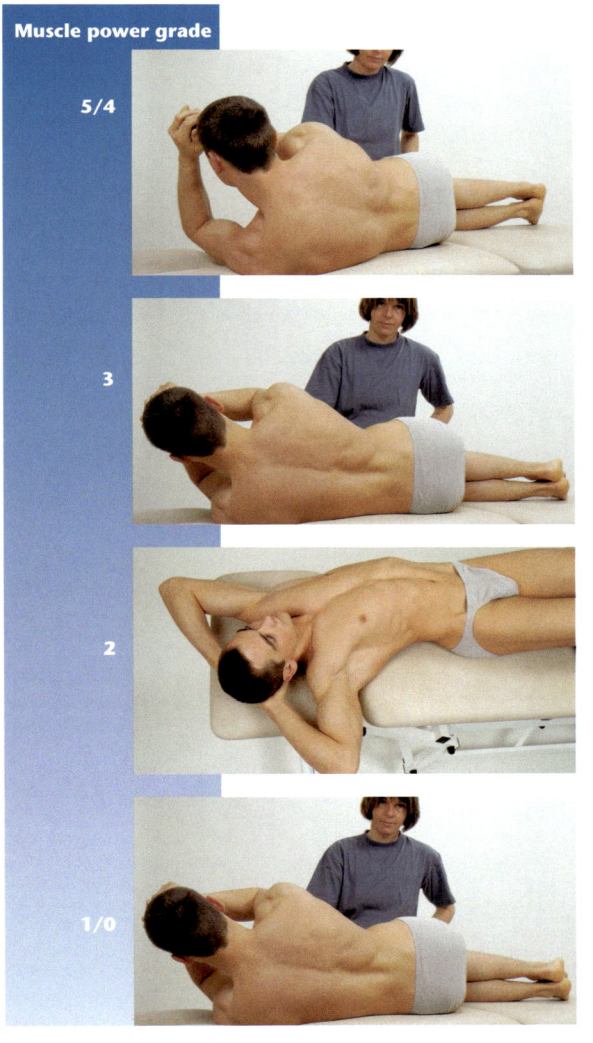

Muscle power grade

5/4

3

2

1/0

Starting position: The patient lies on his side, with the legs flexed 90° at the hip and knee joints and the hands held in front of the head.

Test procedure: The examiner holds the legs in place.

Instruction: "Lift your upper body off the couch sideways."

Starting position: The patient lies on his side, with the legs flexed 90° at the hip and knee joints hands and the hands held in front of the head.

Test procedure: The examiner holds the legs in place.

Instruction: "Lift your upper body off the couch as far as possible."

Starting position: The patient lies on his back with the arms folded behind his head and the knees up.

Test procedure: The examiner looks at the movement of the patient's trunk.

Instruction: "Push your upper body along the couch, moving the right side of your ribcage towards the right side of your pelvis."

Starting position: The patient lies on his side, with the legs flexed 90° at the hip and knee joints hands and the hands held in front of the head.

Test procedure: The examiner holds the legs in place.

Instruction: "Try to lift your upper body off the couch sideways.

⚠ Problems/comments

- It is not possible to palpate the quadratus lumborum muscle.
- The activity of the quadratus lumborum cannot be isolated from the activity of the other muscles of the back and abdomen.

Stretch tests

Quadratus lumborum muscle

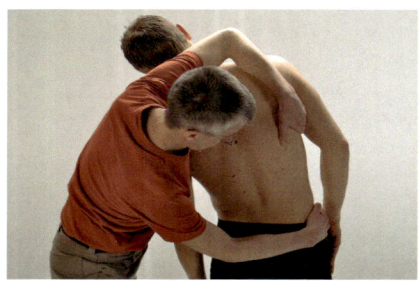

Method

The therapist holds the patient's pelvis in place and moves the trunk into maximum lateral flexion of the lumbar spine to the opposite side.

Finding

Muscle shortening is present if the movement cannot be performed to its full extent and if the endpoint feels soft and elastic.

The patient reports a stretching sensation along the course of the muscle.

Rectus abdominis muscle

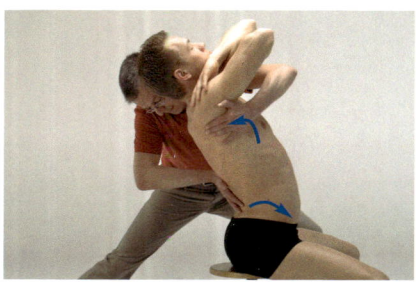

Method

The therapist moves the patient's trunk into maximum extension of the lumbar spine by tilting the pelvis forwards and guiding the upper body backwards and upwards.

Finding

Muscle shortening is present if the movement cannot be performed to its full extent and if the endpoint feels soft and elastic.

The patient reports a stretching sensation along the course of the muscle.

Rectus abdominis muscle, obliquus internus (right) and internus abdominis (left) muscles

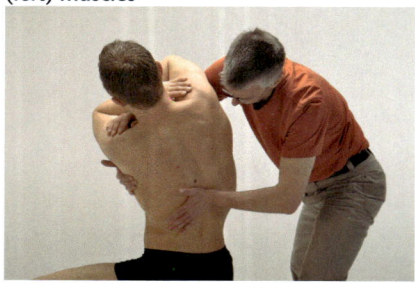

Method

The therapist moves the patient's trunk into maximum extension, lateral flexion to the left and rotation to the right.

Finding

Muscle shortening is present if the movement cannot be performed to its full extent and if the endpoint feels soft and elastic.

The patient reports a stretching sensation along the course of the muscle.

Note

If there is any capsular mobility impairment, the endpoint feels pathologically hard and elastic when the two sides are compared.

4 Trunk

Ventral musculature, thoracic

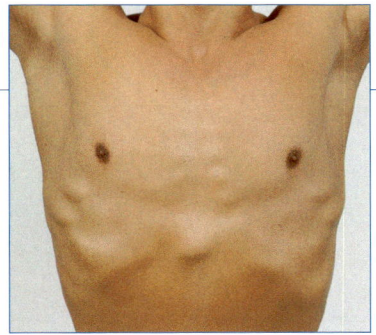

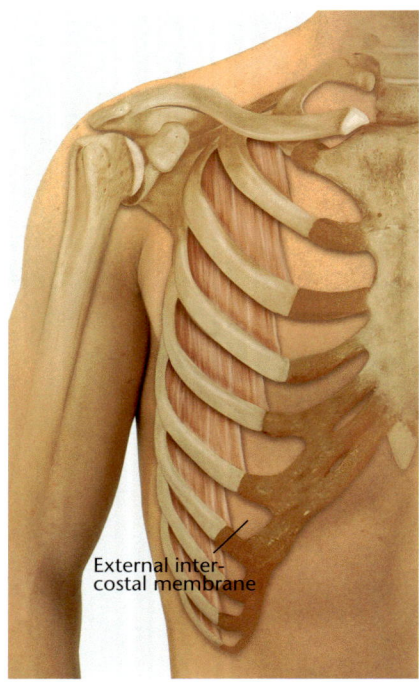

External inter-
costal membrane

External intercostal muscles

The external intercostal muscles widen the intercostal spaces; that is, they elevate the rib above each space. This also results in widening of the pleural cavities, and consequently increased inspiration. This enlarges the sagittal and, in particular, the transverse diameter of the thorax, with elevation of the sternum and the ventral ribs. For the functional relationship between the intercostals and the diaphragm, see under the latter.

Origin	Lower margins of ribs 1–11
Insertion	Upper margins of ribs 2–12
Innervation	Intercostal nerves, T1–T11
Special features	In the anterior, parasternal part of the intercostal space, the muscle is replaced by an external intercostal membrane made of connective tissue

Functions

 Synergists

 Antagonists

Costovertebral and sternocostal joints

Rotation of the ribs around the rib neck axes (elevation of the ventral ribs)
Internal intercostal muscles (intercartilaginous part)
Scalene muscles
Serratus posterior superior muscle

Internal intercostal muscles (interosseous part)
Transversus thoracis muscle
Transversus abdominis muscle
Oblique abdominal muscles

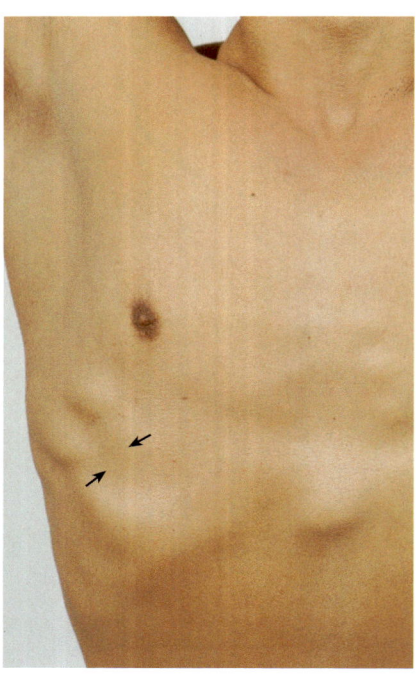

Muscle function testing

Muscle power grade	
4/5	

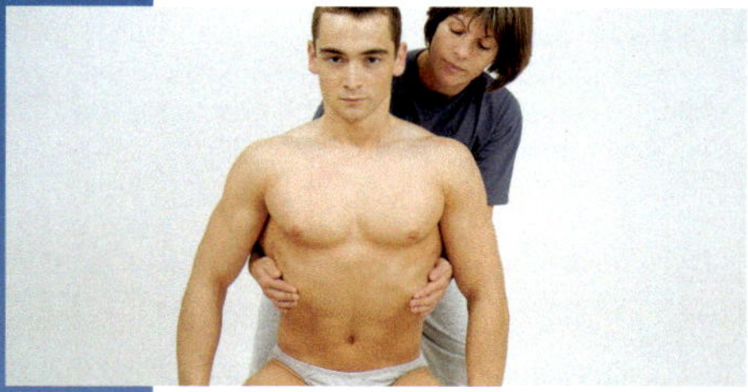

Starting position: The patient sits upright.

Test procedure: Placing one hand on each side of the patient's ribcage, the examiner holds the lower costal arch in place and applies pressure in the direction of expiration.

Instruction: "Breathe in deeply against my resistance."

2/3

Starting position: The patient sits upright.

Test procedure: The examiner looks at the movement of the thorax.

Instruction: "Breathe in deeply."

1/0

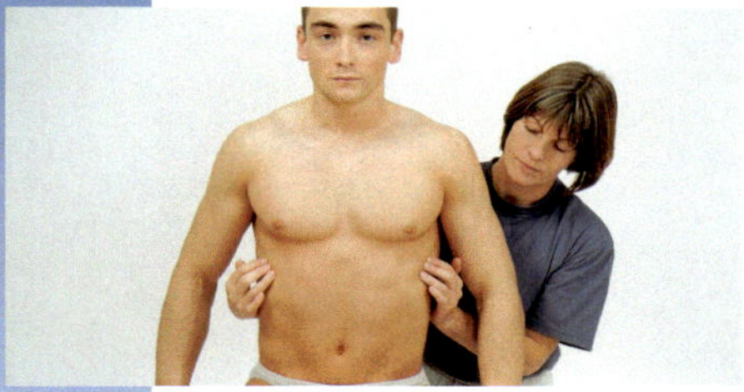

Starting position: The patient sits upright.

Test procedure: The examiner palpates the external intercostal muscles.

Instruction: "Breathe in as deeply as possible."

 Problems/comments

- The serratus posterior superior muscle has the same function as the external intercostals.

- The scalenes are the main muscles to contract on gentle inspiration; the external intercostals come into play only when the inspiration becomes more forced.

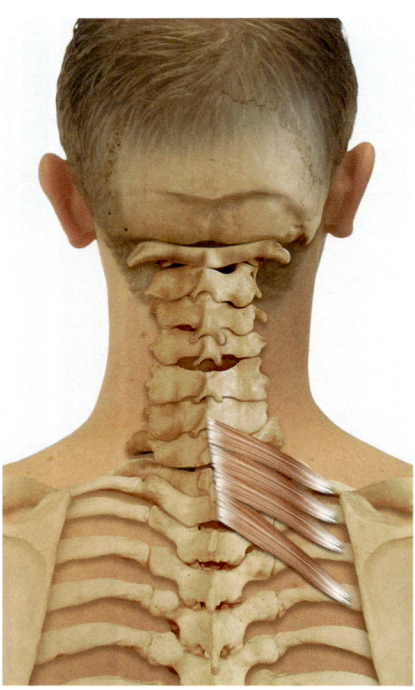

Serratus posterior superior muscle

The serratus posterior superior muscle runs from the last two cervical and first two thoracic vertebrae, extending broadly downwards and laterally to the second to fifth ribs past the costal angle. Consequently, it can elevate the ribs and, eventually, the entire ventral ribcage during inspiration. Its extensor effect on the spine is negligible.

Origin	Spinous processes of cervical vertebrae 6 and 7
	Spinous processes of thoracic vertebrae 1 and 2
Insertion	Ribs 2–5, laterally of the costal angle
Innervation	Ventral branches of spinal nerves C6–T2

Functions

 Synergists

 Antagonists

Costovertebral and sternocostal joints II to V

Rotation of the ribs around the rib neck axes (elevation of the ventral ribs)
External intercostal muscles
Internal intercostal muscles (intercartilaginous part)
Scalene muscles

Internal intercostal muscles (interosseous part)
Transversus thoracis muscle
Transversus abdominis muscle

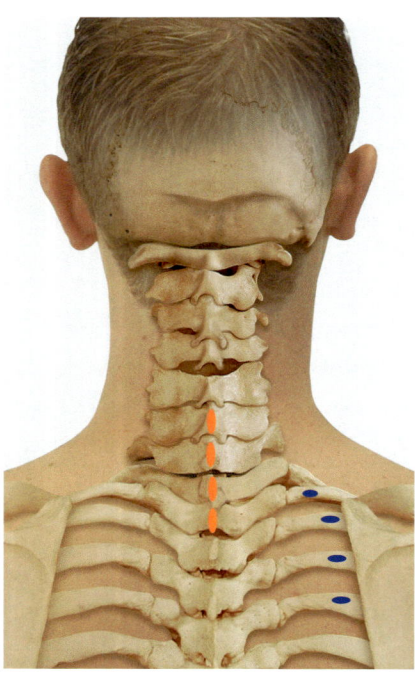

Muscle function testing

Muscle power grade	
5/4	
3/2/1/0	

Starting position: The patient sits upright.

Test procedure: Placing one hand on each side of the patient's ribcage, the examiner holds the lower costal arch in place and applies pressure in the direction of expiration.

Instruction: "Breathe in deeply against my resistance."

Starting position: The patient sits upright.

Test procedure: The examiner looks at the movement of the thorax.

Instruction: "Breathe in deeply."

 Problems/comments

- The serratus posterior superior muscle functions together with the external intercostals as a rib elevator. It is therefore not possible to apply direct resistance to ribs 2 to 5 in order to distinguish its activity from that of the intercostals.

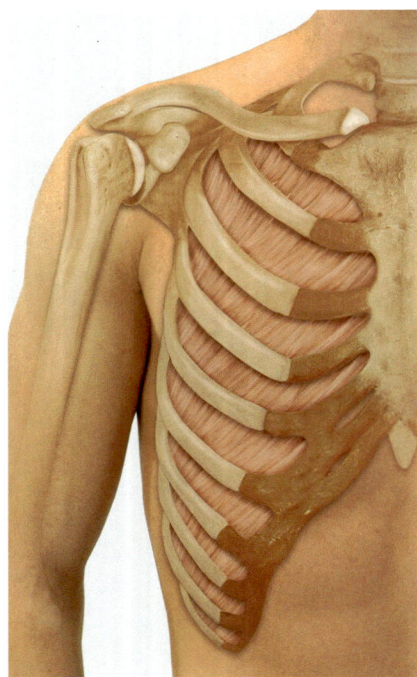

Internal intercostal muscles

The internal intercostal muscles produce an expiratory effect with the parts located between the bony parts of the ribs; that is, they pull each rib up against the one above it. However, the parts located between the costal cartilages have a converse and, therefore, inspiratory effect (see under the external intercostal muscles). However, to keep the list of synergists and antagonists clear and comprehensible, only their expiratory function is considered here.

Origin	Costal cartilage and costal sulcus of ribs 1–11
Insertion	Upper margins of ribs 2–12
Innervation	Intercostal nerves, T1–T11
Special features	In the posterior, paravertebral part of the intercostal space, the muscle is replaced by an internal intercostal membrane made of connective tissue

Functions

 Synergists

 Antagonists

Costovertebral and sternocostal joints

Rotation of the ribs around the rib neck axes (depression of the ventral ribs)
Transversus thoracis muscle
Transversus abdominis muscle
Obliqui abdominis muscles
Rectus abdominis muscle

External intercostal muscles
Internal intercostal muscles (intercartilaginous part)
Scalene muscles
Serratus posterior superior muscle

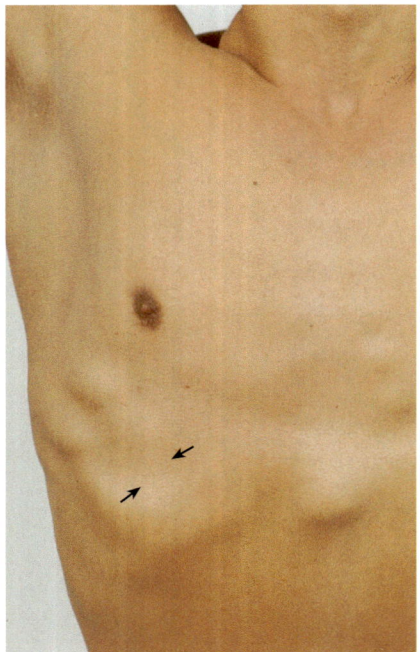

Muscle function testing

Muscle activity

Starting position: The patient sits upright.

Test procedure: The examiner looks at the movement of the thorax.

Instruction: "Breathe out deeply."

 Problems/comments

- The abdominal muscles help with this function.
- The innermost intercostal muscles and the serratus posterior inferior muscle have the same function.
- The internal intercostals are not palpable due to their position.

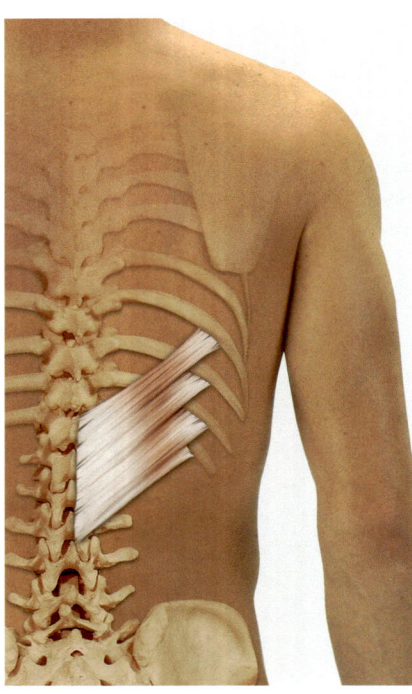

Serratus posterior inferior muscle

The serratus posterior inferior muscle is a muscle which acts during inspiration to prevent the dorsal circumference of the thoracic aperture (i.e. the false ribs) from folding upwards and ventrally into the thorax under the pull of the diaphragm.

Origin	Spinous processes of thoracic vertebrae 11 and 12
	Spinous processes of lumbar vertebrae 1 and 2
Insertion	Ribs 9–12, lower margin
Innervation	Ventral branches of spinal nerves T11–L2

Functions

 Synergists

Costovertebral and sternocostal joints II to V

Stabilization of the dorsal circumference of the lower thoracic aperture during inspiration
Quadratus lumborum muscle
Iliocostalis lumborum muscle

 Antagonists

Diaphragm

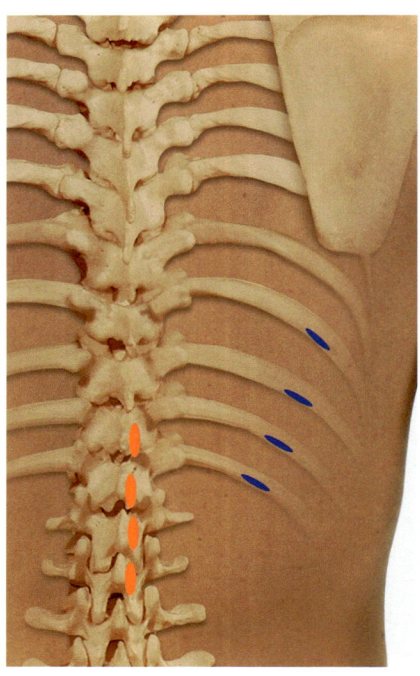

Muscle function testing

Muscle activity

Starting position: The patient sits upright.

Test procedure: The examiner looks at the movement of the thorax.

Instruction: "Breathe out deeply."

 Problems/comments

- Like the internal intercostals, the serratus posterior inferior muscle is a rib depressor.
- The abdominal muscles help with this function.

Diaphragm

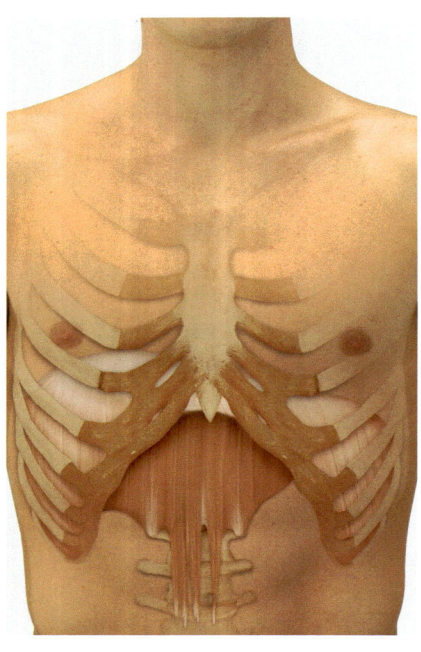

The diaphragm has the highest volumetric efficiency of all the human respiratory muscles. Its muscular component, whose parts are named after the origins of its muscle fibers, depresses the central tendon, thus increasing the volume of the chest above the diaphragm. This reduces the pressure in the thorax, and consequently also in the pleural cavities and the lungs. When the glottic aperture is open, respiratory air flows into the lungs. At the same time, the organs in the thorax and upper abdomen are displaced downwards and the circumference of the abdomen increases. The antagonistic abdominal muscles have a corresponding effect in that they pull the ribs (from which they originate) directly downwards, while simultaneously forcing the organs of the upper abdomen upwards against the diaphragm. In its inspiratory function, however, the diaphragm is dependent on the muscular stabilization of the ribcage by the intercostal muscles and the stabilization of the lower thoracic aperture, in particular, by the serratus posterior inferior and quadratus lumborum muscles. If the diaphragm acts by itself, without this stabilization, it produces paradoxical breathing: an attempt at inspiration causes the diaphragm to contract the flaccid thorax, the lower thoracic aperture is constricted and the patient breathes out.

Origin	Sternal part: posterior surface of the xiphoid process, deep lamina of the rectus sheath
	Costal part: inner surface of the cartilages of ribs 7–12
	Lumbar part: medial and lateral crus of the diaphragm, originating left and right, respectively, of the anterior longitudinal ligament in front of the first three lumbar vertebrae and from the medial and lateral arcuate ligament
Insertion	Central tendon of the diaphragm
Innervation	Right and left phrenic nerve, C3–C6

Functions

 Synergists

 Antagonists

Inspiration

External intercostal muscles	Internal intercostal muscles (interosseous part)
Internal intercostal muscles (intercartilaginous part)	Transversus thoracis muscle
	Obliquus abdominis externus muscle
Scalene muscles	Obliquus abdominis internus muscle
Serratus posterior superior muscle	Transversus abdominis muscle
	Rectus abdominis muscle

4 Trunk

Pelvic floor muscles

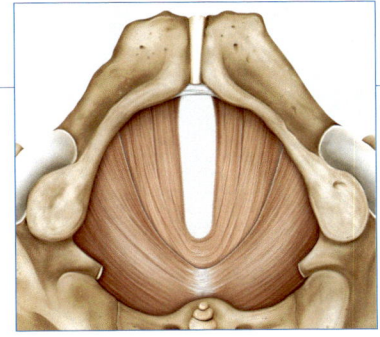

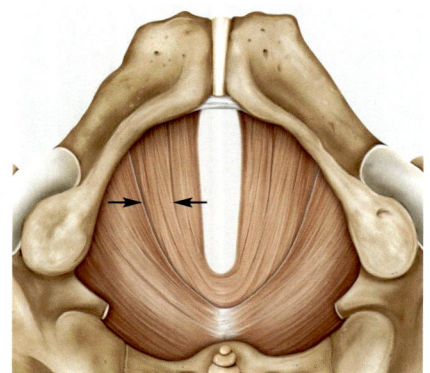

Pubococcygeus muscle

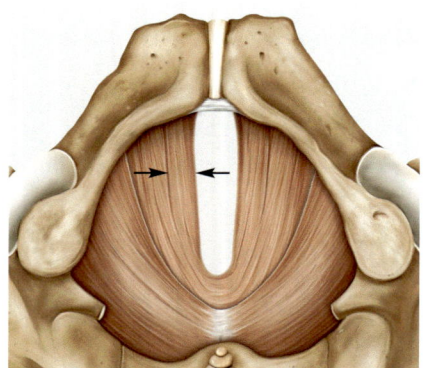

Puborectalis muscle

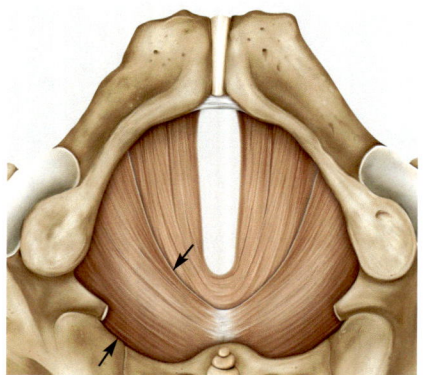

Iliococcygeus muscle

Levator ani muscle

The levator ani muscle forms the muscular pelvic floor; with its tone, which increases with the intraperitoneal pressure (e.g. when a person sneezes or coughs), it can contain the abdominal and pelvic viscera. It forms the basic muscular framework into which the rectum, urethra and vagina in women and which fixes them into place. Its parts are referred to by individual muscle names, depending on their position, as listed below, going outwards:
pubococcygeus muscle
puborectalis muscle
iliococcygeus muscle

Origin	From the pubic bone via the tendinous arch, which is anchored to the obturatorius muscle fascia, up to the ischial spine
Insertion	Tendinous center of the perineum, anchoring itself between the outlets for the organ openings in the midline, with individual fibers also found in the walls of the rectum and vagina, as well as dorsally at the coccygeal bone
Innervation	Direct branches from the sacral plexus, S3–S4

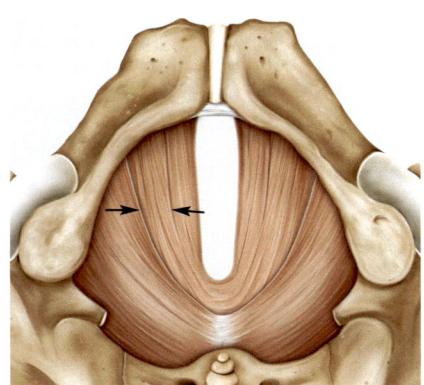

Pubococcygeus muscle

The pubococcygeus (pubococcygeal) muscle is situated laterally between the pubic and the coccygeal bones. It links up laterally with the puborectalis and iliococcygeus muscles but, in contrast with them, does reach the coccygeal bone. The fibers, which run close to the midline between the coccygeal bone and the posterior circumference of the rectum, can antagonize the ventral pull of the pubovaginalis and puborectalis muscles, and thus help to open the rectum and vagina.

Origin	Pubic bone, inner surface, tendinous arch of the levator ani muscle
Insertion	Coccygeal bone, tendinous center of the perineum
Innervation	Direct branches from the sacral plexus, S3–S4
Special features	The pubococcygeus muscle, together with the levator prostatae (or pubovaginalis muscle), the puborectalis muscle and the iliococcygeus muscle, forms the levator ani muscle

Pubovaginalis muscle

The pubovaginalis (pubovaginal) muscle can constrict from the side and from dorsally, which pulls it forwards and upwards towards the pubic bone. From the course of this muscle, it would seem likely that this causes the vagina to be brought into a more horizontal position. In this position, it works like a valve, in that an increase in intraperitoneal pressure or the weight of the unborn child compress it from above, causing it to close.

Puboprostaticus muscle

The puboprostaticus (puboprostatic) muscle is the male counterpart of the pubovaginalis muscle and should be capable of elevating the prostate slightly.

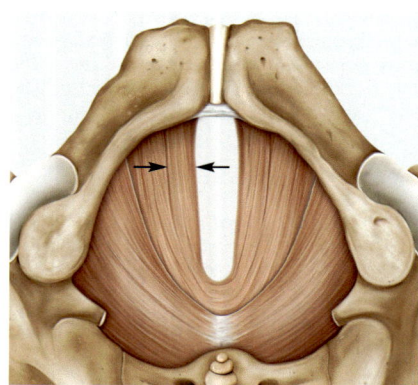

Puborectalis muscle

The puborectalis (puborectal) muscle surrounds the lower rectum from the dorsal side; by tightening, it can pull the rectum forwards against the firm connective tissue layer of the perineum, thus compressing it. This is thought to be the most important closing mechanism for the anus. The tone of this muscle relaxes to allow defecation.

Origin	Pubic bone, inner surface
Insertion	Surrounds the rectum like a snake
Innervation	Coccygeal plexus, S3–S4
Special features	The puborectalis, together with the pubococcygeus muscle, the levator prostatae or pubovaginalis muscle and the iliococcygeus muscle, forms the levator ani muscle
	The puborectalis forms the levator "gate" through which the intestine can pass

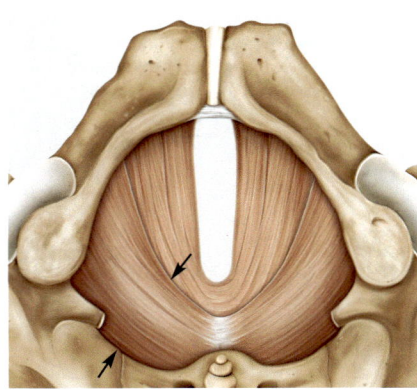

Iliococcygeus muscle

The iliococcygeus (iliococcygeal) muscle has no direct link with the outlets for the organs and connects with the pubococcygeus towards the lateral side. Its fibers run almost transversely from the tendinous arch to the lower sacral vertebrae and the coccygeal bone.

Origin	Tendinous arch of the levator ani muscle
Insertion	Coccygeal bone
Innervation	Direct branches from the sacral plexus, S3–S4
Special features	The iliococcygeus muscle, together with the pubococcygeus muscle, the puborectalis muscle and the levator prostatae or pubovaginalis muscle, forms the levator ani muscle.

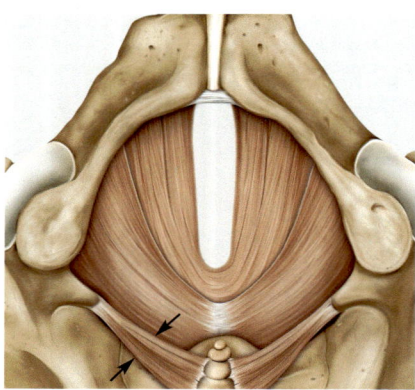

Ischiococcygeus muscle

The ischiococcygeus (ischiococcygeal) muscle also has no direct link with the organ outlets. Its fibers run in the same direction as those of the iliococcygeus muscle, but further dorsally than the latter, over a broad area from the lateral sacral and coccygeal bones to the ischial spine. This muscle is not regarded as part of the levator ani, but it also contributes towards the muscular tone of the pelvic floor.

Origin	Ischial spine, inner surface
Insertion	Sacral bone, lower lateral margins Coccygeal bone
Innervation	Direct branches from the sacral plexus, S3–S4

Sphincter ani externus muscle

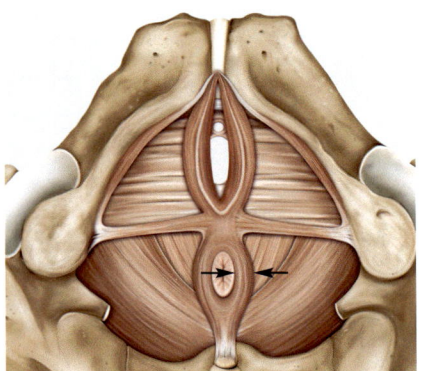

Female pelvic floor

The sphincter ani externus muscle (external sphincter muscle of the anus) is divided into three parts according to its position relative to the skin of the anal region. The subcutaneous part is a ring-shaped cutaneous muscle which encircles the anal opening, similar to the orbicular muscles of the face. The muscle fibers of the superficial (very powerful) part and the deep part lie alongside the anus and can thus constrict it between them. Consequently, they do not surround the anus like a ring but are both attached to the anococcygeal ligament and the perineum. This muscle tightens in the anal reflex and on voluntary closure of the anus in fecal urgency. Its retentive function is of particular importance if there has been loss of tone of the sphincter ani internus muscle. Thus, these two muscles do not, strictly speaking, act as synergists.

Origin	Dermis surrounding the anus
	Tendinous center of the perineum
Insertion	Dermis and subcutis surrounding the anus
	Anococcygeal ligament
Innervation	Pudendal nerve, S2–S4

Male anal region, image: Foto-Archiv KVM

Transversus perinei profundus muscle

The transversus perinei profundus muscle (deep transverse muscle of the perineum) secures the levator "gate" of the levator ani muscle, which might offer a weak point for hernias without this retaining structure. In men, it forms the sphincter urethrae muscle (urethral sphincter) at the aperture admitting the membranous part of the urethra. In women, this structure of circular muscle fibers surrounding the urethra is less pronounced.

Origin	Inferior pubic ramus, ischial ramus
Insertion	Muscle sheet around the apertures for the urethra (in men) and the urethra and vagina (in women)
Innervation	Pudendal nerve or perineal nerve or dorsal nerve of the clitoris or penis, S2–S4

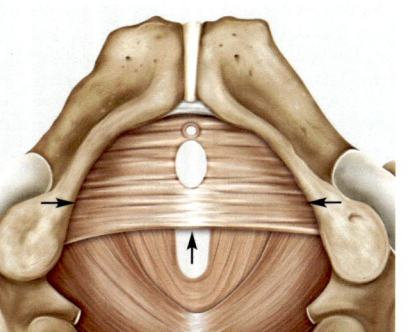

Female pelvic floor

Male anal region, image: Foto-Archiv KVM

Transversus perinei superficialis muscle

The transversus perinei superficialis muscle (superficial transverse muscle of the perineum), like the transversus perinei profundus, can be capable of fixing the perineal raphe, and thus the root of the penis or the posterior part of the vulva, to the perineum at the midline. This muscle is often present only in rudimentary form.

Origin	Superficial branch of the transversus perinei profundus muscle, ischial tuberosity, ischial ramus
Insertion	Tendinous center of the perineum
Innervation	Pudendal nerve or perineal nerve, S2–S4
Special features	The transversus perinei superficialis muscle is not always present.

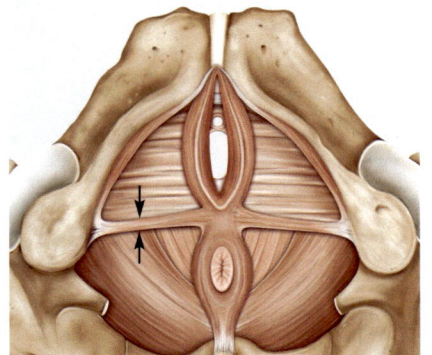

Female pelvic floor

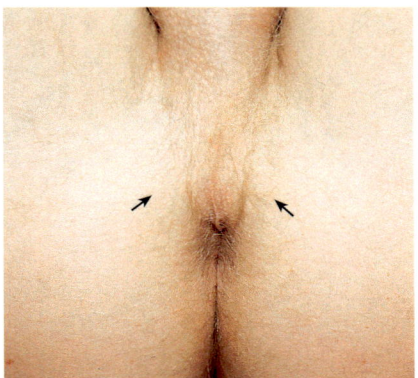

Male anal region, image: Foto-Archiv KVM

Female pelvic floor

Ischiocavernosus muscle

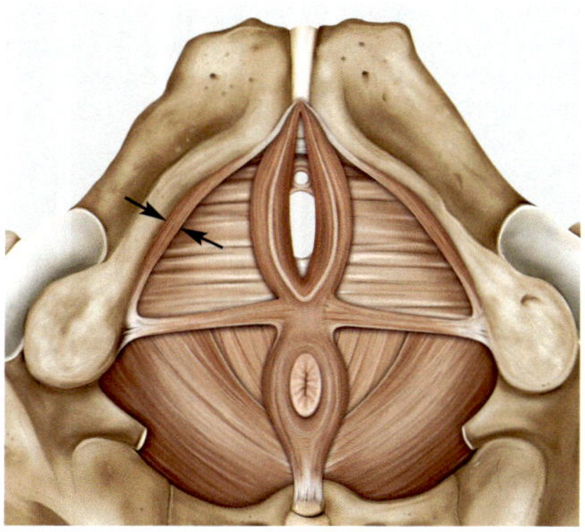

In the flaccid penis, contraction of this muscle produces slight retraction, while it pulls the erect member is pulled towards the abdominal wall. While the spongy body is being filled with blood, the ischiocavernosus muscle (also known as the erector muscle of the penis/clitoris) can compress the deep part of the spongy body of the penis or clitoris against the pubic bone (to which it is attached), thus squeezing out this part of the spongy body into the empty anterior part. Consequently, the posterior part can fill with blood again like a sponge, which is then pumped forward once more with the next wave of contraction of the ischiocavernosus muscle. This can increase the pressure required for erection.

Origin   Ischial ramus

Insertion   Tunica albuginea of the spongy body of the clitoris or penis

Innervation   Pudendal nerve, S2–S4

Bulbospongiosus muscle

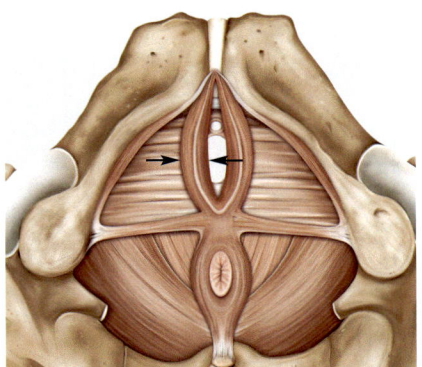

Female pelvic floor

In men, the bulbospongiosus muscle surrounds the bulb of the penis and can compress it, thus also compressing the spongy body. This pumps blood from the posterior bulb of the penis into the glans, enhancing erection. However, its contraction can also propel the contents of the long male urethra towards the outer opening, for instance in the final phase of urination or during ejaculation. In women, this muscle runs laterally of the vaginal opening to the perineal raphe; it can constrict the introitus of the vagina and the raphe slightly, and thus also pulls the anus forward a little.

Origin	Tendinous center of the perineum
Insertion	In women: spongy body of the clitoris In men: inferior urogenital fascia and dorsum of the penis
Innervation	Pudendal nerve or perineal nerve, S2–S4

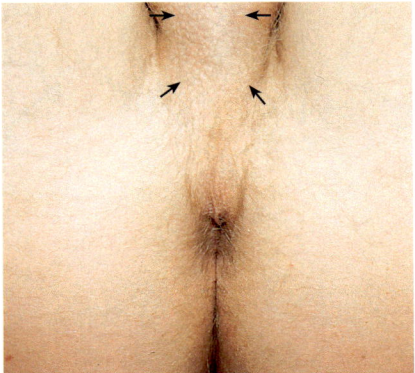

Male anal region, image: Foto-Archiv KVM

Innervation of muscles of the trunk															
Nerve	**Innervated muscles**	**Segments of origin**													
		T1	T2	T3	T4	T5	T6	T7	T8	T9	T10	T11	T12	L1	L2
Phrenic nerve C3–C5															
	Diaphragm C3–C5															
Intercostal nerves		■	■	■	■	■	■	■	■	■	■	■	■		
	Serratus posterior superior muscle	■	■	■	■										
	Rectus abdominis muscle					■	■	■	■	■	■	■	■		
	Obliquus externus abdominis muscle					■	■	■	■	■	■	■	■		
	Obliquus internus abdominis muscle							■	■	■	■	■	■		
	Transversus abdominis muscle					■	■	■	■	■	■	■	■		
	Serratus posterior inferior muscle														
	External and internal intercostal muscles	■	■	■	■										
Subcostal nerve (12th intercostal nerve)										■	■	■	■		
	Pyramidalis muscle												■		
	Quadratus lumborum muscle												■		
Iliohypogastric nerve													■	■	
	Obliquus externus abdominis muscle													■	
	Obliquus internus abdominis muscle													■	
	Transversus abdominis muscle													■	
Ilioinguinal nerve															
	Obliquus internus abdominis muscle													■	
	Transversus abdominis muscle													■	
Genitofemoral nerve														■	■
	Transversus abdominis muscle													■	■
	Cremaster muscle													■	■

The deep muscles of the back are innervated by the dorsal branches of the spinal nerves

The pelvic floor muscles are innervated by the pudendal nerve (S1–S4)

5 Neck

Ventral musculature

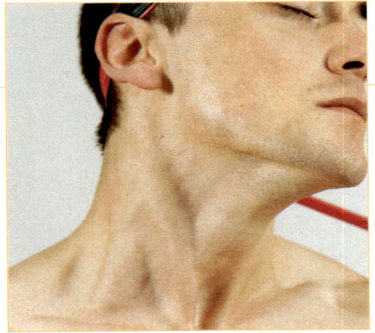

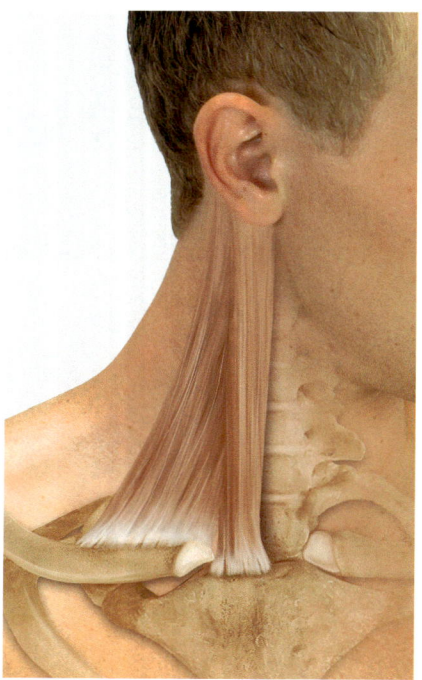

Sternocleidomastoideus muscle

The sternocleidomastoideus (sternocleidomastoid) muscle inclines the head and cervical spine to the same side and, when contracting on one side only, rotates both to the opposite side. The rotatory component is cancelled out when both sides contract. Its flexion and extension effect depends on the position of the head. When the head is flexed, the muscle flexes the head and cervical spine even further; when the head is extended, it extends both. The actions on the sternum and the clavicle are negligible.

Origin	Sternal head: manubrium of the sternum
	Clavicular head: medial part of the clavicle
Insertion	Mastoid process
Innervation	Accessory nerve (cranial nerve XI), cervical plexus, C2

Functions

 Synergists   Antagonists

Atlanto–occipital joint

Flexion (with head flexed)
Rectus capitis anterior muscle, longus capitis muscle, scalenus anterior muscle, suprahyoid muscles, infrahyoid muscles

Deep muscles of the back of the neck which pull on the head
Sternocleidomastoideus (with head extended)
Levator scapulae muscle
Trapezius muscle (descending part)

Extension (with head extended)
Semispinalis capitis muscle
Longissimus capitis muscle
Splenius capitis muscle
Levator scapulae muscle
Trapezius muscle (descending part)

Rectus capitis anterior muscle
Longus capitis muscle
Suprahyoid muscles
Infrahyoid muscles

Atlanto–axial joint

Rotation to the opposite side
Trapezius muscle (descending part)
All the muscles which act as antagonists on the same side become synergists when contracting on the opposite side

Longissimus capitis muscle
Splenius capitis muscle
Rectus capitis posterior major muscle
Obliquus capitis inferior muscle
All the muscles which act as antagonists on the same side become synergists when contracting on the opposite side

Intervertebral discs and joints (cervical spine)

Flexion (with head flexed)
Longus colli muscle, longus capitis muscle, scalenus anterior muscle, suprahyoid muscles, infrahyoid muscles

Deep muscles of the back of the neck
Levator scapulae muscle
Trapezius muscle (descending part)

Extension (with head extended)
Semispinalis capitis muscle, longissimus capitis muscle, splenius capitis muscle, levator scapulae muscle, trapezius muscle (descending part)

Rectus capitis anterior muscle, longus capitis muscle, longus colli muscle, trapezius muscle (descending part), scalenus anterior muscle

Intervertebral discs and joints (cervical spine), atlanto–occipital and atlanto–axial joints

Lateral inclination to the same side
Splenius capitis muscle, longissimus capitis muscle, rectus capitis lateralis muscle (atlanto-occipital joint only), scalenus muscles, trapezius muscle (descending part), levator scapulae muscle

All the muscles which act as antagonists on the same side become synergists when contracting on the opposite side

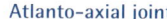

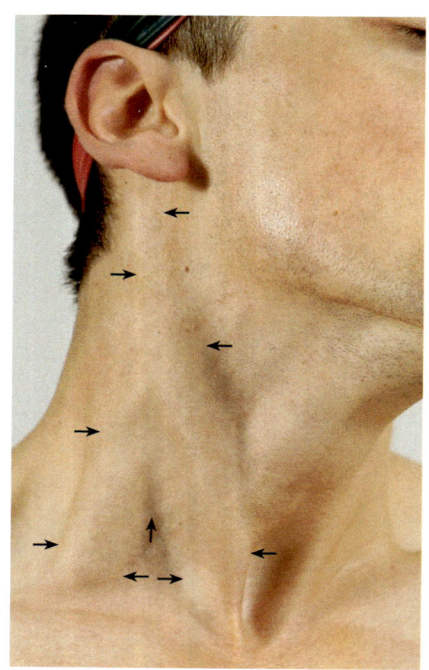

Muscle function testing

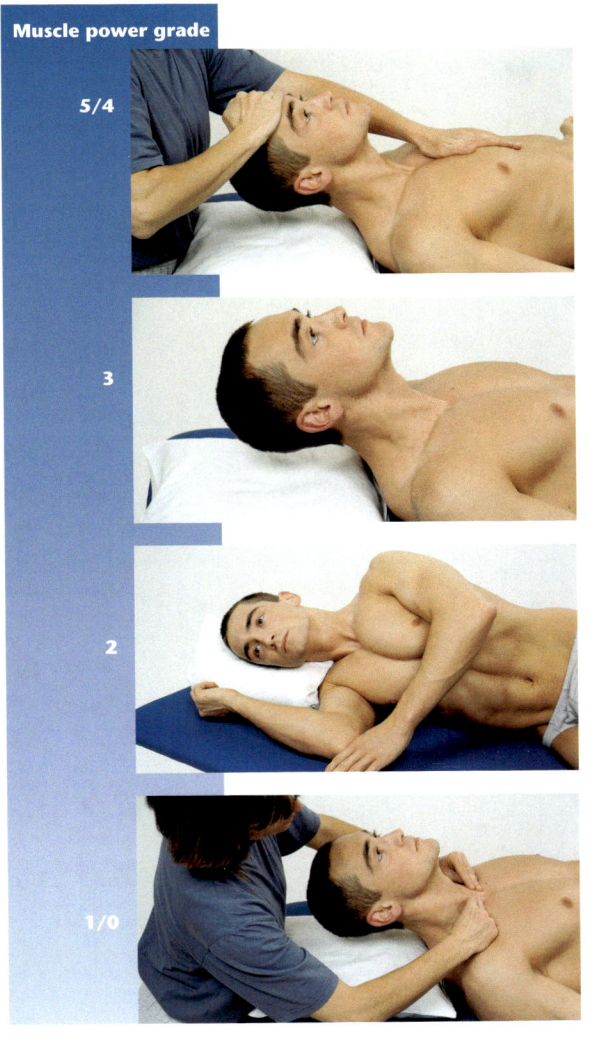

Muscle power grade

5/4

3

2

1/0

BILATERAL TEST

Starting position: The patient lies on his back.

Test procedure: The examiner holds the sternum in place with one hand and applies downward pressure on the patient's forehead with the other.

Instruction: "Raise your head against the resistance of my hand and hold this position."

BILATERAL TEST

Starting position: The patient lies on his back.

Test procedure: The examiner looks at the movement of the patient's head.

Instruction: "Raise your head from the couch."

BILATERAL TEST

Starting position: The patient lies on his side; the head is supported to shoulder width.

Test procedure: The examiner looks at the movement of the patient's head.

Instruction: "Push your head forwards along the head support. Your chest should stay on the couch."

BILATERAL TEST

Starting position: The patient lies on his back.

Test procedure: The examiner palpates the sternocleidomastoideus muscles on both sides of the neck.

Instruction: "Try to raise your head from the couch."

 Clinical relevance

- Unilateral contracture of the sternocleidomastoideus muscle causes torticollis.
- Contraction of the sternocleidomastoideus muscle can intensify the lordosis of the cervical spine by pushing the head forwards relative to the cervical spine and flexing the lower cervical spine relative to the thoracic spine.
- The sternocleidomastoideus muscle is one of the accessory respiratory muscles.

! **Problems/comments**

- The short muscles of the back of the neck (extensors) help with this movement.
- The sternocleidomastoideus test may also be done unilaterally with added rotation of the head to the opposite side.

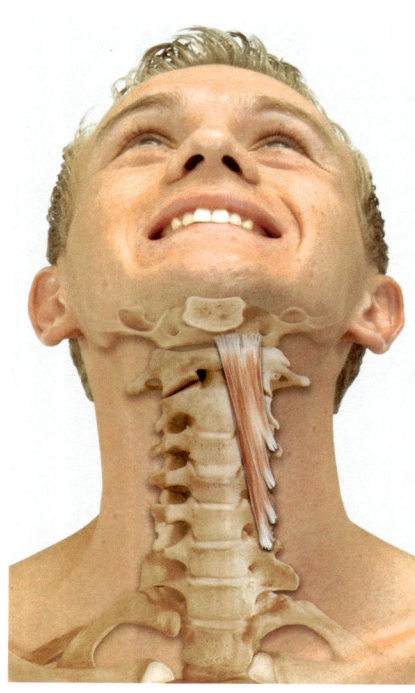

Longus capitis muscle

The longus capitis muscle (long muscle of the head) flexes the cervical spine and thus counteracts the lordotic effect of the muscles of the back of the neck. It also flexes the head against resistance.

Origin   Anterior tubercles of the transverse processes of the cervical vertebrae 3–6

Insertion   Basilar apophysis of the occipital bone

Innervation   Direct branches from the cervical plexus, C1–C4

Functions

 Synergists

 Antagonists

Atlanto–occipital joint

Flexion
Sternocleidomastoideus (with head flexed)
Rectus capitis anterior muscle
Syprahyoid muscles
Infrahyoid muscles

Deep muscles of the back of the neck which pull on the head
Sternocleidomastoideus (with head extended)
Trapezius muscle (descending part)
Levator scapulae muscle

Intervertebral discs and joints (cervical spine)

Flexion
Sternocleidomastoideus (with head flexed)
Longus colli muscle
Scalenus anterior muscle
Suprahyoid muscles
Infrahyoid muscles

Deep muscles of the back of the neck
Sternocleidomastoideus (with head extended)
Trapezius muscle (descending part)
Levator scapulae muscle

Rectus capitis anterior muscle

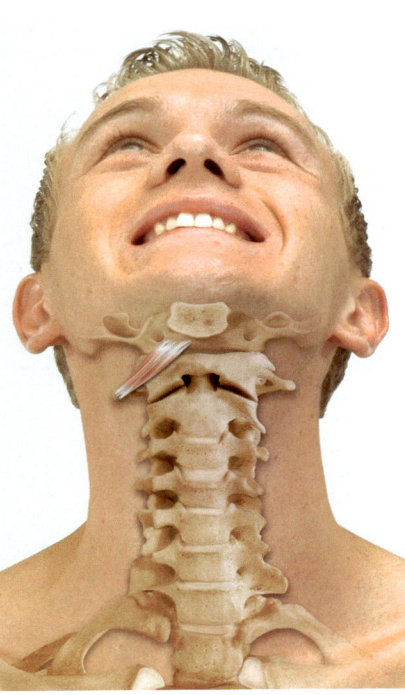

The rectus capitis anterior muscle effects minimal head flexion in the atlanto-occipital joint. However, far more efficient muscles with greater leverage exist for this function. The main task of this muscle, as with all the rectus capitis muscles, is to stabilize the atlanto-occipital joint when the head is moved by other muscles or during flinging movements. In this context, the muscle works flexibly with all the other muscles in this region.

Origin	Transverse process of the atlas
Insertion	Basilar apophysis of the occipital bone
Innervation	Direct branches from the cervical plexus, C1–C4

Functions

 Synergists

Antagonists

Atlanto-occipital joint

Flexion
Sternocleidomastoideus (with head flexed)
Longus capitis muscle
Scalenus anterior muscle
Syprahyoid muscles
Infrahyoid muscles

Deep muscles of the back of the neck which pull on the head
Trapezius muscle (descending part)
Levator scapulae muscle

Longus colli muscle

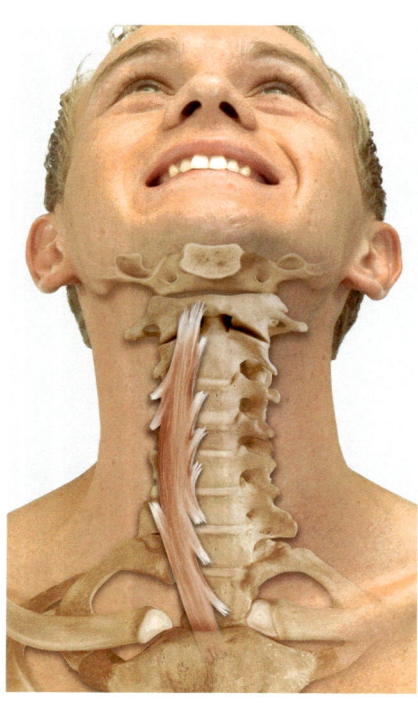

The longus colli muscle (long muscle of the neck) flexes the cervical spine and thus, like the longus capitis muscle, counteracts the lordotic effect of the muscles of the back of the neck.

Origin	Anterior tubercles of the transverse processes of the upper cervical vertebrae, bodies of the last cervical and first thoracic vertebrae
Insertion	Anterior tubercle of the atlas Transverse processes of the lower cervical vertebrae Bodies of the upper cervical vertebrae
Innervation	Anterior branches of spinal nerves C3–C6
Special features	The longus colli muscle is also known as the longus cervicis muscle

Functions

 Synergists

Antagonists

Intervertebral discs and joints (cervical spine)

Flexion

Sternocleidomastoideus (with head flexed)
Longus capitis muscle
Scalenus anterior muscle
Suprahyoid muscles
Infrahyoid muscles

Deep muscles of the back of the neck
Sternocleidomastoideus (with head extended)
Levator scapulae muscle
Trapezius muscle (descending part)

Muscle function testing

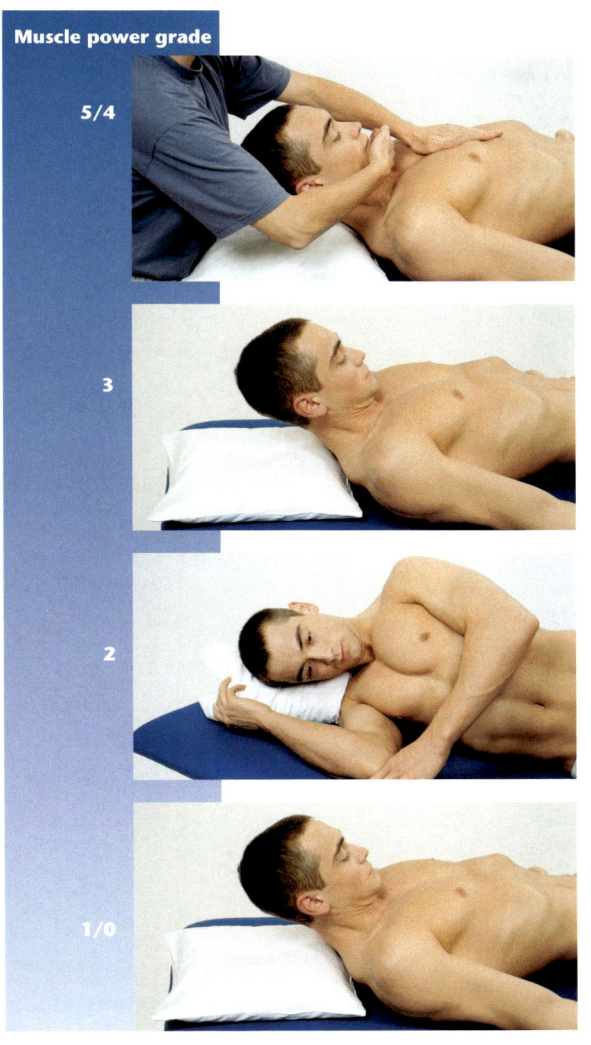

Muscle power grade

5/4

3

2

1/0

Starting position: The patient lies on his back.

Test procedure: The examiner holds the sternum in place with one hand and with the other applies pressure below the chin in the direction of extension of the cervical spine.

Instruction: "Raise your head and bring your chin down to your breastbone against the resistance of my hand."

Starting position: The patient lies on his back.

Test procedure: The examiner looks at the movement of the patient's head.

Instruction: "Raise your head from the couch and bring your chin down to your breastbone."

Starting position: The patient lies on his side.

Test procedure: The examiner looks at the movement of the patient's head.

Instruction: "Bend the spine of your neck by pulling your chin towards your breastbone."

Starting position: The patient lies on his back.

Test procedure: The examiner looks at the movement of the patient's head.

Instruction: "Try to raise your head from the couch."

 Problems/comments

- It is not possible to differentiate functionally between the longus capitis, rectus capitis and longus colli muscles.

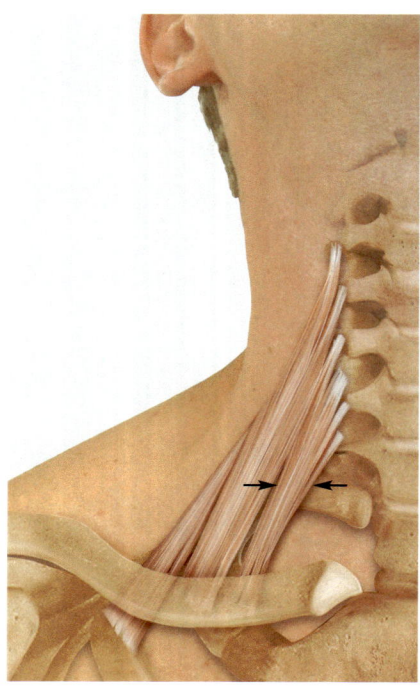

Scalenus anterior muscle

The scalenus anterior (anterior scalene) muscle inclines the cervical spine to the same side and, if the first rib is its fixed point, rotates it to the opposite side. The rotatory and lateral inclination components are cancelled out when both sides contract; if this is the case, the muscle flexes the cervical spine. When the cervical spine is stabilized, it lifts the first rib and thus has an inspiratory effect.

Origin	Anterior tubercles of the transverse processes of cervical vertebrae 3–6
Insertion	Tubercle of the scalenus anterior muscle of the first rib
Innervation	Ventral branches of spinal nerves C5–C8
Special features	The scalenus anterior muscle forms the anterior boundary of the interscalene triangle. The subclavian vein runs in front of the muscle, while the subclavian artery and the brachial plexus run behind this muscle to the arm

Functions

 Synergists

Antagonists

Intervertebral discs and joints (cervical spine)

Lateral inclination to the same side
Sternocleidomastoideus muscle
Scalenus medius muscle
Scalenus posterior muscle
Trapezius muscle (descending part)
Levator scapulae muscle
All the deep back muscles of the
region (excluding the spinal and intraspinal
muscles)

All the muscles which act as antagonists on the
same side become synergists when contracting
on the opposite side

Rotation to the opposite side
Sternocleidomastoideus muscle
Trapezius muscle (descending part)
Rotatores cervicis muscles
Multifidus cervicis muscle
All the muscles which act as antagonists on the
same side become synergists when contracting
on the opposite side

Splenius capitis muscle
Splenius cervicis muscle
Longissimus capitis muscle
Rectus capitis posterior major muscle
Obliquus capitis inferior muscle
All the muscles which act as antagonists on the
same side become synergists when contracting
on the opposite side

Flexion (bilateral contraction)
Sternocleidomastoideus muscle (with head
flexed)
Longus capitis muscle
Longus colli muscle
Suprahyoid muscles
Infrahyoid muscles

Sternocleidomastoideus muscle (with head
extended)
Trapezius muscle (descending part)
Levator scapulae muscle
All the deep back muscles of the region

Scalenus medius muscle

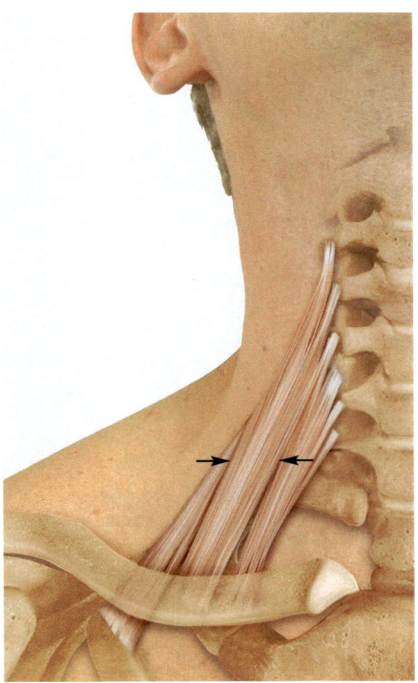

The scalenus medius (middle scalene) muscle inclines the cervical spine to the side and, as one of the accessory respiratory muscles, aids inspiration.

Origin	Anterior tubercles of the transverse processes of cervical vertebrae 2–7
Insertion	First rib, dorsally of the sulcus of the subclavian artery
Innervation	Ventral branches of spinal nerves C4–C8
Special features	The scalenus anterior muscle forms the posterior boundary of the interscalene triangle. The subclavian artery and the brachial plexus run in front of this muscle to the arm

Functions

 Synergists

Intervertebral discs and joints (cervical spine)

Lateral inclination to the same side
Sternocleidomastoideus muscle
Scalenus anterior muscle
Scalenus posterior muscle
Trapezius muscle (descending part)
Levator scapulae muscle
All the deep back muscles of the region (excluding the spinal and intraspinal muscles)

Antagonists

All the muscles which act as antagonists on the same side become synergists when contracting on the opposite side

Scalenus posterior muscle

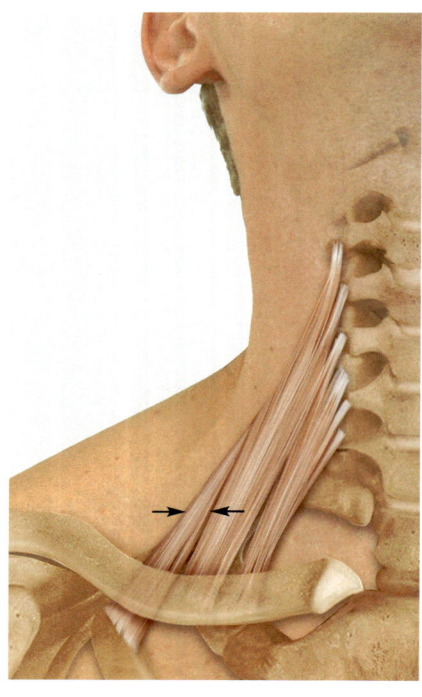

The scalenus posterior (posterior scalene) has a corresponding effect to that of the scalenus medius muscle. Both incline the cervical spine to the side and, as accessory respiratory muscles, aid inspiration.

Origin	Posterior tubercles of the transverse processes of cervical vertebrae 5 and 6
Insertion	Upper margin of rib 2
Innervation	Ventral branches of spinal nerves C7–C8

Functions

 Synergists

 Antagonists

Intervertebral discs and joints (cervical spine)

Lateral inclination to the same side
Sternocleidomastoideus muscle
Scalenus anterior muscle
Scalenus medius muscle
Trapezius muscle (descending part)
Levator scapulae muscle
All the deep back muscles of the region (excluding the spinal and intraspinal muscles

All the muscles which act as antagonists on the same side become synergists when contracting on the opposite side

Muscle function testing

Muscle power grade

5/4

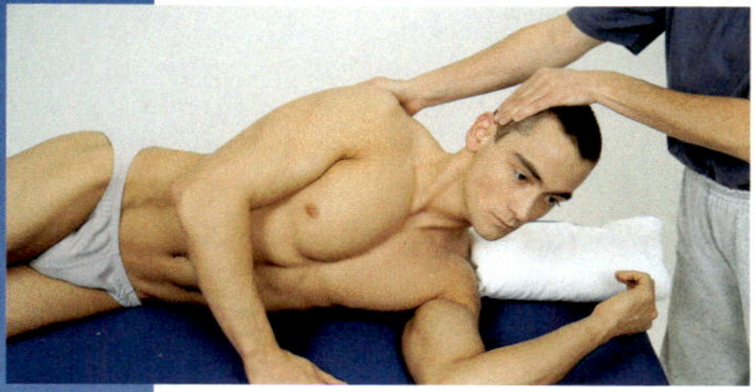

Starting position: The patient lies on his side.

Test procedure: The examiner holds the patient's shoulder in place with one hand and with the other applies downward pressure to the side of the head.

Instruction: "Raise your head against the resistance of my hand and hold this position."

3

Starting position: The patient lies on his side.

Test procedure: The examiner holds the patient's shoulder in place with one hand.

Instruction: "Raise your head from the couch."

2

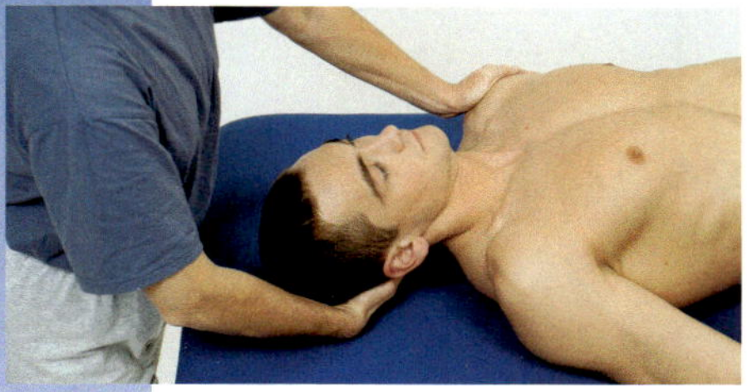

Starting position: The patient lies on his back.

Test procedure: The examiner holds the patient's opposite shoulder in place and takes the weight of the patient's head.

Instruction: "Bring your right ear close to your right shoulder."

1/0

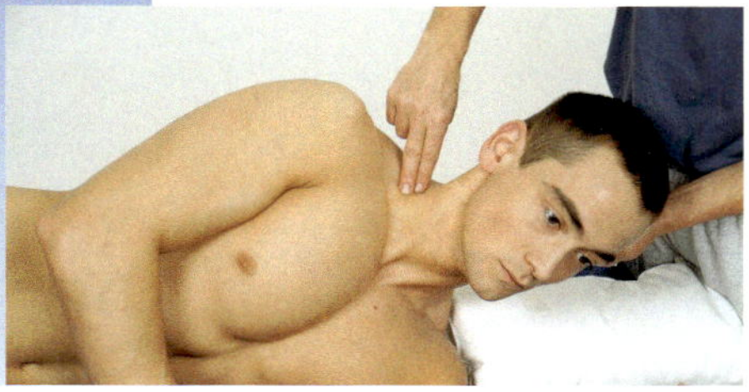

Starting position: The patient lies on his side.

Test procedure: The examiner palpates the scalenus muscles.

Instruction: "Try to raise your head from the couch."

Clinical relevance

- The brachial plexus and the subclavian artery run between the scalenus anterior and scalenus medius muscles (the interscalene triangle).
- The brachial plexus may become stretched over the first rib or trapped between the two scalenus muscles (e.g. when carrying a heavy object), causing scalenus syndrome.

Problems/comments

- The sternocleidomastoideus muscle can take over the function of the scalenus muscles.
- The scalenus muscles are aided in their function by the anterior intertransversarii muscles.

319

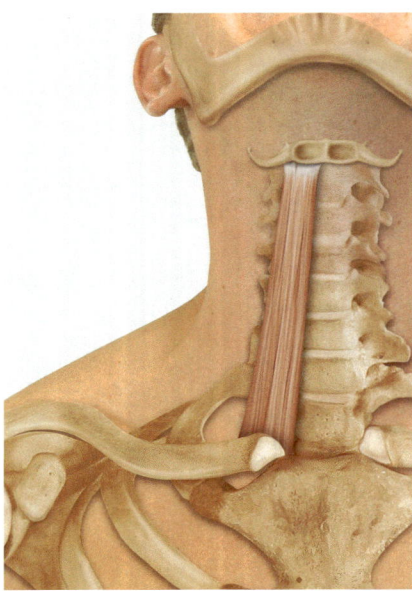

Sternohyoideus muscle

The sternohyoideus (sternohyoid) muscle depresses the hyoid bone (tongue bone). If the suprahyoid muscles between the hyoid bone and the mandible are contracted, this muscle, like all the infrahyoid muscles except the sternothyroideus muscle, acts together with the digastricus and mylohyoideus muscles to open the lower jaw. This function is not included in the list of synergists and antagonists because the muscle does not act directly on the temporomandibular joint.

Origin	Medial section of the clavicle Posterior sternoclavicular ligament Dorsal surface of the manubrium of the sternum
Insertion	Body of the hyoid bone
Innervation	Cervical ansa, C1–C4

Functions

 Synergists

 Antagonists

Hyoid bone

Depressing the hyoid bone

Sternothyroideus muscle
Thyrohyoideus muscle (with larynx fixed)
Omohyoideus muscle (weak)

Digastricus muscle
Geniohyoideus muscle (with mandible fixed)
Mylohyoideus muscle (with mandible fixed)

Intervertebral discs and joints (cervical spine) and atlanto–occipital joint

Flexion (indirect)

Sternocleidomastoideus muscle (with head flexed)
Longus capitis muscle
Longus colli muscle (cervical spine only)
Scalenus anterior muscle (cervical spine only)
Suprahyoid muscles
Sternothyroideus muscle
Thyrohyoideus muscle (with larynx fixed)
Omohyoideus muscle (weak)

Deep muscles of the back of the neck which pull on the head
Sternocleidomastoideus muscle (with head extended)
Trapezius muscle (descending part)
Levator scapulae muscle

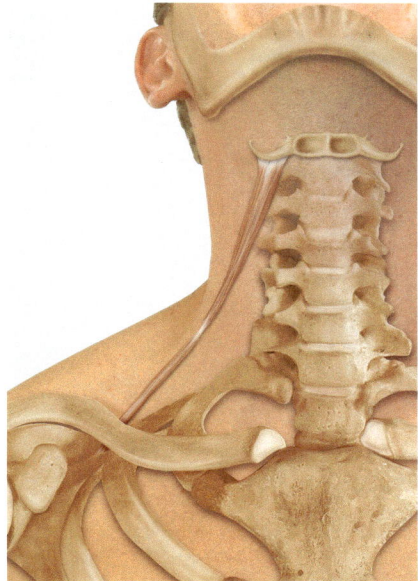

Omohyoideus muscle

The omohyoideus (omohyoid) muscle depresses the hyoid bone and pulls it slightly in the dorsal direction. Its effect on the shoulder blade is slight and may thus be ignored. In contrast, the effect of the omohyoideus muscle on the jugular vein is essential. It keeps this vein open against negative pressure by tightening it with the pretracheal fascia, which is attached to the vein.

Origin	Superior margin of the scapula
	Superior transverse ligament of the scapula
Insertion	Body of the hyoid bone
Innervation	Cervical ansa, C1–C4
Special features	This muscle possesses an intermediate tendon

Functions

 Synergists

 Antagonists

Hyoid bone

Depressing the hyoid bone

Sternohyoideus muscle	Digastricus muscle
Sternothyroideus muscle	Geniohyoideus muscle (with mandible fixed)
Thyrohyoideus muscle (with the larynx fixed)	Mylohyoideus muscle (with mandible fixed)

Intervertebral discs and joints (cervical spine) and atlanto-occipital joint

Flexion (indirect)

Sternocleidomastoideus muscle (with head flexed)	Deep muscles of the back of the neck which pull on the head
Longus capitis muscle	Sternocleidomastoideus muscle (with head extended)
Longus colli muscle (cervical spine only)	Trapezius muscle (descending part)
Scalenus anterior muscle (cervical spine only)	Levator scapulae muscle
Suprahyoid muscles	
Sternothyroideus muscle	
Thyrohyoideus muscle (with the larynx fixed)	
Sternohyoideus muscle	

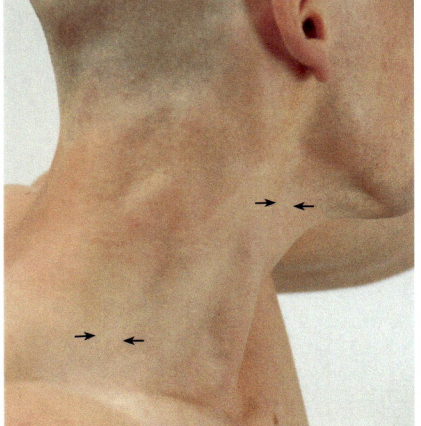

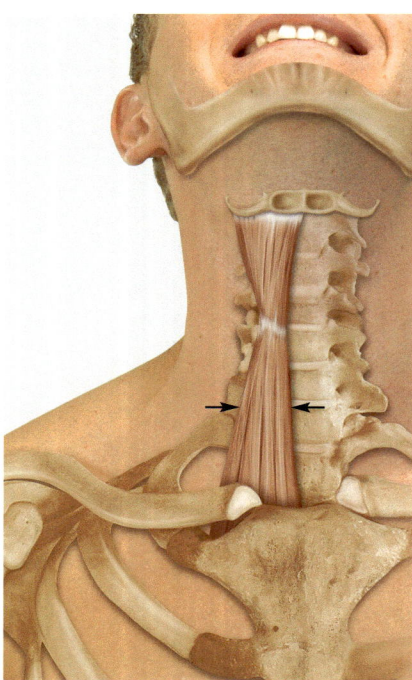

Sternothyroideus muscle

The sternothyroideus (sternothyroid) muscle can depress the larynx, thus bringing it back into its original position after swallowing. The epiglottis can open, because it is moved downwards by the hyoid bone and thus by the adipose body of the larynx. This ensures that the full laryngeal volume needed for voice production is restored.

Origin	Dorsal surface of the manubrium of the sternum Dorsal surface of the costal cartilage of rib 1
Insertion	Oblique line of the thyroid cartilage
Innervation	Cervical ansa, C1–C4

Functions

 Synergists   Antagonists

Larynx

Depressing the larynx
Elastic longitudinal stretching of the trachea

Thyrohyoideus muscle
(indirectly: stylopharyngeus muscle)
(indirectly: suprahyoid muscles)

Intervertebral discs and joints (cervical spine) and atlanto-occipital joint

Flexion (indirect)
Sternocleidomastoideus muscle (with head flexed)
Longus capitis muscle
Longus colli muscle (cervical spine only)
Scalenus anterior muscle (cervical spine only)
Suprahyoid muscles
Sternohyoideus muscle
Thyrohyoideus muscle (with larynx fixed)
Omohyoideus muscle (weak)

Deep muscles of the back of the neck which pull on the head
Sternocleidomastoideus muscle (with head extended)
Trapezius muscle (descending part)
Levator scapulae muscle

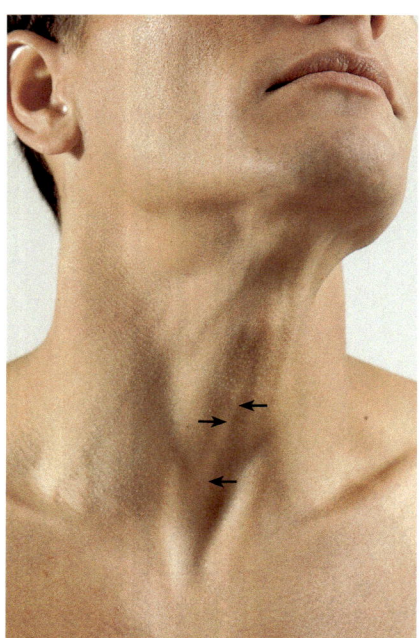

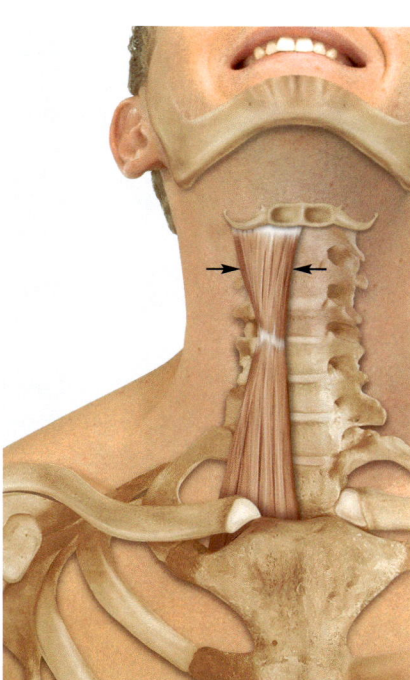

Thyrohyoideus muscle

The thyrohyoideus (thyrohyoid) muscle lifts the larynx towards the hyoid bone, thus helping press the epiglottis against the pre-epiglottal adipose body, depressing the latter and covering the opening of the larynx during swallowing. Consequently, the thyrohyoideus muscle is an important muscle in the act of swallowing. However, if the larynx is fixed at the sternum by the sternothyroideus muscle, the thyrohyoideus, together with the sternohyoideus, can also depress the hyoid bone. If the suprahyoid musculature between the hyoid bone and the mandible is contracted, this muscle, like all the infrahyoid muscles, acts together with the digastricus and the mylohyoideus muscles as an opener of the lower jaw. This function is not included in the list of synergists and antagonists because the muscle does not act directly on the temporomandibular joint.

Origin	Oblique line of the thyroid cartilage
Insertion	Greater horn of the hyoid bone
Innervation	Cervical ansa via the hypoglossal nerve (cranial nerve XII), C1–C2

Functions

 Synergists

 Antagonists

Larynx

Raising the larynx (indirectly with the hyoid bone fixed)
Stylopharyngeus muscle
Suprahyoid muscles

Sternothyroideus muscle
Omothyroideus muscle (weak)
(elastic longitudinal stretching of the trachea)

Hyoid bone

Depressing the hyoid bone (with larynx fixed)
Sternohyoideus muscle
Sternothyroideus muscle
Omohyoideus muscle (weak)

Digastricus muscle
Geniohyoideus muscle (with mandible fixed)
Mylohyoideus muscle (with mandible fixed)

Intervertebral discs and joints (cervical spine) and atlanto-occipital joint

Flexion (indirect)
Sternocleidomastoideus muscle (with head flexed)
Longus capitis muscle
Longus colli muscle (cervical spine only)
Scalenus anterior muscle (cervical spine only)
Suprahyoid muscles
Sternohyoideus muscle
Sternothyroideus muscle
Omohyoideus muscle (weak)

Deep muscles of the back of the neck which pull on the head
Sternocleidomastoideus muscle (with head extended)
Trapezius muscle (descending part)
Levator scapulae muscle

Muscles tested together

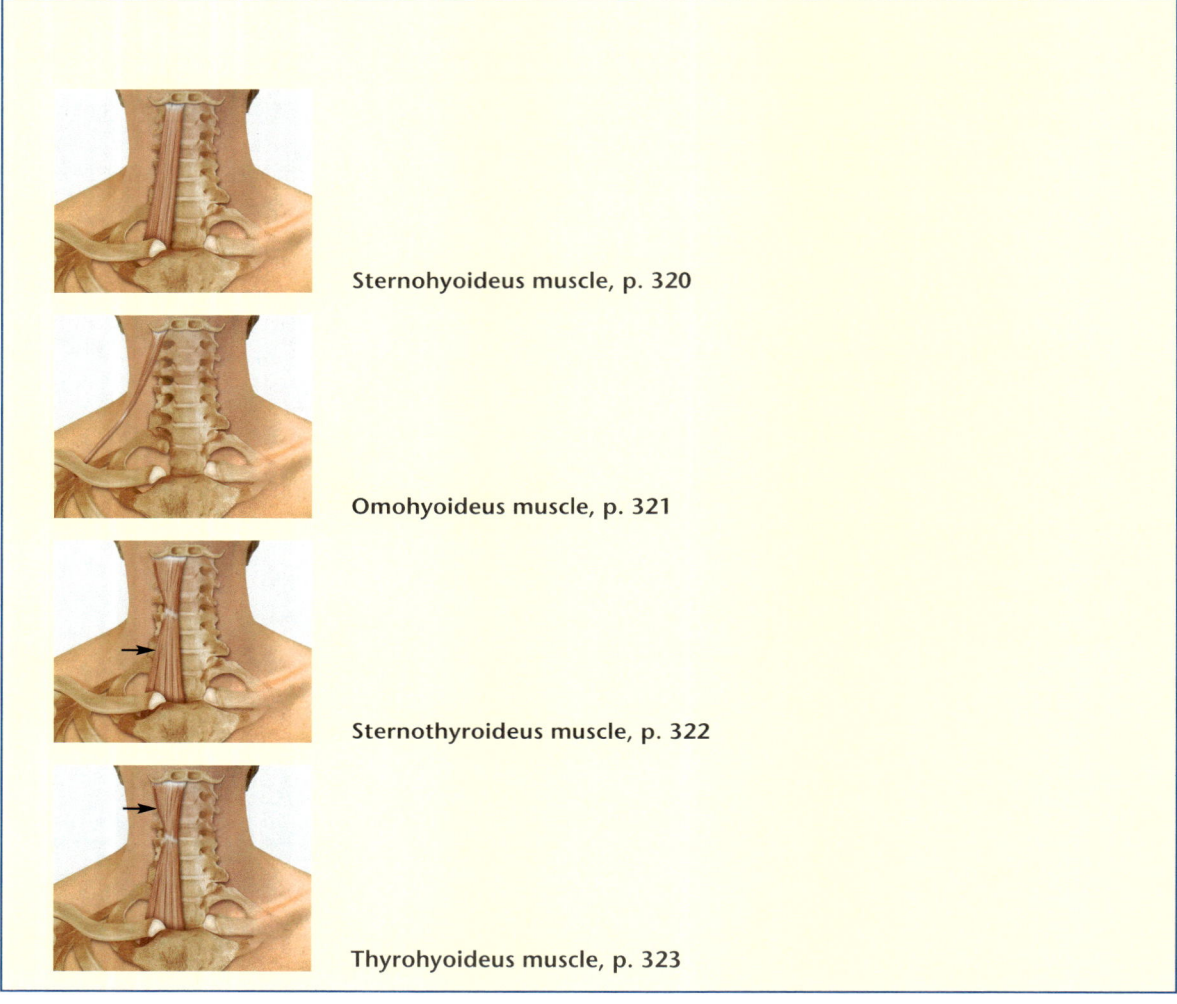

Sternohyoideus muscle, p. 320

Omohyoideus muscle, p. 321

Sternothyroideus muscle, p. 322

Thyrohyoideus muscle, p. 323

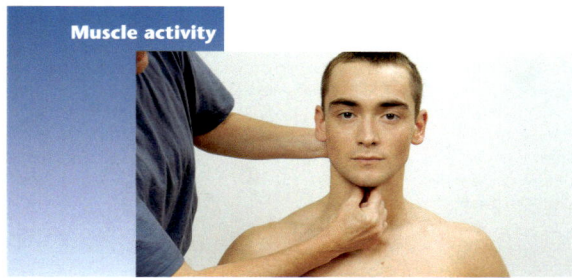

Muscle activity

Muscle function testing

Starting position: The patient sits and looks straight ahead with his face relaxed.

Test procedure: The examiner palpates the hyoid bone with thumb and index finger and follows the progress of that bone in the caudal direction.

Instruction: "Please swallow."

 Clinical relevance

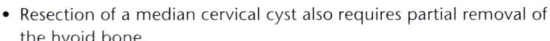 Problems/comments

- Resection of a median cervical cyst also requires partial removal of the hyoid bone.

- It is not possible to differentiate functionally between the individual infrahyoid muscles (lower musculature of the hyoid bone).

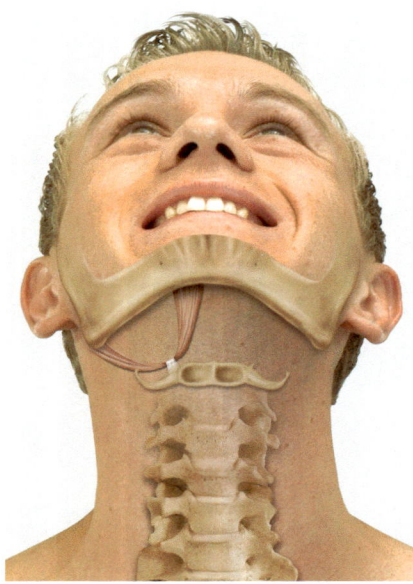

Digastricus muscle

The digastricus (digastric) muscle can pull the mandible downwards against the fixed hyoid bone and is, therefore, an important muscle in the act of actively opening the mouth (e.g. against resistance). During swallowing, it lifts the hyoid bone from the mandible (if the latter is fixed by the muscles of mastication) and from the mastoid process. Its posterior belly can also displace the hyoid bone in the dorsal direction. Since the hyoid bone forms no true joints with the rest of the skeleton, this muscle can only be said to displace this bone relative to the soft tissues of the neck.

Origin	Posterior belly: mastoid process of the temporal bone
Insertion	Anterior belly: lower margin of the mandible
Innervation	Anterior belly: mandibular branch of the trigeminal nerve (cranial nerve V) Posterior belly: facial nerve (cranial nerve VII)
Special features	The intermediate tendon of the digastricus muscle attaches to the hyoid bone via the stylohyoideus muscle

Functions

 Synergists   Antagonists

Hyoid bone

Lifting the hyoid bone
Mylohyoideus muscle (with mandible fixed)
Stylohyoideus muscle
Geniohyoideus muscle (with mandible fixed)

Sternohyoideus muscles
Thyrohyoideus muscle (with larynx fixed)
Omohyoideus muscle (weak)

Displacement of the hyoid bone in the dorsal direction (posterior belly)
Stylohyoideus muscle

Geniohyoideus muscle
Mylohyoideus muscle

Displacement of the hyoid bone in the ventral direction (anterior belly)
Geniohyoideus muscle
Mylohyoideus muscle

Digastricus muscle (anterior belly)

Stylohyoideus muscle
Digastricus muscle (posterior belly)

Temporomandibular joint

Depressing the lower jaw (with hyoid bone fixed)
Geniohyoideus muscle
Mylohyoideus muscle
Pterygoideus lateralis muscle

Temporalis muscle
Masseter muscle
Pterygoideus medialis muscle

Intervertebral discs and joints (cervical spine) and atlanto–occipital joint

Flexion (indirect)
Sternocleidomastoideus muscle (with head flexed)
Longus capitis muscle
Longus colli muscle (cervical spine only)
Scalenus anterior muscle (cervical spine only)
Infrahyoid muscles
Mylohyoideus muscle
Stylohyoideus muscle
Geniohyoideus muscle

Autochthonous muscles of the back of the neck which pull on the head
Sternocleidomastoideus muscle (with head extended)
Trapezius muscle, descending part
Levator scapulae muscle

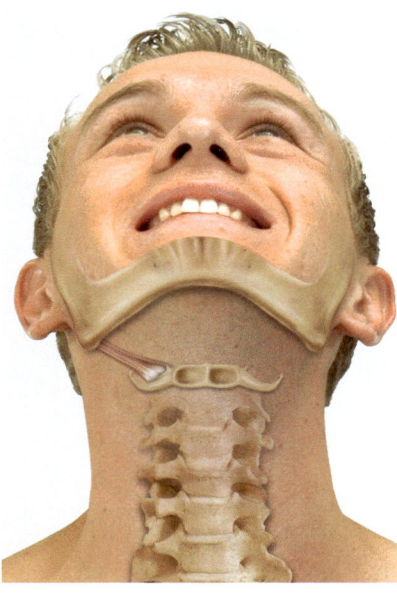

Stylohyoideus muscle

The stylohyoideus (stylohyoid) muscle lifts the hyoid bone and displaces it backwards. In doing so, it causes the muscular floor of the mouth to lengthen. Shortly before its insertion at the hyoid bone, it grasps the intermediate tendon of the digastricus muscle, thus attaching it to the hyoid bone. It is also indirectly involved in the action of the digastricus muscle on the hyoid bone.

Origin	Styloid process of the temporal bone
Insertion	Lateral margin of the hyoid bone
Innervation	Facial nerve (cranial nerve VII)
Special features	The muscle attaches the intermediate tendon of the digastricus muscle to the hyoid bone

Functions

 Synergists Antagonists

Hyoid bone

Lifting the hyoid bone
Digastricus muscle
Mylohyoideus muscle (with mandible fixed)
Geniohyoideus muscle (with mandible fixed)

Sternohyoideus muscles
Thyrohyoideus muscle (with larynx fixed)
Omohyoideus muscle (weak)

Displacement of the hyoid bone in the dorsal direction (posterior belly)
Digastricus muscle (posterior belly)

Geniohyoideus muscle
Mylohyoideus muscle
Digastricus muscle (anterior belly)

Intervertebral discs and joints (cervical spine) and atlanto–occipital joint

Flexion (indirect)
Sternocleidomastoideus muscle (with head flexed)
Longus capitis muscle
Longus colli muscle (cervical spine only)
Scalenus anterior muscle (cervical spine only)
Infrahyoid muscles
Mylohyoideus muscle
Digastricus muscle
Geniohyoideus muscle

Deep muscles of the back of the neck which pull on the head
Sternocleidomastoideus muscle (with head extended)
Trapezius muscle, descending part
Levator scapulae muscle

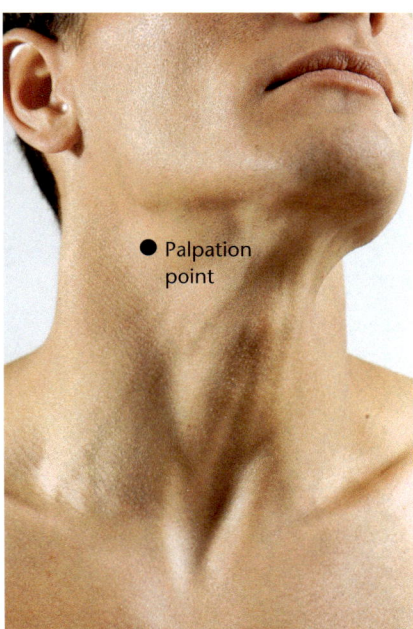

● Palpation point

Mylohyoideus muscle

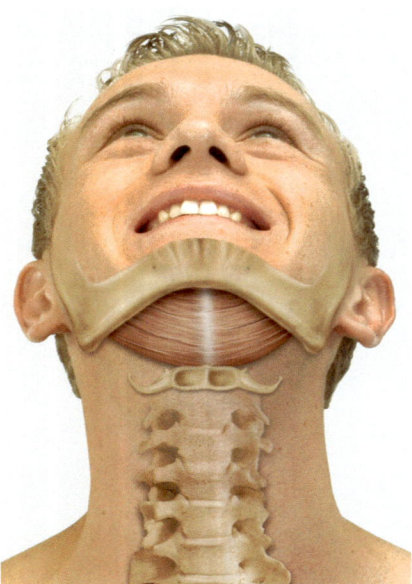

The mylohyoideus (mylohyoid) muscle, depending on its fixed point, can pull the mandible downwards against the fixed hyoid bone or, as it does during swallowing, lift the hyoid bone from the mandible (if the latter is fixed by the muscles of mastication) and displace it in the dorsal direction, thus helping to press the tongue against the palate. In either case, it lifts and tightens the floor of the mouth, which also helps to press the tongue against the roof of the mouth during swallowing. Since the hyoid bone forms no true joints with the rest of the skeleton, this muscle can only be said to displace this bone relative to the soft tissues of the neck.

Origin	Mylohyoid line on the inner surface of the mandible
Insertion	Cranial margin of the body of the hyoid bone
Innervation	Mylohyoid nerve from the mandibular branch of the trigeminal nerve (cranial nerve V)
Special features	The muscles on both sides fan out into a median raphe, which runs from the mandible to the hyoid bone and forms the muscular floor of the mouth

Functions

 Synergists   Antagonists

Hyoid bone

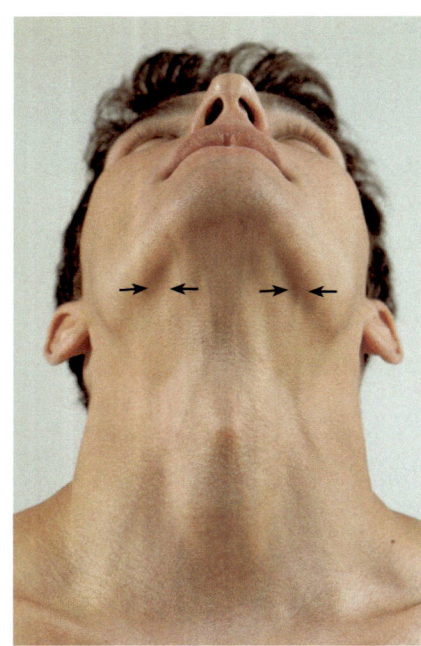

Lifting the hyoid bone
Digastricus muscle (with mandible fixed)
Stylohyoideus muscle
Geniohyoideus muscle (with mandible fixed)

Sternohyoideus muscles
Thyrohyoideus muscle (with larynx fixed)
Omohyoideus muscle (weak)

Displacement of the hyoid bone in the ventral direction
Geniohyoideus muscle
Digastricus muscle (anterior belly)

Stylohyoideus muscle
Digastricus muscle (posterior belly)

Temporomandibular joint

Depressing the lower jaw (with hyoid bone fixed)
Digastricus muscle
Geniohyoideus muscle
Pterygoideus lateralis muscle

Temporalis muscle
Masseter muscle
Pterygoideus medialis muscle

Intervertebral discs and joints (cervical spine) and atlanto–occipital joint

Flexion (indirect)
Sternocleidomastoideus muscle (with head flexed)
Longus capitis muscle
Longus colli muscle (cervical spine only)
Scalenus anterior muscle (cervical spine only)
Infrahyoid muscles
Stylohyoideus muscle
Digastricus muscle
Geniohyoideus muscle

Deep muscles of the back of the neck which pull on the head
Sternocleidomastoideus muscle (with head extended)
Trapezius muscle, descending part
Levator scapulae muscle

Geniohyoideus muscle

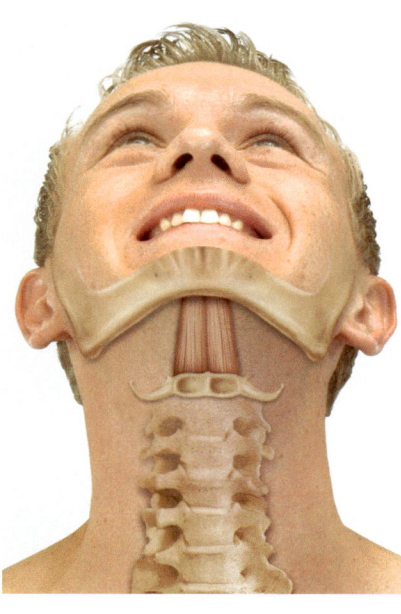

The geniohyoideus (geniohyoid) muscle, depending on its fixed point, can pull the mandible downwards against the fixed hyoid bone or, as it does during swallowing, lift the hyoid bone from the mandible (if the latter is fixed by the muscles of mastication) and displace it in the ventral direction. During swallowing, it shortens the muscular floor of the mouth, simultaneously widening the pharynx. Since the hyoid bone forms no true joints with the rest of the skeleton, this muscle can only be said to displace this bone relative to the soft tissues of the neck.

Origin	Mental spine
Insertion	Anterior surface of the body of the hyoid bone
Innervation	Cervical ansa, C1–C2 via the hypoglossal nerve (cranial nerve XII)

Functions

 Synergists   Antagonists

Hyoid bone

Lifting the hyoid bone
Digastricus muscle (with mandible fixed)
Stylohyoideus muscle
Mylohyoideus muscle (with mandible fixed)

Sternohyoideus muscles
Thyrohyoideus muscle (with larynx fixed)
Omohyoideus muscle (weak)

Displacement of the hyoid bone in the ventral direction
Mylohyoideus muscle
Digastricus muscle (anterior belly)

Stylohyoideus muscle
Digastricus muscle (posterior belly)

Temporomandibular joint

Depressing the lower jaw
Digastricus muscle
Mylohyoideus muscle
Pterygoideus lateralis muscle

Temporalis muscle
Masseter muscle
Pterygoideus medialis muscle

Intervertebral discs and joints (cervical spine) and atlanto–occipital joint

Flexion (indirect)
Sternocleidomastoideus muscle (with head flexed)
Longus capitis muscle
Longus colli muscle (cervical spine only)
Scalenus anterior muscle (cervical spine only)
Infrahyoid muscles
Mylohyoideus muscle
Digastricus muscle
Stylohyoideus muscle

Deep muscles of the back of the neck which pull on the head
Sternocleidomastoideus muscle (with head extended)
Trapezius muscle, descending part
Levator scapulae muscle

Muscles tested together

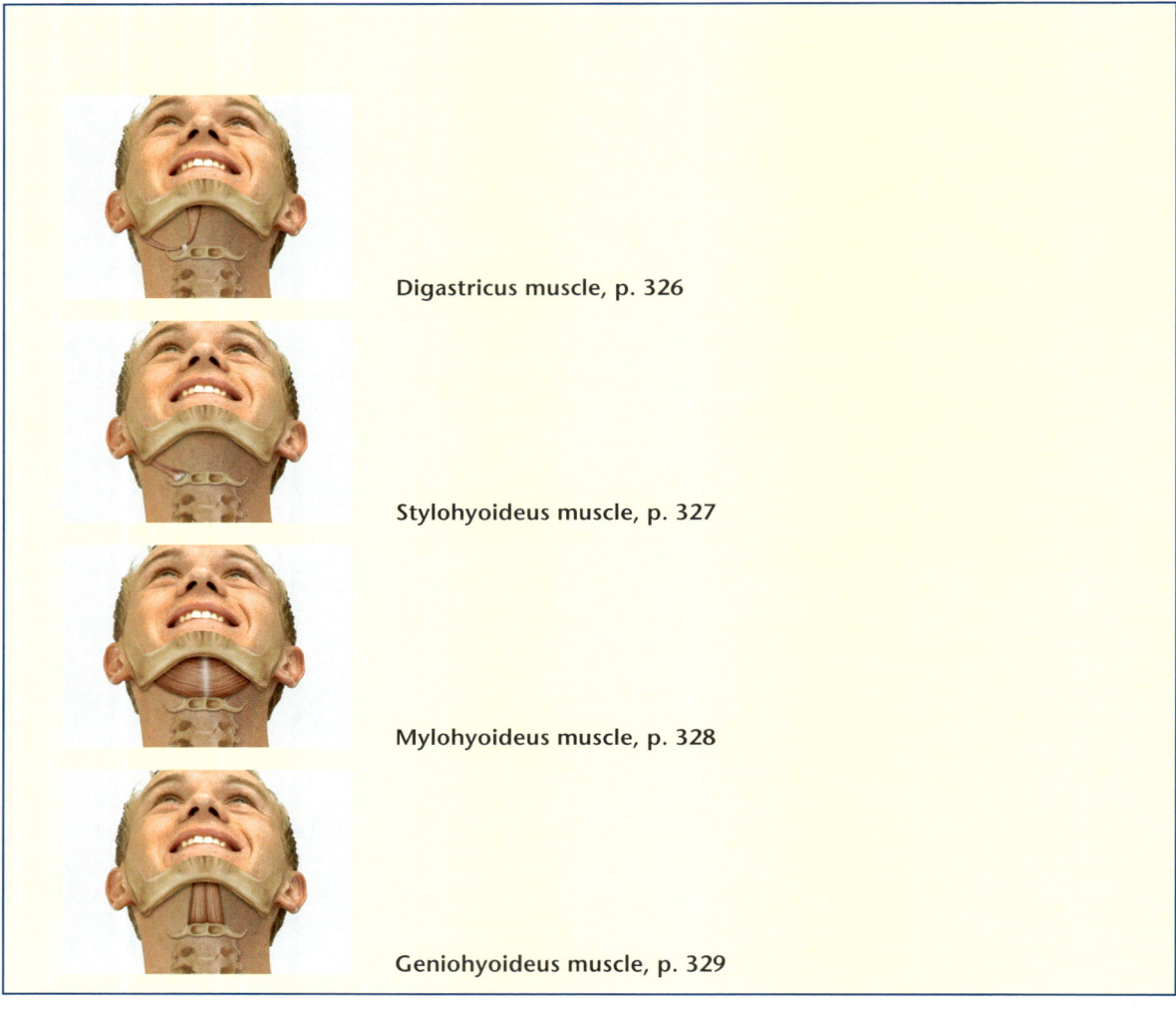

Digastricus muscle, p. 326

Stylohyoideus muscle, p. 327

Mylohyoideus muscle, p. 328

Geniohyoideus muscle, p. 329

Muscle function testing

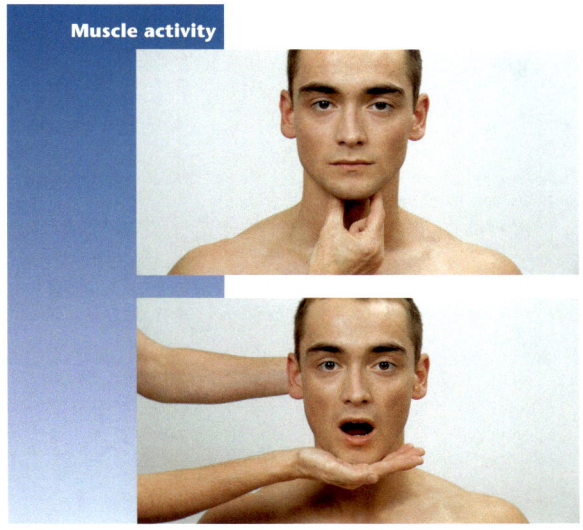

Muscle activity

Starting position: The patient sits and looks straight ahead with his face relaxed.

Test procedure: The examiner palpates the hyoid bone with thumb and index finger and follows the progress of that bone in the cranial direction.

Instruction: "Please swallow."

Alternative:

Starting position: The patient sits and looks straight ahead with his face relaxed.

Test procedure: The examiner holds the back of the head in place with one hand and places the other under the chin, applying upward resistance (as if to close the mouth).

Instruction: "Open your mouth by pressing against my hand with your lower jaw."

 **Clinical relevance**

- The suprahyoid muscles (upper musculature of the hyoid bone) are active during chewing, swallowing and also during speech.
- Ossification of the stylohyoid ligament causes the hyoid bone to be immobilized within the neck.

 Problems/comments

- It is not possible to differentiate functionally between the individual suprahyoid muscles.

Stretch tests

Scalenus anterior and scalenus medius muscles

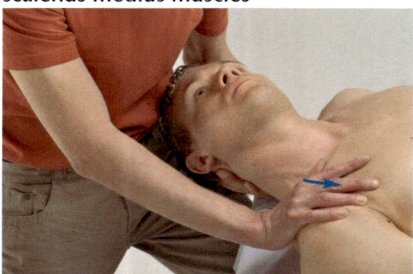

Method

The therapist applies gentle traction and moves the cervical spine and head into extension, lateral flexion to the opposite side and rotation to the same side.
The patient is asked to breathe out. The therapist now moves the patient's first and second ribs in the caudal direction as far as possible.

Finding

Muscle shortening is present if the movement cannot be performed to its full extent and if the endpoint feels soft and elastic.
The patient reports a stretching sensation along the course of the muscle.

Sternocleidomastoideus muscle

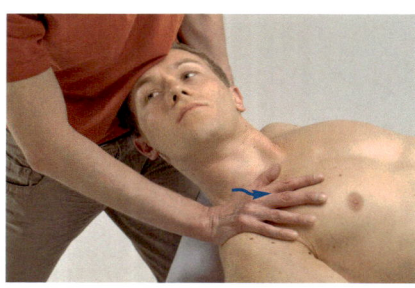

Method

The therapist applies gentle traction and moves the patient's cervical spine and head into flexion, maximum rotation to the same side and lateral flexion to the opposite side.
He pushes the sternum and the sternal end of the clavicle in the caudal and dorsal directions as far as possible.

Finding

Muscle shortening is present if the movement cannot be performed to its full extent and if the endpoint feels soft and elastic.
The patient reports a stretching sensation along the course of the muscle.

Supra- and infrahyoid muscles (both sides)

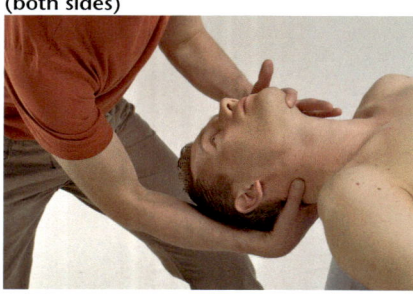

Method

The therapist moves the patient's head and cervical spine into maximum extension, with the patient's mouth closed.

Finding

If the movement is restricted, the patient should open his mouth.
If greater extension of the cervical spine is now possible, this means that shortening of the hyoid muscles is present.
The endpoint feels soft and elastic.
The patient reports a superficial stretching sensation in the ventral region of the neck.

Innervation of the muscles of the neck										
Nerve	**Innervated muscles**	**Alternative designation**								
Trigeminal nerve		Cranial nerve V								
	Digastricus muscle, anterior belly									
	Mylohyoideus muscle									
Facial nerve		Cranial nerve VII								
	Digastricus muscle, posterior belly									
	Stylohyoideus muscle									
Accessory nerve		Cranial nerve XI								
	Sternocleidomastoideus muscle									
		C1	C2	C3	C4	C5	C6	C7	C8
Cervical plexus		X	X	X	X				
	Sternocleidomastoideus muscle	X	X						
	Longus capitis muscle	X	X	X	X				
	Longus colli muscle		X	X	X	X	X		
	Sternohyoideus muscle	X	X	X	X				
	Omohyoideus muscle	X	X	X	X				
	Sternothyroideus muscle	X	X	X	X				
	Thyrohyoideus muscle	X	X						
	Geniohyoideus muscle	X	X						
Ventral branch									
	Rectus capitis anterior muscle	X							
	Rectus capitis lateralis muscle	X							
	Scalenus anterior muscle					X	X	X	X
	Scalenus medius muscle				X	X	X	X	X
	Scalenus posterior muscle							X	X
Greater occipital nerve			X						
Lesser occipital nerve			X	X					

6 Head

Muscles of facial expression

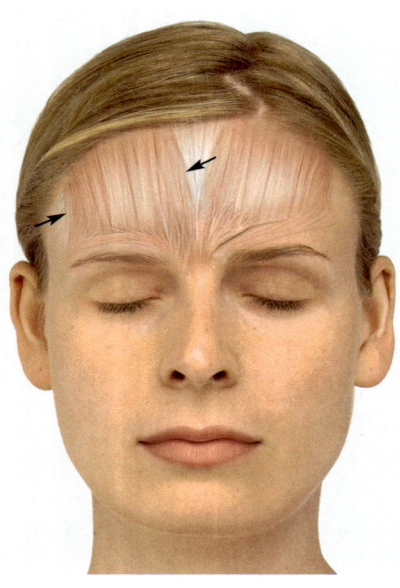

Epicranius muscle

The epicranius (epicranial) muscle can raise the eyebrows, producing deep lines across the forehead. It is thus a significant antagonist of the orbicularis oculi muscle and opens the palpebral fissure together with the levator palbebrae superioris muscle. This takes place via the frontal part of the occipitofrontalis (occipitofrontal) muscle, with the help of its occipital part. Through its contraction, the latter fixes the epicranial aponeurosis as the fixed point for the frontal part. The temporoparietalis (temporoparietal) muscle is irregular and rudimentary.

Origin

Occipitofrontalis muscle:
Occipital part: short, tendinous fibers from the supreme arcuate line of the occipital bone
Frontal part: the medial fibers project from the procerus muscle, the later fibers join with those of the corrugator supercilii and orbicularis oculi muscles
Temporoparietalis muscle: the skin of the temple, temporal fascia

Insertion

Occipitofrontalis muscle:
Occipital part: epicranial aponeurosis (galea aponeurotica)
Frontal part: epicranial aponeurosis, ventrally of the coronal suture
Temporoparietalis muscle: epicranial aponeurosis

Innervation

Occipitofrontalis muscle:
Occipital part: posterior auricular nerve from the facial nerve (cranial nerve VII)
Frontal part: temporal branches of the facial nerve (cranial nerve VII)
Temporoparietalis muscle: temporal branches of the facial nerve (cranial nerve VII)

Special features

The occipitofrontalis muscle and the generally inconspicuous temporoparietalis muscle are together referred to as the epicranius muscle.

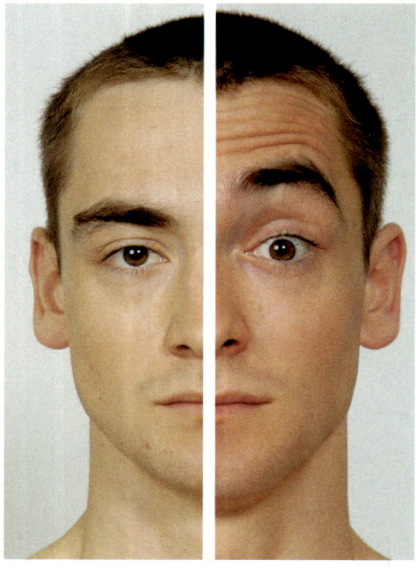

Muscle function testing

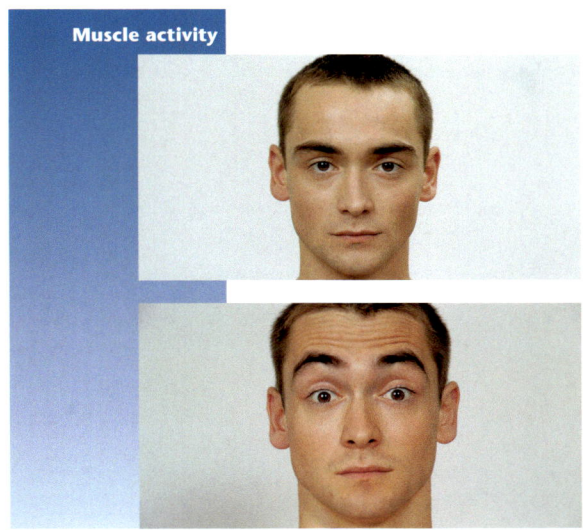

Muscle activity

Starting position: The patient looks straight ahead with his face relaxed.

Test procedure: The examiner looks at the patient's face.

Instruction: "Raise your eyebrows and wrinkle your forehead."

 Clinical relevance

- In contrast to peripheral facial nerve paralysis, the function of the occipital part of the epicranius muscle is preserved in central facial nerve paralysis.

 Problems/comments

- This muscle activity can take place only through the joint action of the frontal and occipital parts of the epicranius muscle.

Corrugator supercilii muscle

The corrugator supercilii (superciliary corrugator) muscle draws the medial sides of the eyebrows towards the middle and downwards, thus producing vertical lines between them and over the root of the nose.

Origin   Nasal part of the frontal bone

Insertion   Epicranial aponeurosis (galea aponeurotica)
                        Skin above the middle section of the eyebrow

Innervation   Temporal branches of the facial nerve (cranial nerve VII)

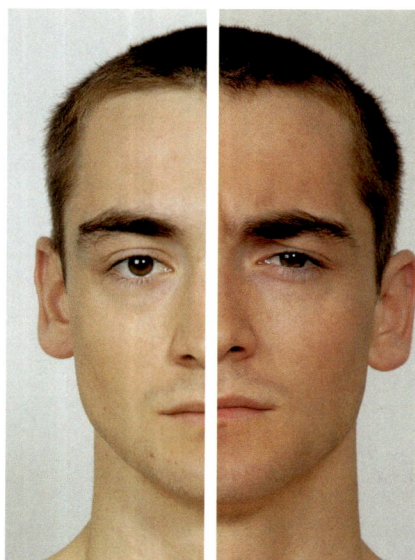

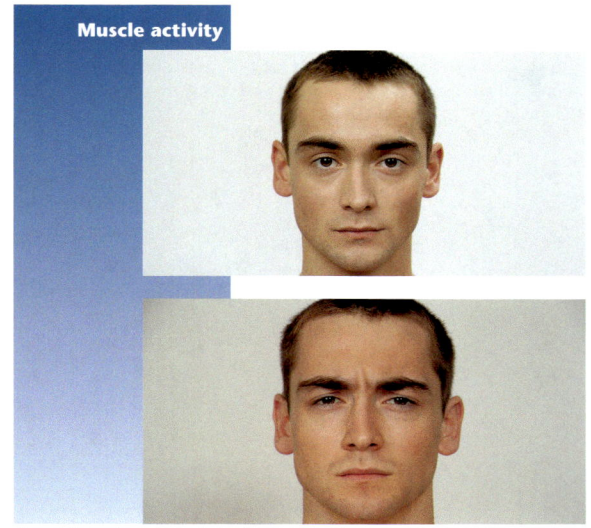

Muscle activity

Muscle function testing

Starting position: The patient looks straight ahead with his face relaxed.

Test procedure: The examiner looks carefully at the patient's face.

Instruction: "Pull your eyebrows together towards the nose."

 **Clinical relevance**

- Bright light causes contraction of the corrugator supercilii muscle, as a protective reaction.

Problems/comments

- This movement generally take place with the aid of the depressor supercilii muscle.

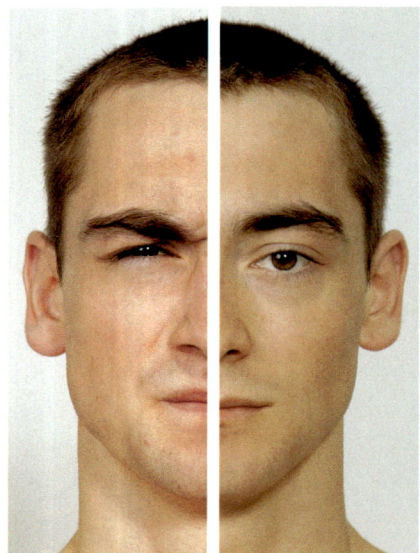

Procerus muscle

The procerus muscle, together with the corrugator supercilii muscle, pulls the skin of the medial side of the eyebrows down towards the root of the nose, thus producing deep transverse lines above the nose.

Origin   Lower part of the nasal bone
                              Upper part of the nasal cartilage

Insertion   Skin of the forehead, between the eyebrows

Innervation   Buccal branches of the facial nerve (cranial nerve VII)

Muscle function testing

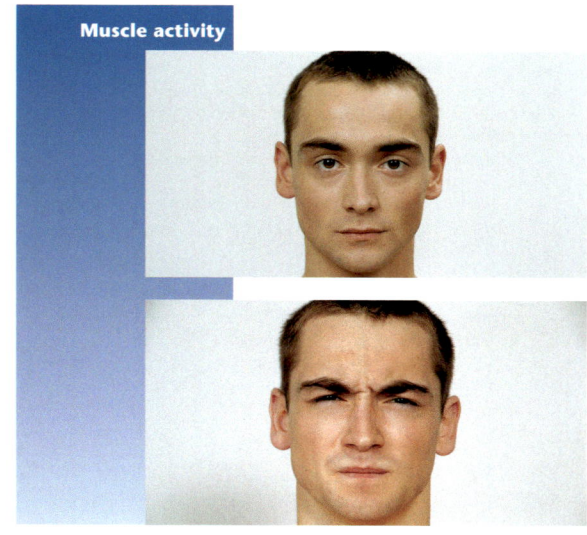

Muscle activity

Starting position: The patient looks straight ahead with his face relaxed.

Test procedure: The examiner looks at the patient's face.

Instruction: "Pull your eyebrows downwards."

 **Clinical relevance**

- The procerus muscle gives the face an angry expression.

 Problems/comments

- The procerus muscle may be missing. In some cases, the patient may also be unable to activate it voluntarily.

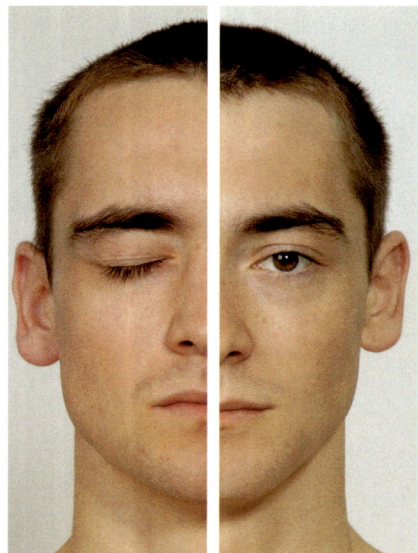

Orbicularis oculi muscle

The orbicularis oculi muscle (orbicular muscle of the eye), with its two parts, the palpebral and the orbital, can, together with the corrugator supercilii muscle, cause narrowing of the palpebral fissure; it thus acts as an antagonist to the levator palpebrae superioris muscle and to the tarsalis superior and tarsalis inferior muscles. During laughter, this muscle produces the typical laugh lines which radiate outwards from the corner of the eye.

Origin	Medial part of the orbit (nasal part of the frontal process of the maxilla, anterior lacrimal crest and medial palpebral ligament)
Insertion	Palpebral part: skin of the upper and lower eyelids Orbital part: fanning out broadly to attach to the skin of the orbit, the forehead and the cheeks
Innervation	Temporal and zygomatic branches of the facial nerve (cranial nerve VII)

Muscle function testing

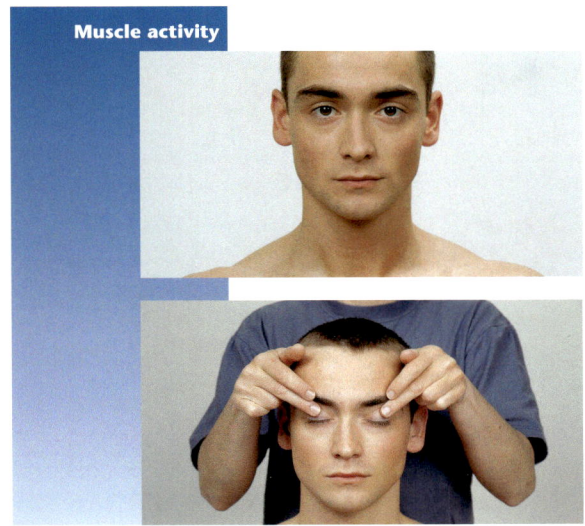

Muscle activity

Starting position: The patient looks straight ahead with his face relaxed.

Test procedure: The examiner applies resistance to the closed eyelids.

Instruction: "Keep your eyes shut."

Clinical relevance

- In central facial nerve paralysis, the orbicularis oculi muscle, which is innervated centrally and bilaterally, may not be particularly affected.
- Patients with paralysis of the orbicularis oculi muscle cannot close the affected eye (lagophthalmos). When the individual tries to close the affected eye, the eyeball may be seen to roll upwards (Bell's phenomenon).

Problems/comments

- When the eyes are tightly shut, it is mainly the orbital part of the orbicularis oculi muscle that is activated; if they are lightly shut, the palpebral part tends to be the one that is activated.
- The lines in the lateral angle of the eye, which are produced by the action of the orbicularis oculi muscle, can become permanent with age, when they are described as "crow's feet."

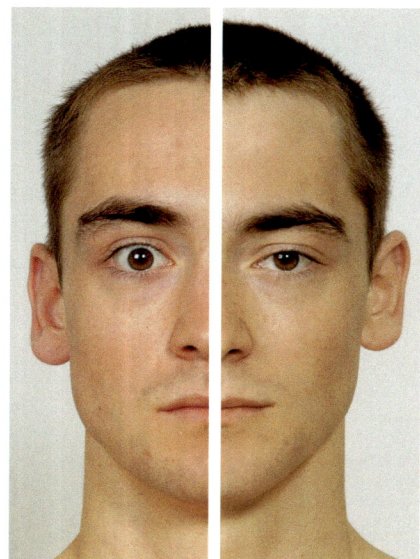

Levator palpebrae superioris muscle

According to its innervation and origin, the levator palpebrae superioris muscle (levator muscle of the upper eyelid) belongs with the muscles of the orbit. As it is tested together with the muscles of facial expression, it will be addressed here with those muscles. It lifts the upper eyelid, thus controlling the width of the palpebral fissure, and is involved in blinking. However, the fine adjustment of the retraction of the upper and lower eyelids is controlled by the tarsalis superior and tarsalis inferior muscles and by the sympathetic nervous system.

Origin
Caudal surface of the lesser wing of the sphenoid bone
Cranially and ventrally of the optic canal

Insertion
Tarsal cartilage (tarsus) and skin of the upper eyelid

Innervation
Superior branch of the oculomotor nerve (cranial nerve III)

Muscle function testing

Muscle activity

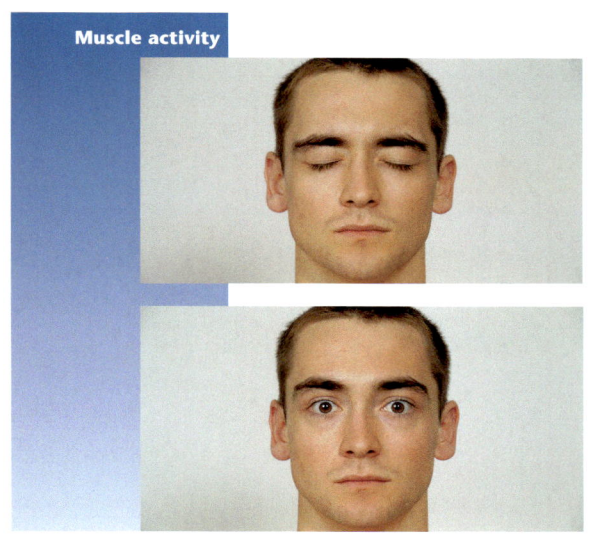

Starting position: The patient faces straight ahead with his face relaxed and eyes closed.

Test procedure: The examiner looks at the patient's face.

Instruction: "Open your eyes, leaving your forehead relaxed."

℞ Clinical relevance

- Paralysis of the levator palpebrae superioris muscle causes the upper eyelid to droop (ptosis).

- The levator palpebrae superioris muscle consists of transversely striated skeletal muscle innervated by the oculomotor nerve. A smooth muscle, the tarsalis superior, is innervated via post-ganglionic, sympathetic fibers from the superior cervical ganglion. Nerve block of this ganglion, for example to treat Sudeck's atrophy (reflex sympathetic dystrophy), results in drooping of the ipsilateral upper eyelid.

- The levator palpebrae superioris muscle is permanently activated in the waking state, unless the eyes are closed.

! Problems/comments

- The patient may also activate the occipitofrontalis muscle.

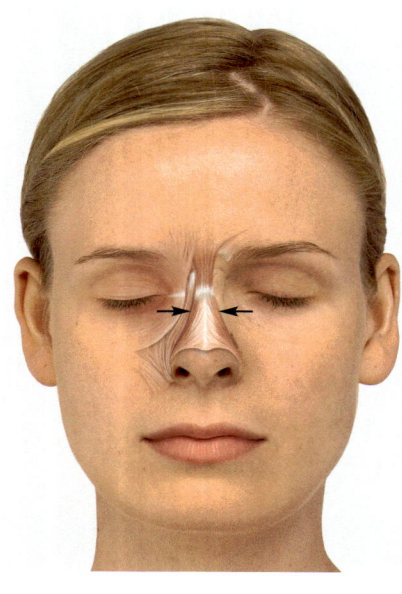

Nasalis muscle

The upper part of the nasalis (nasal) muscle (alar part) dilates and flares the nostril. In doing so, it can help to reduce respiratory work. The lowest parts of this muscle and a few fibers which run to the septum (transverse part) can also compress the nostril and pull the tip of the nose slightly downwards.

Origin	Maxilla Alveolar jugum of the lateral incisor and the canine
Insertion	Wing of the nose, margin of the nostril Lateral nasal cartilage Aponeurosis of the dorsum of the nose
Innervation	Buccal branches of the facial nerve (cranial nerve VII)

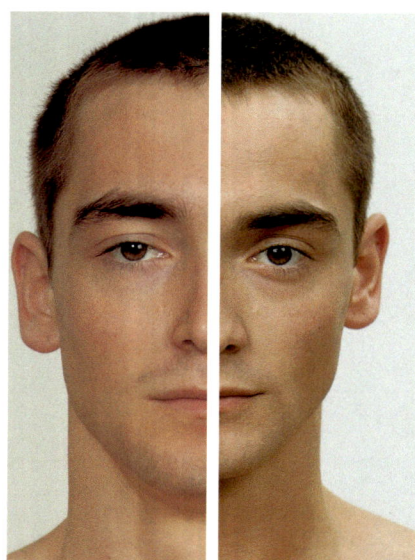

Muscle function testing

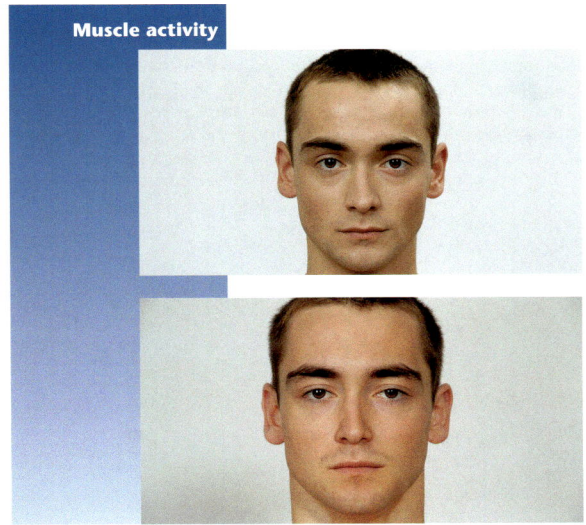

Muscle activity

Starting position: The patient looks straight ahead with his face relaxed.

Test procedure: The examiner looks at the patient's face.

Instruction: "Pull the wings of your nose downwards."

 Problems/comments

- The patient may not be able to activate the nasalis muscle in isolation.
- If the patient is unable to perform this movement voluntarily, it may help to ask him to breathe deeply and slowly through the nose.

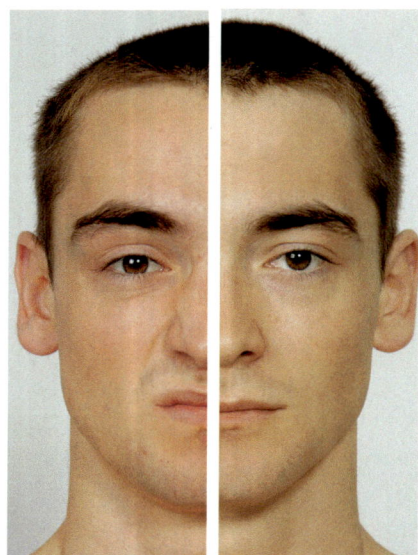

Levator labii superioris alaeque nasi muscle

The levator labii superioris alaeque nasi muscle (levator muscle of the upper lip and nasal wing), acting jointly with the levator labii superioris muscle, lifts the upper lip. As a few of its fibers are also inserted at the lateral circumference of the nostril, it can lift the free edge of the nose slightly, particularly during inspiration. This stops the wings of the nose from collapsing as a result of the negative pressure in the nasal vestibule and reduces resistance to airflow in this part of the nasal cavity. This function is aided by the nasalis muscle.

Origin	Frontal process of the maxilla
	Muscle mass of the orbicularis oculi muscle
Insertion	Wings of the nose, upper lip
	Lateral and dorsal circumference of the nostril
Innervation	Zygomatic branches of the facial nerve (cranial nerve VII)

Muscle function testing

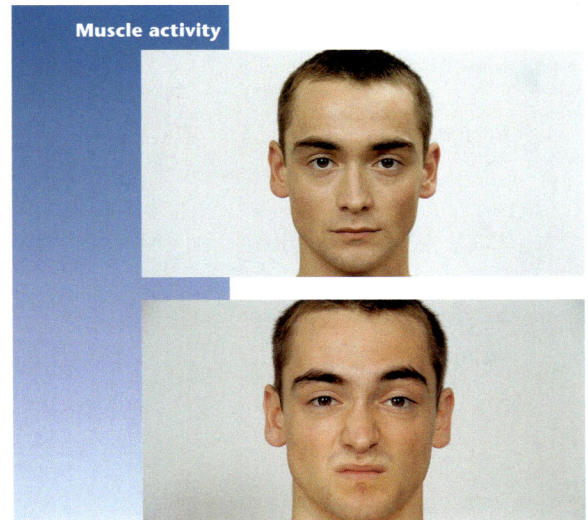

Starting position: The patient looks straight ahead with his face relaxed.

Test procedure: The examiner looks at the patient's face.

Instruction: "Lift the wings of your nose."

 **Clinical relevance**

- The action of the levator labii superiors alaeque nasi is enhanced in infant respiratory distress (nasal flaring).

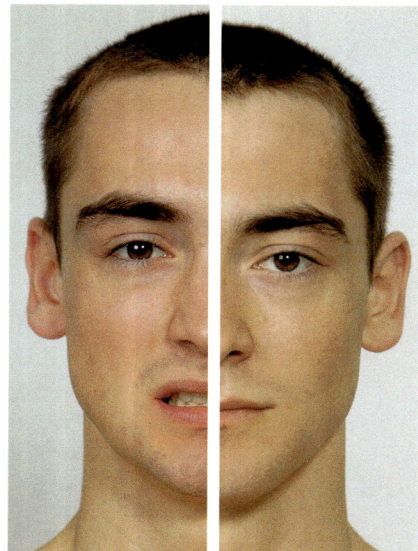

Levator labii superioris muscle

The levator labii superioris muscle (levator muscle of the upper lip) lifts the upper lip and deepens the nasolabial lines. This exposes the anterior teeth and gums of the upper jaw. This muscle tightens when a person smiles broadly or laughs.

Origin

Infraorbital margin of the maxilla
Frontal process of the maxilla

Insertion

Upper lip
Orbicularis oris muscle

Innervation

Zygomatic branches of the facial nerve (cranial nerve VII)

Muscle function testing

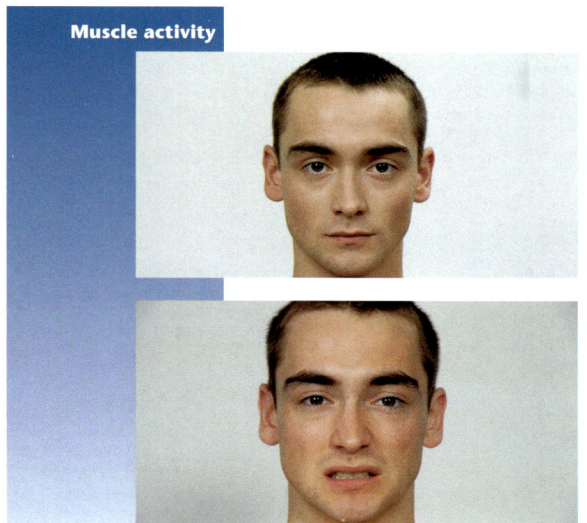

Muscle activity

Starting position: The patient looks straight ahead with his face relaxed.

Test procedure: The examiner looks at the patient's face.

Instruction: "Pull up your upper lip."

 Problems/comments

- The levator labii superioris muscle is aided in its function by the levator labii superioris alaeque nasi and the zygomaticus minor muscles.

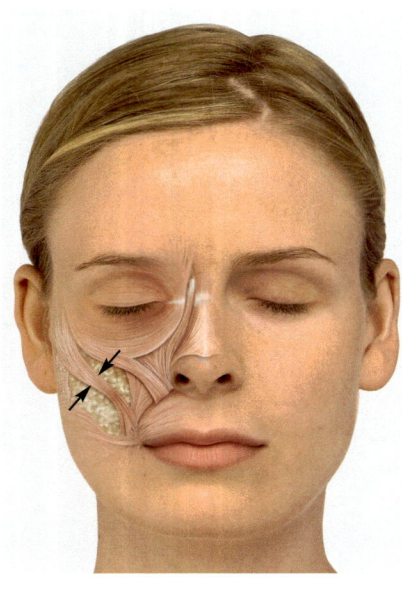

Zygomaticus major muscle

The zygomaticus major muscle (greater zygomatic muscle) lifts the angles of the mouth and deepens the nasolabial lines, for example when a person laughs or smiles. In doing so, it works in conjunction with the levator anguli oris muscle.

Origin	Central sections of the zygomatic bone, ventrally of the temporozygomatic suture Parotid Fascia
Insertion	Skin of the angle of the mouth, lips
Innervation	Zygomatic branches of the facial nerve (cramal nerve VII)

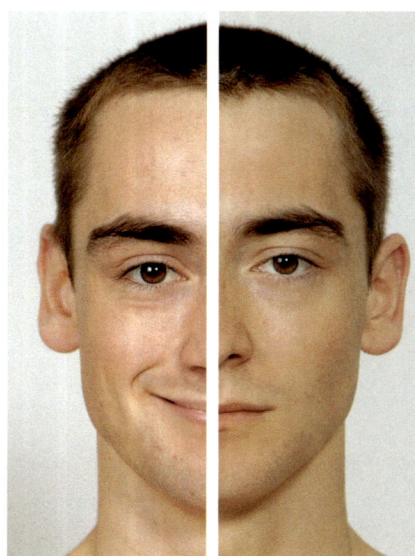

Zygomaticus minor muscle

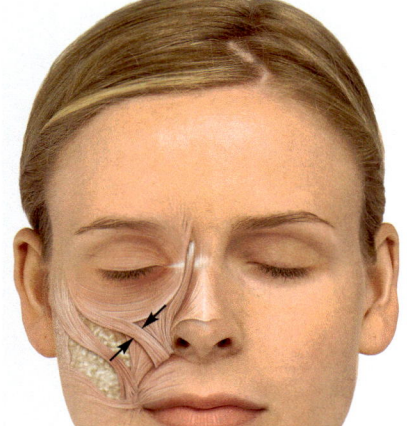

The zygomaticus minor muscle (lesser zygomatic muscle) acts as a lifter of the upper lip, in comparable fashion to the levator labii superioris.

Origin	Medial sections of the zygomatic bone, dorsally of the temporozygomatic suture Parotid fascia
Insertion	Lateral sections of the upper lip
Innervation	Zygomatic branches of the facial nerve (cranial nerve VII)

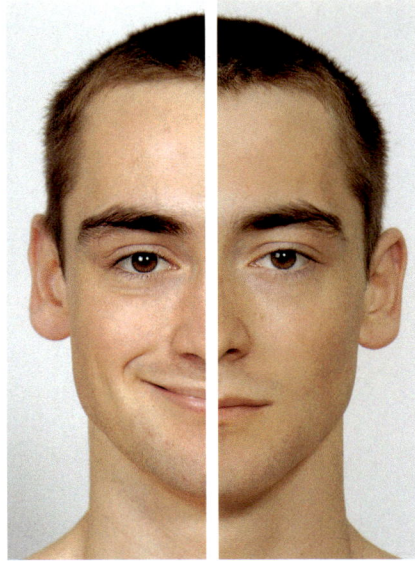

The following muscles are tested jointly:

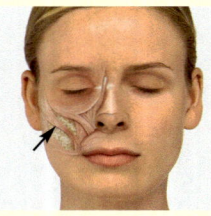

Zygomaticus major muscle, p. 352

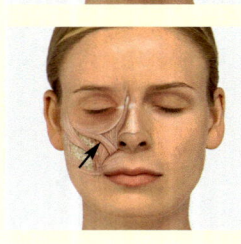

Zygomaticus minor muscle, p. 353

Muscle function testing

Muscle activity

Starting position: The patient looks straight ahead with his face relaxed.

Test procedure: The examiner looks at the patient's face.

Instruction: "Smile."

 **Clinical relevance**

- In facial nerve paralysis (peripheral or central), weakness of the ipsilateral muscles manifests as drooping of the lips.

 Problems/comments

- Smiling and laughing takes place with the aid of the risorius muscle.
- The zygomaticus major and zygomaticus minor muscles work together. The nasolabial lines show up prominently when both these muscles contract.

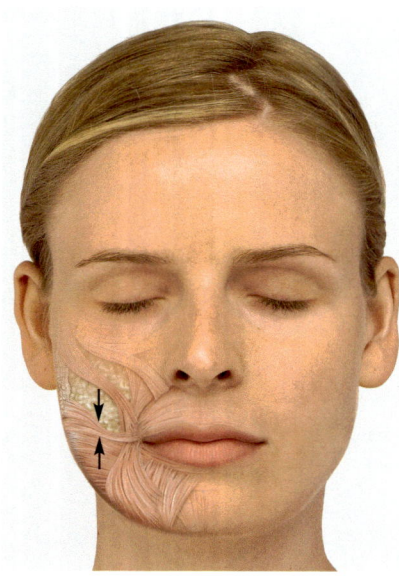

Risorius muscle

The risorius muscle aids the levator anguli oris muscle when a person laughs, by pulling the angle of the mouth to the side and thereby producing dimples in the cheeks.

Origin	Parotid fascia
Insertion	Upper lip
	Angle of the mouth
Innervation	Buccal branches of the facial nerve (cranial nerve VII)

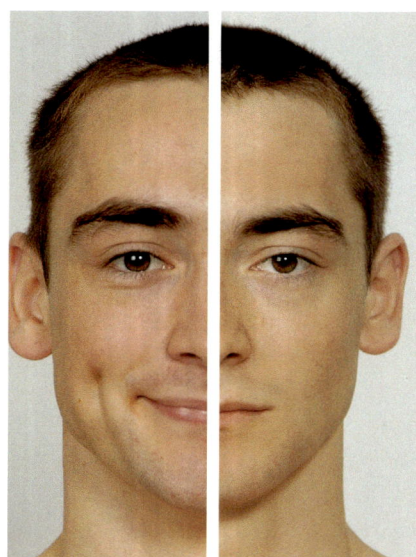

Muscle function testing

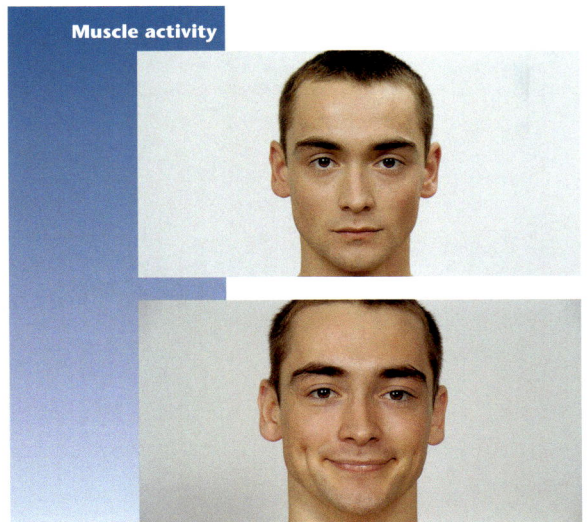

Muscle activity

Starting position: The patient looks straight ahead with his face relaxed.

Test procedure: The examiner looks at the patient's face.

Instruction: "Pull the corners of your mouth apart."

 Problems/comments

- When a person laughs, the risorius muscle works in conjunction with the levator anguli oris muscle.
- The risorius muscle may also be absent.

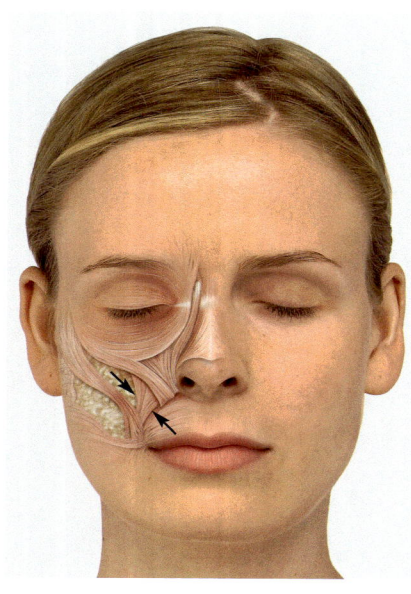

Levator anguli oris muscle

The levator anguli oris muscle (levator muscle of the angle of the mouth) lifts the angles of the mouth, thus emphasizing the nasolabial lines. It is the most important laugh muscle, together with the risorius.

Origin   Infraorbital margin of the maxilla
                    Frontal process of the maxilla
                    Muscle mass of the orbicularis oris

Insertion   Upper lip

Innervation   Zygomatic branches of the facial nerve (cranial nerve VII)

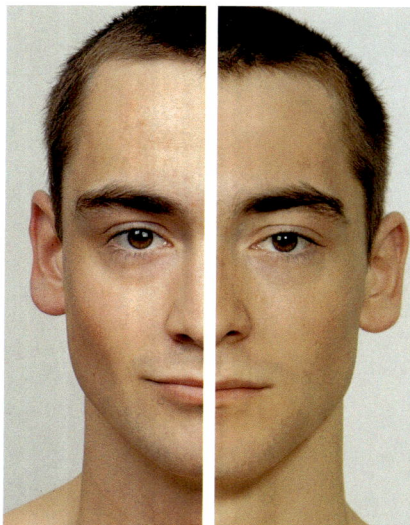

Muscle function testing

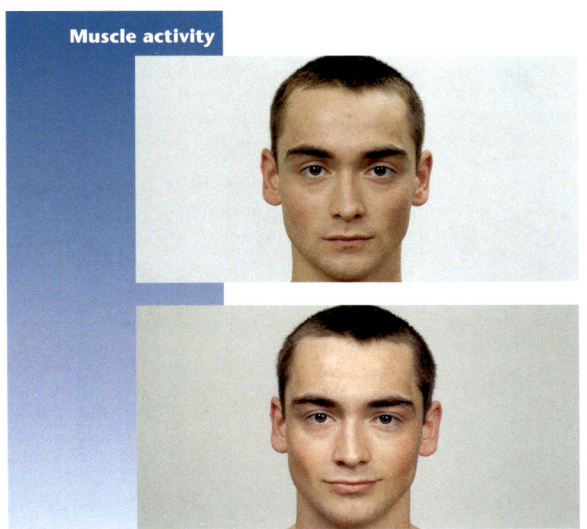

Muscle activity

Starting position: The patient looks straight ahead with his face relaxed.

Test procedure: The examiner looks at the patient's face.

Instruction: "Pull both corners of your mouth upwards."

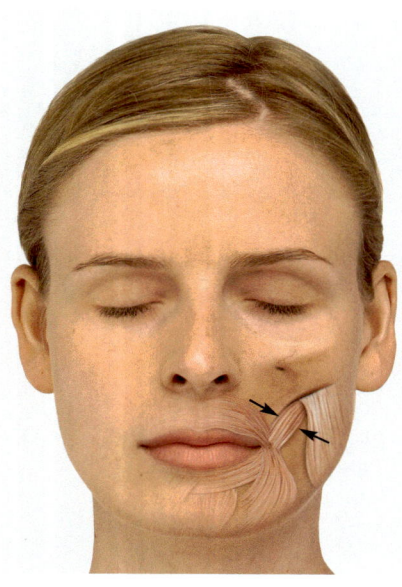

Buccinator muscle

The buccinator muscle can pull back the angle of the mouth, thus acting as an antagonist to the orbicularis oris muscle. When a person chews, however, it acts synergistically with that muscle, which moves food from the lateral oral vestibule back to between the teeth. The tongue also helps with this. As the "trumpet" muscle, it helps to drive air out of the buccal cavity.

Origin	Maxilla: alveolar process in the region of the first molar Mandible: along the alveolar process in the region of the posterior molars Pterygomandibular raphe
Insertion	Angle of the mouth, where part of it joins up with the orbicularis oris muscle
Innervation	Buccal branches of the facial nerve (cranial nerve VII)
Special features	The buccinator muscle forms the muscular basis of the cheek

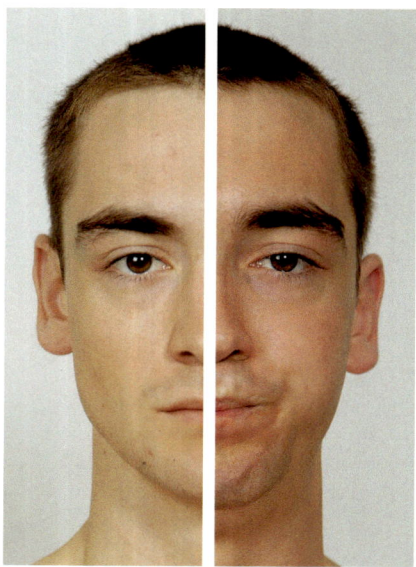

Buccinator muscle, tightened over an inflated cheek

Muscle function testing

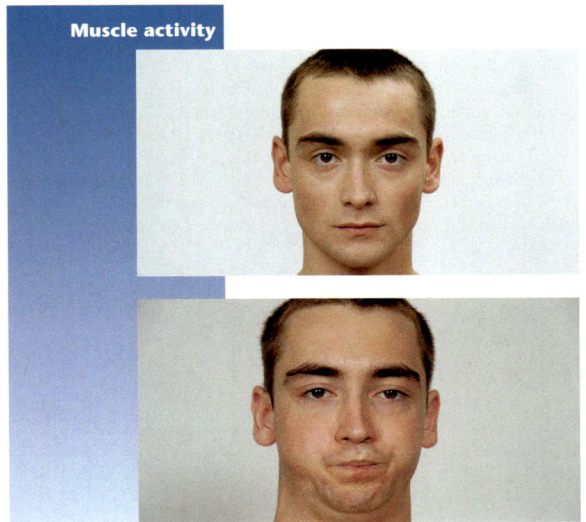

Muscle activity

Starting position: The patient looks straight ahead with his face relaxed.

Test procedure: The examiner looks at the patient's face.

Instruction: "Keeping your lips closed, imagine that you are trying to blow air into the mouthpiece of a trumpet."

 **Clinical relevance**

- In facial nerve paralysis (peripheral or central), weakness of the ipsilateral muscles manifests as drooping of the lips.

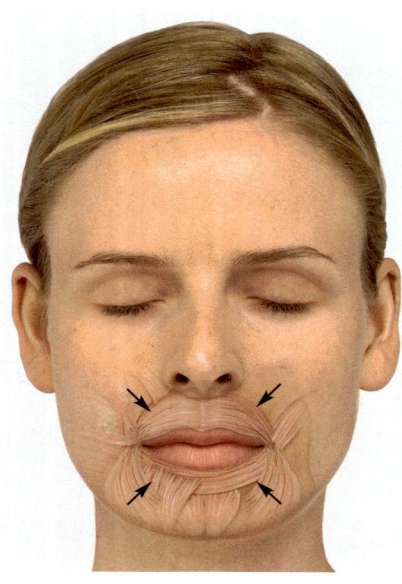

Orbicularis oris muscle

The orbicularis oris muscle (orbicular muscle of the mouth) forms the basis of the motor apparatus of the lips. The section of this muscle ring which lies furthest from oral aperture can reduce the size of that aperture, while protruding the red margin of the lips, as in whistling. If the part which lies at the edge of lip (within the red margin) contracts in isolation, the red margin inclines inwards towards the front teeth, so that less of it is visible. The tone of this muscle is important in retaining saliva, which is discharged from the angle of the mouth if the muscle is paralyzed. All the muscles of facial expression which pull the lips or the angles of the mouth laterally or upwards/downwards (thus widening the oral aperture) act as its antagonists.

Origin	Mandible Maxilla Perioral skin
Insertion	Lips
Innervation	Buccal branches and marginal mandibular branch of the facial nerve (cranial nerve VII)

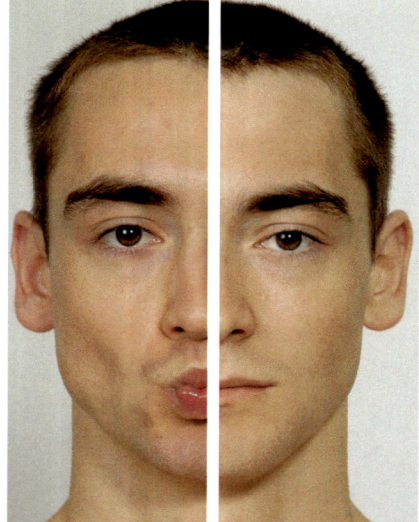

Muscle function testing

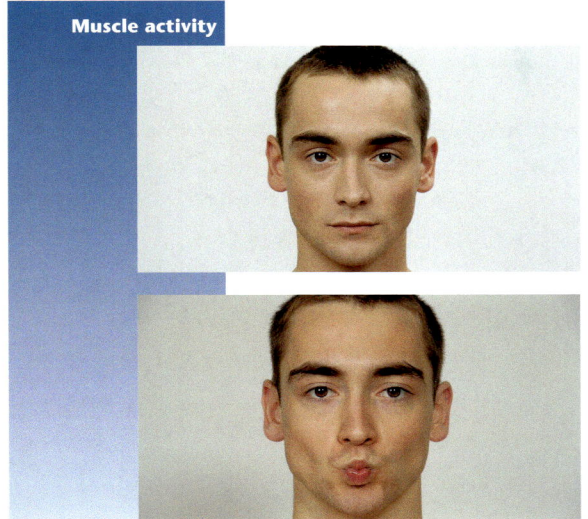

Muscle activity

Starting position: The patient looks straight ahead with his face relaxed.

Test procedure: The examiner looks at the patient's face.

Instruction: "Purse your lips."

 **Clinical relevance**

- In facial nerve paralysis (peripheral or central), weakness of the ipsilateral muscles manifests as drooping of the lips.

 Problems/comments

- Ensure that the lips really are pursed and not just pressed together, as this movement is performed by other muscles.

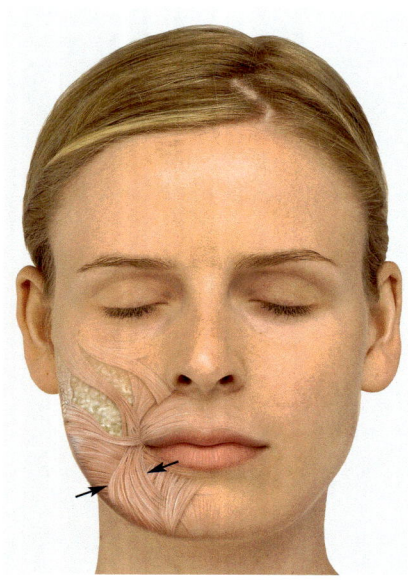

Depressor anguli oris muscle

The depressor anguli oris muscle (depressor muscle of the angle of the mouth) pulls the angle of the mouth downwards, thus flattening out the nasolabial lines.

Origin   Lower edge of the mandible, caudally of the mental foramen

Insertion   Lips
                    Cheek, laterally of the angle of the mouth

Innervation   Marginal mandibular branch of the facial nerve (cranial nerve VII)

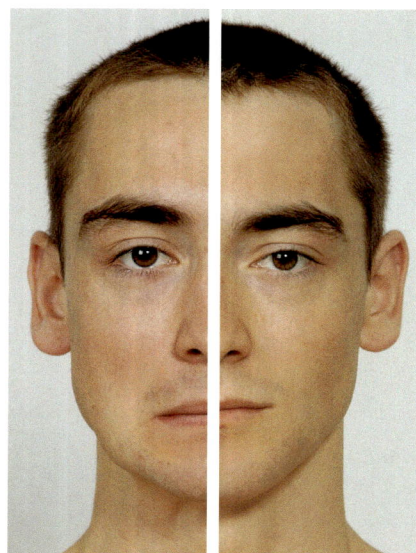

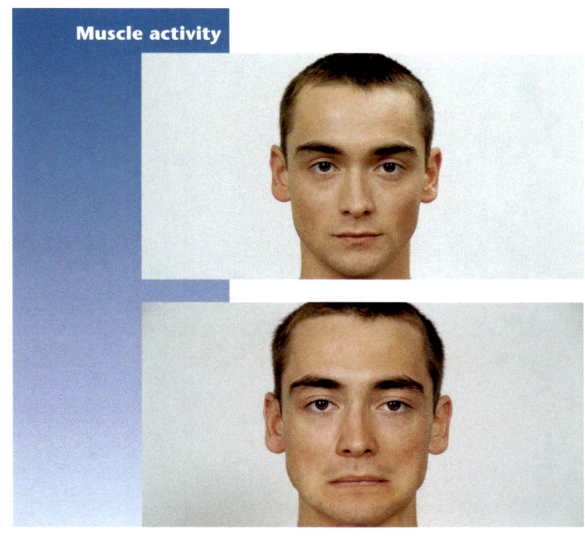

Muscle function testing

Starting position: The patient looks straight ahead with his face relaxed.

Test procedure: The examiner looks at the patient's face and neck.

Instruction: "Pull the corners of your mouth downwards."

 Clinical relevance

- The depressor anguli oris muscle produces a sad expression.

Problems/comments

- The depressor anguli oris and the platysma are jointly involved in this movement.

Depressor labii inferioris muscle

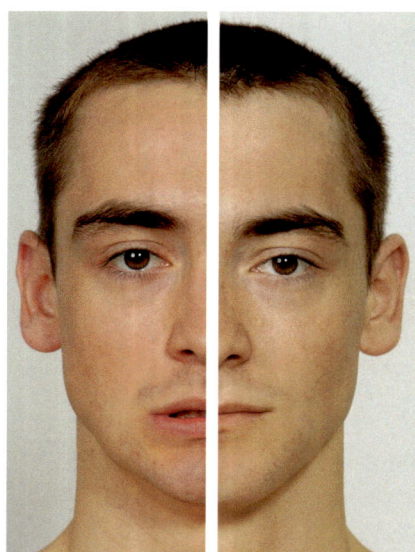

The depressor labii inferioris muscle (depressor muscle of the lower lip) pulls and widens the lower lip downwards, exposing the front teeth of the lower jaw. In doing so, it deepens the line between the lower lip and the chin.

Origin	Base of the mandible, mediocaudally of the mental foramen
Insertion	Lower lip
Innervation	Marginal mandibular branch of the facial nerve (cranial nerve VII)

Muscle function testing

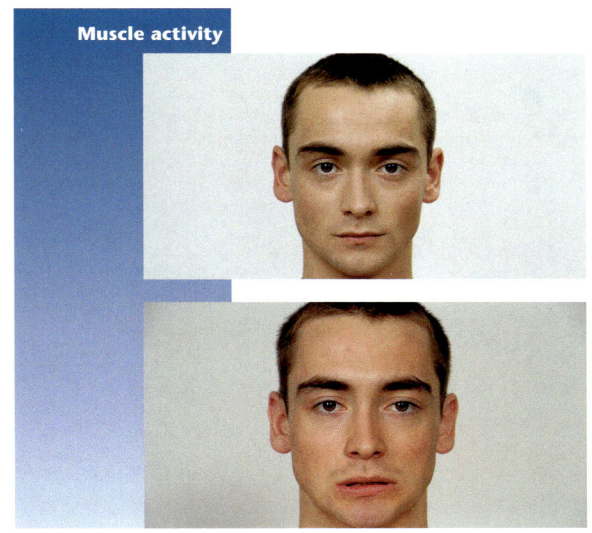

Muscle activity

Starting position: The patient looks straight ahead with his face relaxed.

Test procedure: The examiner looks at the patient's face.

Instruction: "Pull down and widen your lower lip."

Platysma

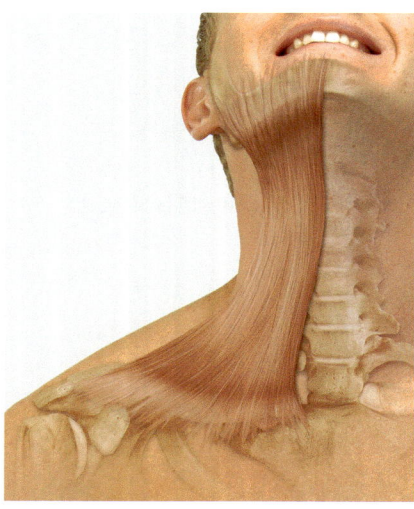

As the cutaneous muscle of the neck, the platysma can tighten the skin of the anterior part of the neck from the lower jaw to the clavicles, particularly as part of the startle reaction or when it is tightened voluntarily; when it tightens in this way, its individual muscle cords often become clearly visible under the skin. The lower lip and the angles of the mouth are also pulled downwards. In humans, this muscle is probably a largely non-functional rudiment. In the great apes, the muscle still has the function of keeping the throat sac toned, enhancing its resonance. Its effects on the cervical spine or the temporomandibular joint are negligible.

Origin   Base of the mandible
                        Parotid fascia

Insertion   Skin caudally of the clavicle
                        Pectoral fascia

Innervation   Cervical branch of the facial nerve (cranial nerve VII)

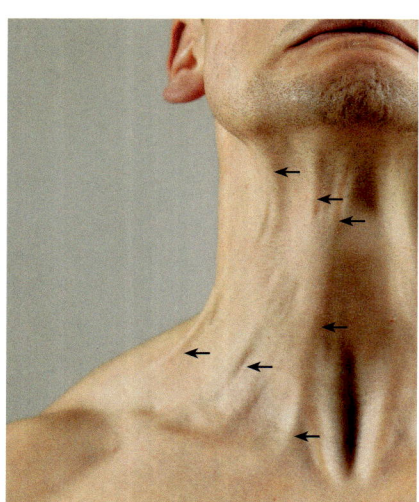

Muscle function testing

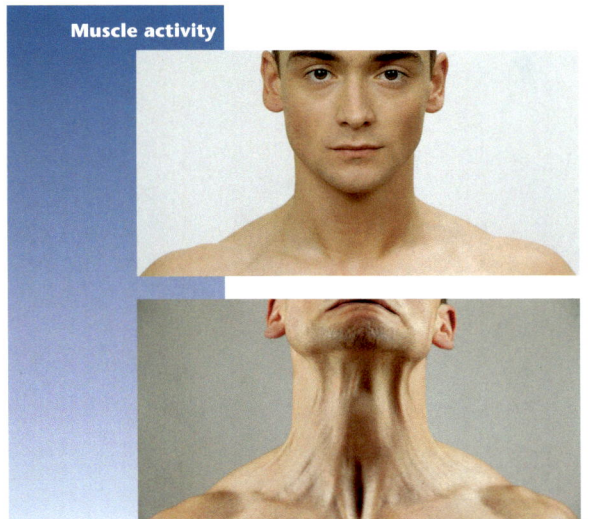

Muscle activity

Starting position: The patient looks straight ahead with his face relaxed.

Test procedure: The examiner looks at the patient's face and neck.

Instruction: "Pull the corners of your mouth and your lower lip hard downwards and sideways, and tighten the skin of your neck."

6 Head

Muscles of mastication

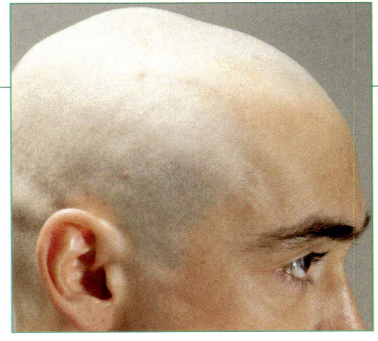

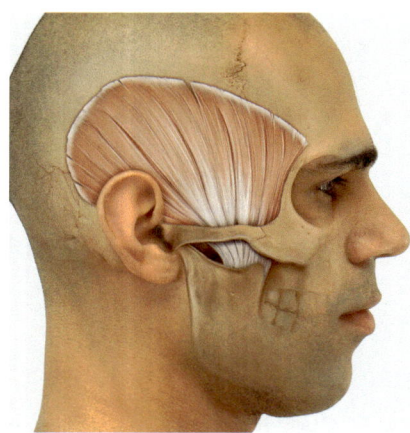

Temporalis muscle

The temporalis (temporal) muscle produces a powerful bite and, at the same time, can pull the mandible back with its horizontal fibers. Its resting tone prevents the mandible from drooping under gravitational pull.

Origin	Temporal fossa Temporal fascia
Insertion	Coronoid process of mandible
Innervation	Deep temporal nerves from the mandibular branch of the trigeminal nerve (cranial nerve V)

Functions

 Synergists Antagonists

Temporomandibular joints

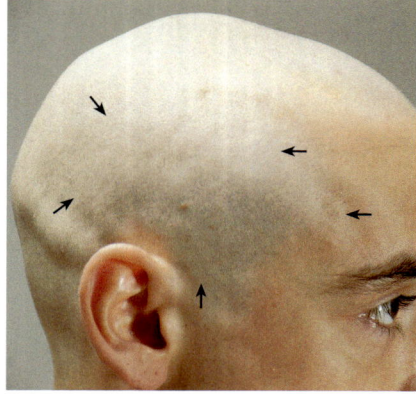

Lifting the mandible
Masseter muscle
Pterygoideus medialis muscle

Digastricus muscle
Mylohyoideus muscle
Geniohyoideus muscle
Pterygoideus lateralis muscle

Pushing the mandible forwards (vertical fibers adjacent to the orbit)
Pterygoideus lateralis muscle
Pterygoideus medialis muscle
Masseter muscle, superficial part

Hyoid muscles
Masseter muscle, deep part
Temporalis muscle (horizontal fibers)

Pulling the mandible back (horizontal fibers lying above the ear)
Hyoid muscles

Pterygoideus lateralis muscle
Pterygoideus medialis muscle
Masseter muscle, superficial part
Temporalis muscle (vertical fibers)

Muscle function testing

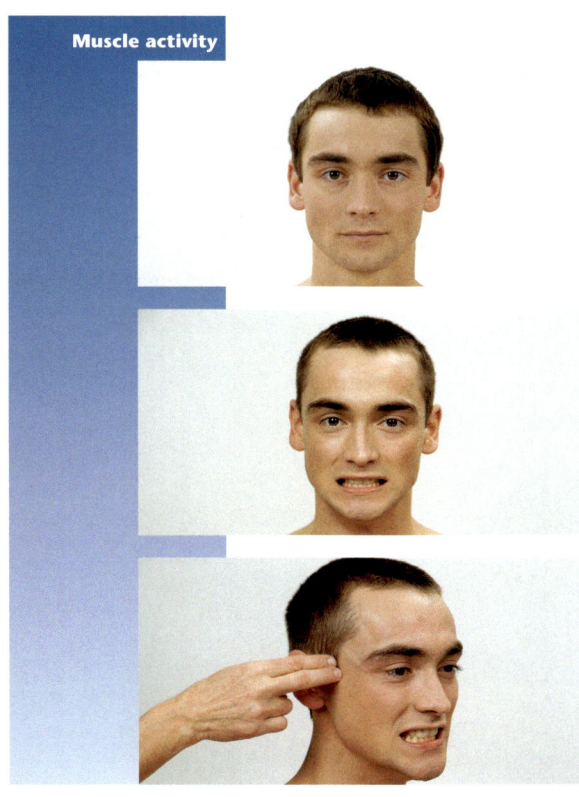

Muscle activity

Starting position: The patient looks straight ahead with his face relaxed.

Test procedure: The examiner looks at the patient's face.
Instruction: "Keeping your lips open, bite your teeth together."

Test procedure: The examiner palpates the temporalis muscle.
Instruction: "Keep your lips open and try to bite your teeth together."

 Clinical relevance

- The temporalis muscle is often involved in temporomandibular syndrome (Costen's syndrome) and in the development of tension headaches.

⚠ **Problems/comments**

- It is difficult to look at the temporalis muscle without its two main synergists, the masseter and pterygoideus medialis muscles.

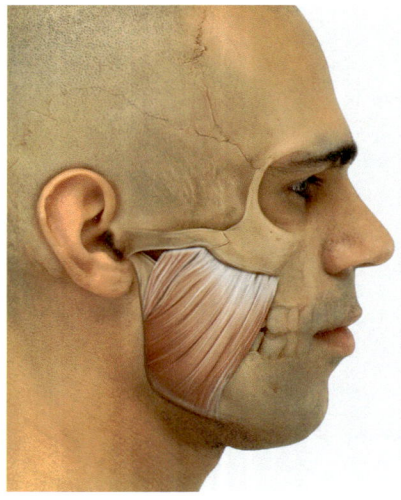

Masseter muscle

The masseter muscle closes the jaws very powerfully. It can also push the mandible forward with its superficial part.

Origin	Superficial part: lower edge and the anterior two thirds of the zygomatic arch Deep part: posterior third and inner surface of the zygomatic arch
Insertion	Superficial part: mandibular angle and masseteric tuberosity Deep part: outer surface of the ramus of the mandible
Innervation	Masseteric nerve from the mandibular branch of the trigeminal nerve (cranial nerve V)

Functions

 Synergists

 Antagonists

Temporomandibular joints

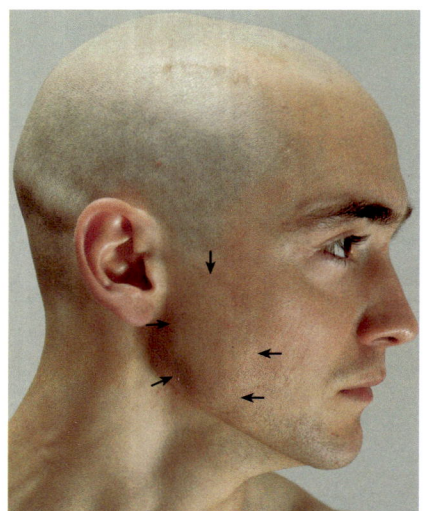

Lifting the mandible
Temporalis muscle
Pterygoideus medialis muscle

Digastricus muscle
Mylohyoideus muscle
Geniohyoideus muscle
Pterygoideus lateralis muscle

Pushing the mandible forwards (superficial part)
Pterygoideus lateralis muscle
Pterygoideus medialis muscle
Temporalis muscle (vertical fibers adjoining the orbit – weak)

Temporalis muscle (horizontal fibers lying above the ear)
Masseter muscle, deep part
Hyoid muscles

Muscle function testing

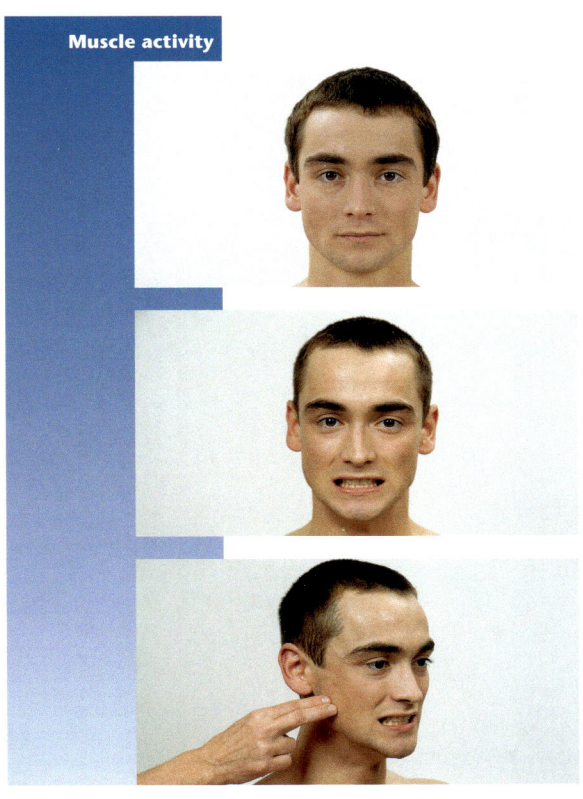

Muscle activity

Starting position: The patient looks straight ahead with his face relaxed.

Test procedure: The examiner looks at the patient's face.
Instruction: "Keeping your lips open, bite your teeth together."

Test procedure: The examiner palpates the masseter muscle.
Instruction: "Keep your lips open and try to bite your teeth together."

⚕ Clinical relevance

- Following a head injury, the masseter muscle can show a spastic reaction pattern. This can cause difficulties opening the mouth and lead to severe grinding of the teeth (bruxism).
- If the masseter muscle is paralyzed, the patient can replace its function with the pterygoideus medialis and temporalis muscles.

❗ Problems/comments

- The temporalis and pterygoideus medialis muscles aid the function of the masseter.

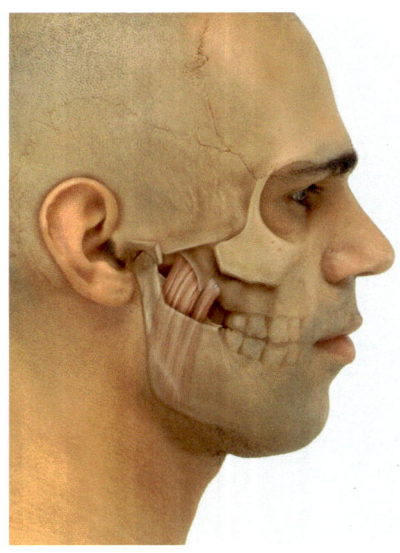

Pterygoideus medialis muscle

The pterygoideus medialis (medial pterygoid) muscle closes the jaws powerfully and pushes the mandible forward slightly.

Origin

Pterygoid fossa of the sphenoid bone
Lateral lamina of the pterygoid process

Insertion

Inner surface of the mandibular angle
Pterygoid tuberosity

Innervation

Medial pterygoid nerve from the mandibular branch of the trigeminal nerve (cranial nerve V3)

Functions

 Synergists

 Antagonists

Temporomandibular joints

Lifting the mandible
Temporalis muscle
Masseter muscle

Digastricus muscle
Mylohyoideus muscle
Geniohyoideus muscle
Pterygoideus lateralis muscle

Pushing the mandible forwards
Pterygoideus lateralis muscle
Masseter muscle, superficial part
Temporalis muscle (vertical fibers adjoining the orbit – weak)

Temporalis muscle (horizontal fibers lying above the ear)

Muscle function testing

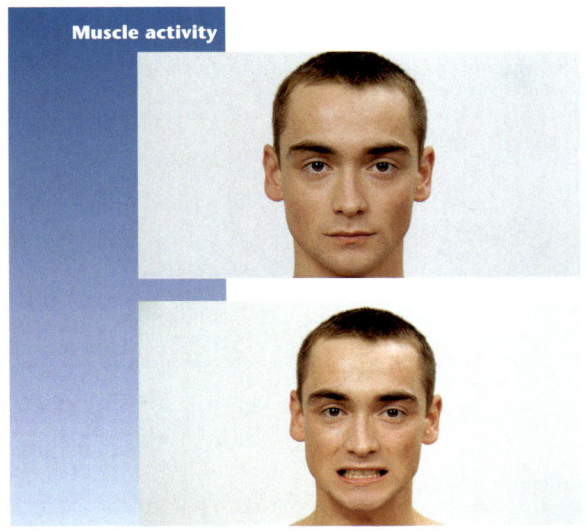

Muscle activity

Starting position: The patient looks straight ahead with his face relaxed.

Test procedure: The examiner looks at the patient's face.

Instruction: "Keeping your lips open, bite your teeth together."

 Problems/comments

• The pterygoideus medialis muscle can also move the mandible sideways (grinding movement while chewing).

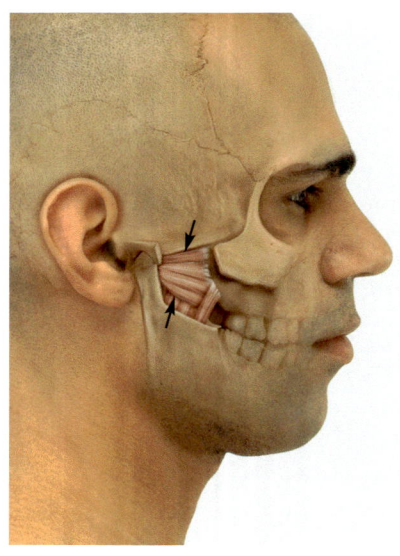

Pterygoideus lateralis muscle

The pterygoideus lateralis (lateral pterygoid) muscle comes forward to meet the lower jaw and can push forward not only the mandible itself (inferior part) but also the temporo-mandibular articular disk (superior part). It, therefore, has a key role in opening the jaw.

Origin	Superior head: temporal surface of the greater wing of the sphenoid bone Inferior head: lateral surface of the lateral lamina of the pterygoid process
Insertion	Superior head: pterygoid fovea of the condylar process, anterior edge of the temporomandibular articular disk Inferior head: pterygoid fovea of the condylar process
Innervation	Lateral pterygoid nerve from the mandibular branch of the trigeminal nerve (cranial nerve V)

Functions

 Synergists

 Antagonists

Temporomandibular joints

Pushing the mandible forwards

Pterygoideus lateralis muscle
Pterygoideus medialis muscle
Masseter muscle, superficial part
Temporalis muscle (vertical fibers adjoining the orbit – weak)

Temporalis muscle (horizontal fibers lying above the ear)
Masseter muscle, deep part

Muscle function testing

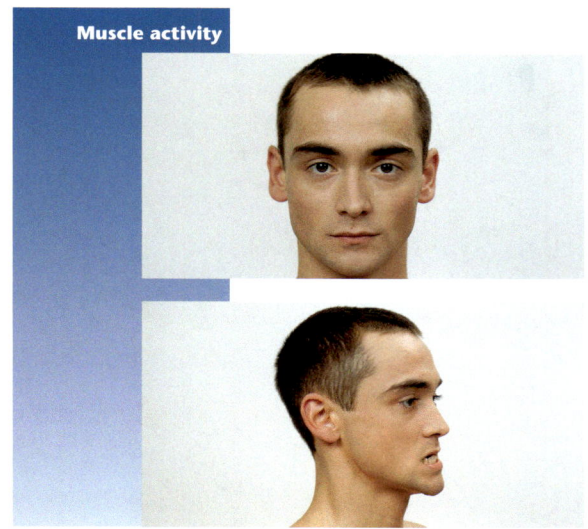

Muscle activity

Starting position: The patient looks straight ahead with his face relaxed.

Test procedure: The examiner looks at the patient's face.

Instruction: "Keeping your lips open, bite your teeth together."

Stretch test

**Masseter, temporalis
and pterygoideus medialis muscles**

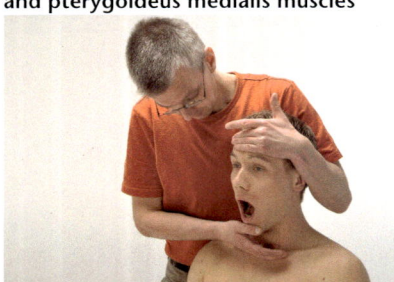

Method

The therapist moves the patient's mandible passively until the mouth is fully open.

Finding

Muscle shortening is present if the mouth cannot be opened fully and the endpoint feels soft and elastic.

The patient reports a stretching sensation along the course of the muscle.

6 Head

Muscles of the tongue

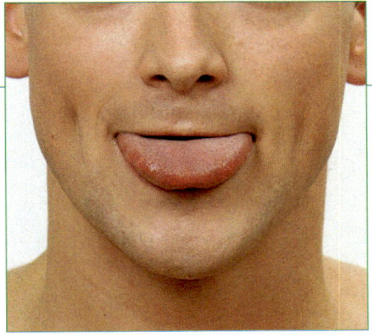

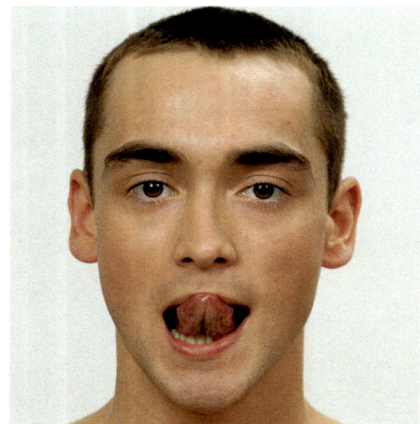

Intrinsic tongue muscles

The so-called intrinsic (internal) muscles alter the shape of the tongue. Their effect depends on which other intrinsic muscles are contracting at the same time. The tissue pressure which builds up during contraction of the tongue's intrinsic muscles acts antagonistically (the water cushion principle).

The longitudinalis superior muscle (superior longitudinal muscle of the tongue) can shorten and widen the tongue and raise its tip.

The longitudinalis inferior muscle (inferior longitudinal muscle of the tongue) can shorten and widen the tongue and lower its tip.

The transversus linguae muscle (transverse muscle of the tongue) can make the tongue narrower, extend it longitudinally and curve the lateral edges of the tongue upwards.

The verticalis linguae muscle (vertical muscle of the tongue) can make the tongue flatter, wider and shorter.

Origin	*Longitudinalis superior muscle*: root of the tongue
	Longitudinalis inferior muscle: root of the tongue
	Transversus linguae muscle: lateral edge of the tongue
	Verticalis linguae muscle: lingual aponeurosis
Insertion	*Longitudinalis superior muscle*: apex of the tongue
	Longitudinalis inferior muscle: apex of the tongue
	Transversus linguae muscle: lateral edge of the tongue
	Verticalis linguae muscle: lower surface of the tongue
Innervation	Hypoglossal nerve (cranial nerve XII)

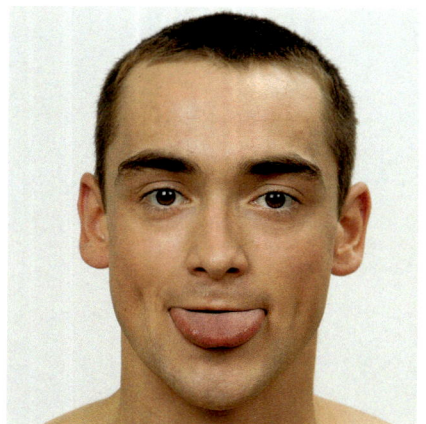

Muscle function testing

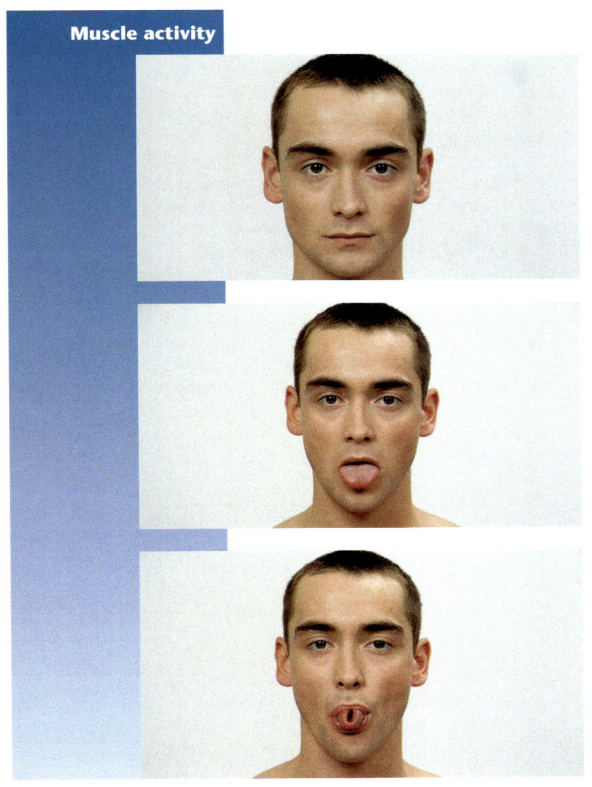

Muscle activity

Starting position: The patient looks straight ahead with his face relaxed.

Test procedure: The examiner looks at the movements of the tongue.
Instruction: "Stick out your tongue and then pull it back in again."

Test procedure: The examiner looks at the movements of the tongue.
Instruction: "Roll up your tongue and then flatten it again."

 **Clinical relevance**

- If the hypoglossal nerve (cranial nerve XII) is damaged, the tip of the tongue deviates toward the damaged side.
- The mobility of the tongue can be restricted by thickening and shortening of the lingual frenulum in progressive systemic sclerosis.

 Problems/comments

- It may be difficult for the patient to perform some of these movements "to order." However, the activity of the intrinsic tongue muscles can sometimes be observed during the transition between various other movements.

Extrinsic tongue muscles

The extrinsic tongue muscles move the whole tongue within the buccal cavity. Depending on the positions of the muscles, they can extend, retract, raise or lower the tongue.

The genioglossus muscle pulls the tongue forwards and downwards. During eating, this action allows the tongue to receive each bite of food with its taste buds. Like all the extrinsic tongue muscles, it displaces the entire tongue into the buccal cavity or out of it if the intrinsic muscles cause the tongue to extend longitudinally at the same time.

The hyoglossus (hyoglossal) muscle pulls the base of the tongue backwards and downwards, thus pushing food towards the esophagus during the swallowing act. This action occurs directly after the swallowing act is initiated by the styloglossus muscle.

The action of the chondroglossus muscle corresponds to that of the hyoglossus.

The styloglossus muscle pulls the base of the tongue backwards and upwards. When it contracts, after food has been chewed, it initiates swallowing by pushing the food against the gums, which triggers the swallowing reflex. This muscle is also important for sucking, because it draws the tongue back into the buccal cavity like the plunger of a syringe.

The palatoglossus (palatoglossal) muscle, together with the palatopharyngeus muscle, contracts the oropharyngeal isthmus during swallowing. It could lower the soft palate, if the latter were not simultaneously raised and tightened by the tensor and levator veli palatini muscles. During swallowing, it thus acts on the tongue and raises it.

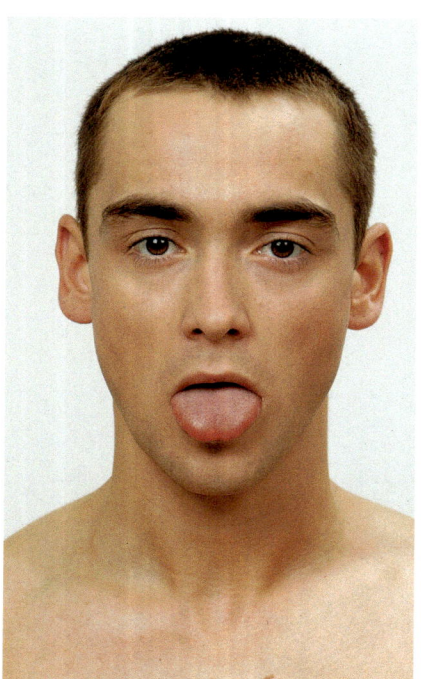

Both the intrinsic tongue muscles and the genioglossus muscle need to contract for the tongue to be stuck out

Origin

Genioglossus muscle: inner surface of the mandible
Hyoglossus muscle: greater horn and body of the hyoid bone
Chondroglossus muscle: lesser horn of the hyoid bone
Styloglossus muscle: styloid process
Palatoglossus muscle: palatine aponeurosis

Insertion

Genioglossus muscle: lingual aponeurosis
Hyoglossus muscle: lingual aponeurosis (lateral)
Chondroglossus muscle: lingual aponeurosis
Styloglossus muscle: lateral edge of the tongue
Palatoglossus muscle: radiating into the intrinsic musculature

Innervation

Genioglossus, hyoglossus, chondroglossus and styloglossus muscles: hypoglossal nerve (cranial nerve XII)
Palatoglossus muscle: glossopharyngeal nerve (cranial nerve IX) and vagus nerve (cranial nerve X)

Muscle function testing

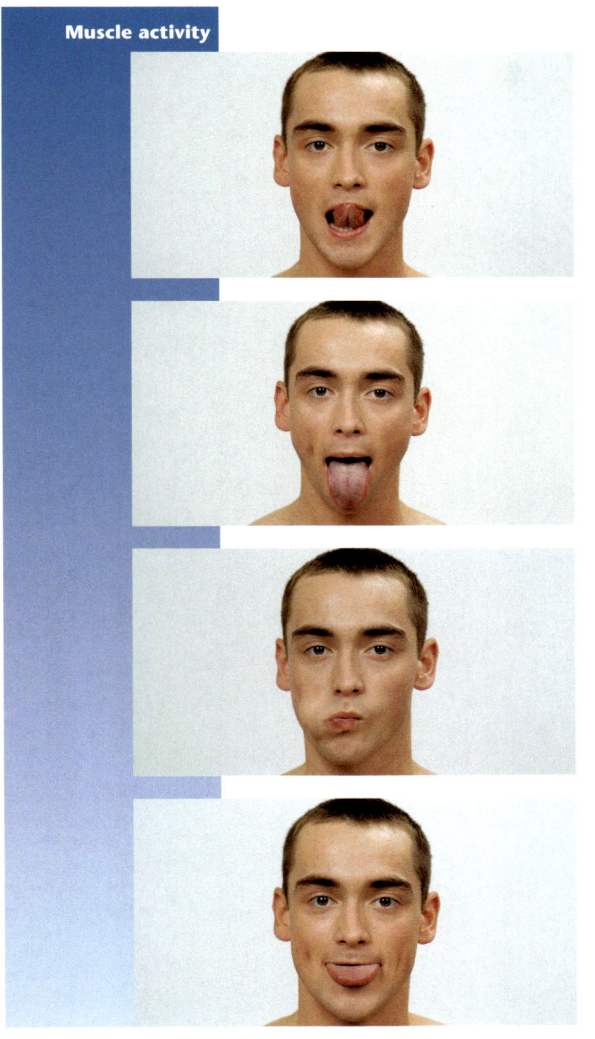

Muscle activity

Starting position: The patient looks straight ahead with his face relaxed.

Test procedure: The examiner looks at the movements of the tongue.

Instruction: "Stick out your tongue, then raise and lower it."

Test procedure: The examiner looks at the movements of the tongue.

Instruction: "Stick out your tongue and move the tip of your tongue downwards."

Test procedure: The examiner looks at the movements of the tongue.

Instruction: "Move the tip of your tongue so that it first touches the inside of your right cheek, and then the inside of your left cheek."

Test procedure: The examiner looks at the movements of the tongue.

Instruction: "Stick out your tongue and make it as wide and flat as possible."

6 Head

Muscles of the eye

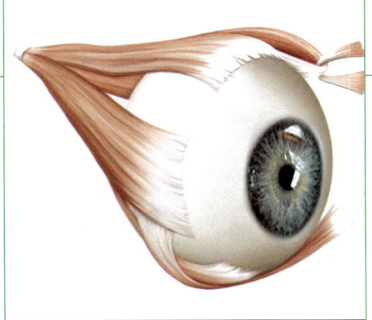

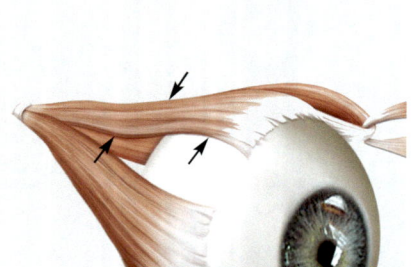

Rectus superior muscle

The rectus superior (superior straight) muscle of the eye lifts and rotates the eyeball in the medial direction. The gaze is thus directed upwards and some 25° towards the nose. The eyeball is also retracted dorsally into the orbit. Depending on the direction of gaze, all the other muscles of the eye may also be its synergists or antagonists. The only synergist listed here is the one which needs to contract in order to move the other eye in the same direction.

Origin	Common tendinous ring
Insertion	Cranially at the eyeball, ventrally of its equator
Innervation	Superior branch of the oculomotor nerve (cranial nerve III)

Functions

 Synergists

Synergist on the other eye
Obliquus inferior muscle

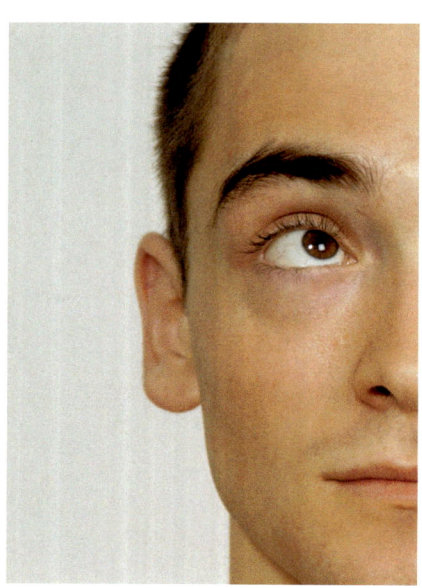

Muscle function testing

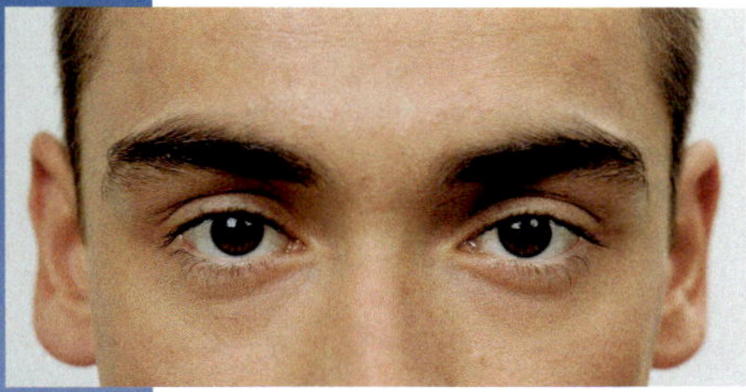

Muscle activity

Starting position: The patient looks straight ahead with his face relaxed.

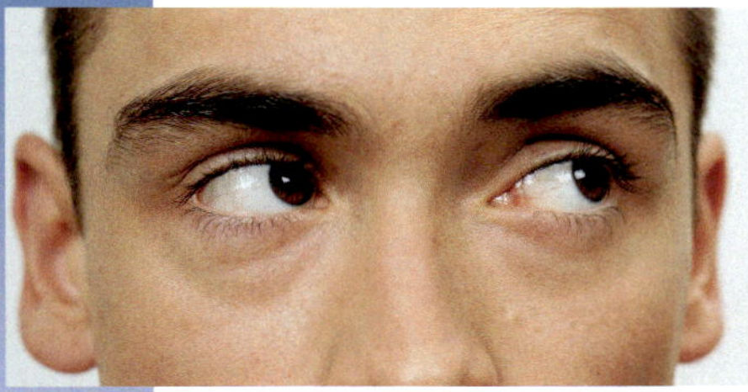

Test of the right rectus superior muscle

Test procedure: The examiner looks at the patient's right eye.

Instruction: "Look up and to the left."

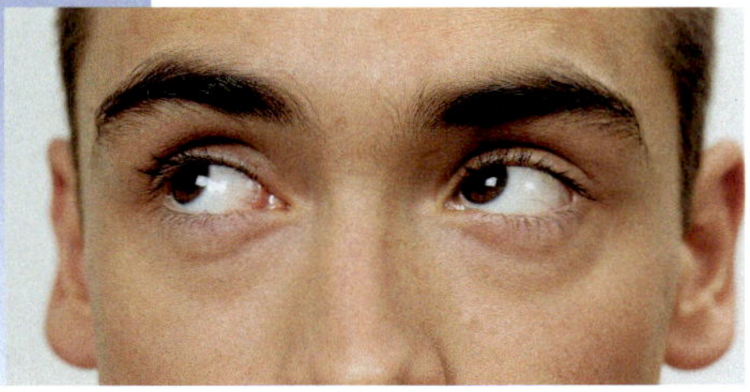

Test of the left rectus superior muscle

Test procedure: The examiner looks at the patient's left eye.

Instruction: "Look up and to the right."

⚕ Clinical relevance

- A deficit in any of the individual nerves of the eye causes double vision. The experienced clinician can use the position of the double images to determine which nerve is affected. The distance between the double images increases if the patient is asked to look in the direction into which the paralyzed muscle takes the eyeball.

❗ Problems/comments

- The left obliquus inferior muscle is simultaneously involved in directing the gaze upwards and to the right.
- The right obliquus inferior muscle is simultaneously involved in directing the gaze upwards and to the left.

Rectus inferior muscle

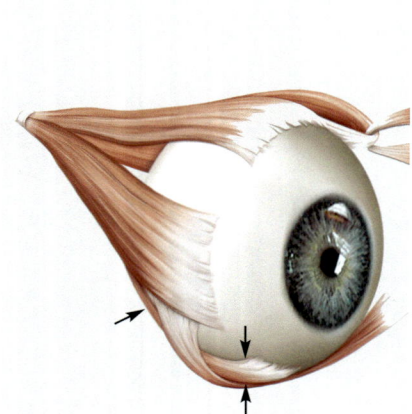

The rectus inferior (inferior straight) muscle of the eye lowers the gaze and directs it medially towards the nose. The eyeball is also retracted dorsally into the orbit. Depending on the direction of gaze, all the other muscles of the eye may also be its synergists or antagonists. The only synergist listed here is the one which needs to contract in order to move the other eye in the same direction.

Origin	Common tendinous ring
Insertion	Caudally at the eyeball, ventrally of its equator
Innervation	Inferior branch of the oculomotor nerve (cranial nerve III)

Functions

 Synergists

Synergist on the other eye
Obliquus superior muscle

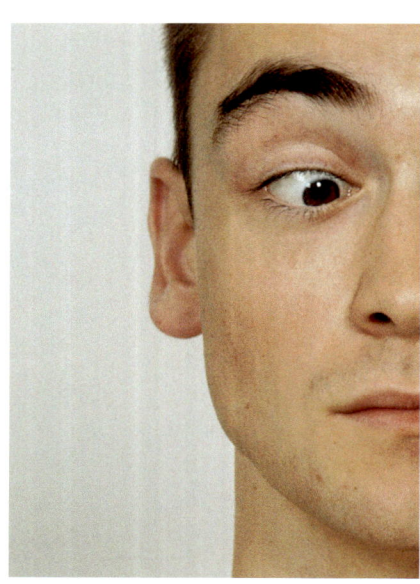

Muscle function testing

Muscle activity

Starting position: The patient looks straight ahead with his face relaxed.

Test of the right rectus inferior muscle

Test procedure: The examiner looks at the patient's right eye.

Instruction: "Look down and to the left."

Test of the left rectus inferior muscle

Test procedure: The examiner looks at the patient's left eye.

Instruction: "Look down and to the right."

 ### Clinical relevance

- A deficit in any of the individual nerves of the eye causes double vision. The experienced clinician can use the position of the double images to determine which nerve is affected. The distance between the double images increases if the patient is asked to look in the direction into which the paralyzed muscle takes the eyeball.

 ### Problems/comments

- The left obliquus superior muscle is simultaneously involved in directing the gaze downwards and to the right.
- The right obliquus superior muscle is simultaneously involved in directing the gaze downwards and to the left.
- The eyelids may be held up with the fingers to enable the direction of gaze to be clearly ascertained.

Trochlea

Insertion

Obliquus superior muscle

The obliquus superior (superior oblique) muscle of the eye rolls the eyeball inwards towards the nose and lowers and abducts the gaze towards the temple. It also pulls the eyeball in the ventral direction within the orbit, thus acting as an antagonist to the extrinsic straight muscles of the eye. Depending on the direction of gaze, all the other muscles of the eye may also be its synergists or antagonists. The only synergist listed here is the one which needs to contract in order to move the other eye in the same direction.

Origin	Orbit, medially of the optic canal
Insertion	Temporally and cranially at the eyeball, dorsally of its equator
Innervation	Trochlear nerve (cranial nerve IV)
Special features	The trochlea of the superior oblique muscle of the eye acts as the fulcrum for the muscle's tendon.

Functions

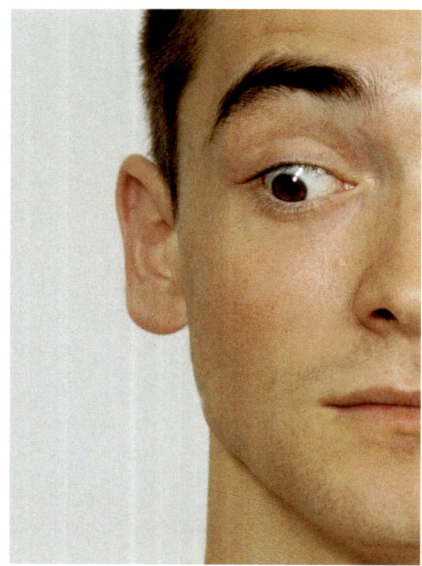 Synergists

Synergist on the other eye
Rectus inferior muscle

Muscle function testing

Starting position: The patient looks straight ahead with his face relaxed.

Test of the right obliquus superior muscle

Test procedure: The examiner looks at the patient's right eye.

Instruction: "Look down and to the right."

Test of the left obliquus superior muscle

Test procedure: The examiner looks at the patient's left eye.

Instruction: "Look down and to the left."

 **Clinical relevance**

- Paralysis of this muscle causes only slight deviation of the eye. Instead, what becomes apparent is that the head is rotated and tilted towards the healthy side to compensate. When the head is tilted towards the paralyzed side, the eyeball deviates upwards when the patient fixes the gaze with the healthy eye (Bielschowsky's sign).

 Problems/comments

- The left rectus inferior muscle is simultaneously involved in directing the gaze downwards and to the left.
- The right rectus inferior muscle is simultaneously involved in directing the gaze downwards and to the right.
- The eyelids may be held up with the fingers to enable the direction of gaze to be clearly ascertained.

Obliquus inferior muscle

The obliquus inferior (inferior oblique) muscle of the eye rolls the eyeball outwards and raises and abducts the gaze towards the temple. It also pulls the eyeball in the ventral direction within the orbit, thus acting as an antagonist to the extrinsic straight muscles of the eye. Depending on the direction of gaze, all the other muscles of the eye may also be its synergists or antagonists. The only synergist listed here is the one which needs to contract in order to move the other eye in the same direction.

Origin   Base of the orbit, laterally of the nasolacrimal canal

Insertion   Temporally at the eyeball, dorsally of its equator

Innervation   Inferior branch of the oculomotor nerve (cranial nerve III)

Functions

 Synergists

Synergist on the other eye
Rectus superior muscle

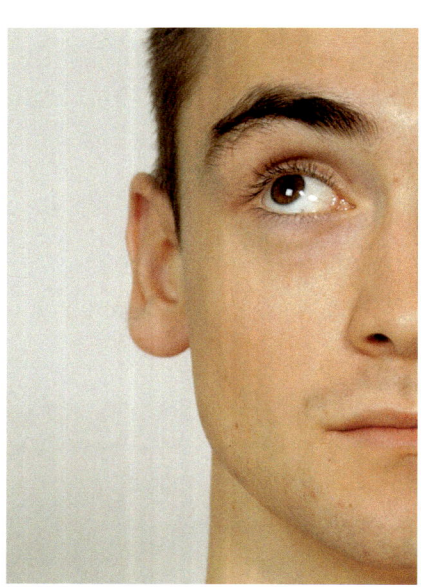

Muscle function testing

Muscle activity

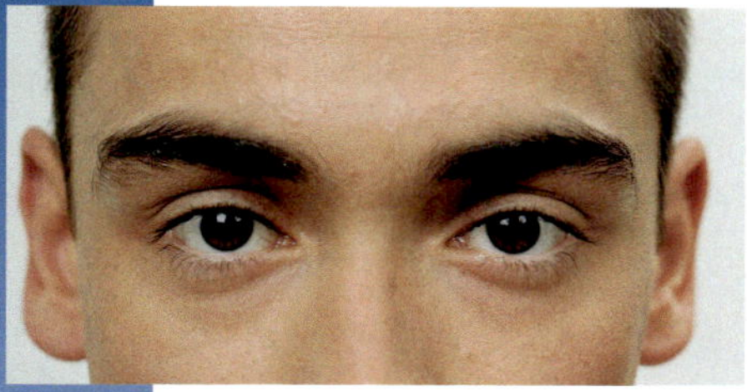

Starting position: The patient looks straight ahead with his face relaxed.

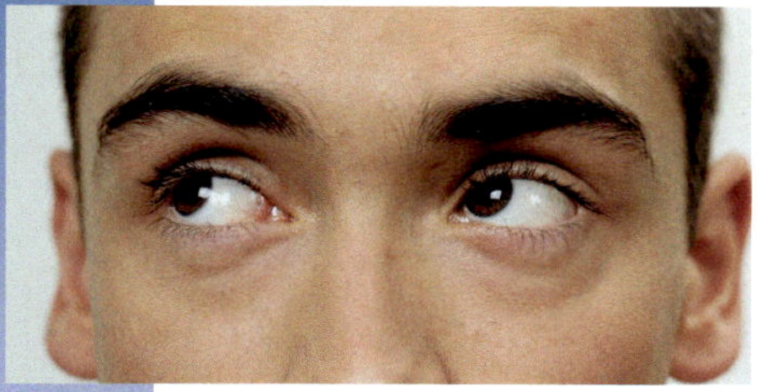

Test of the right obliquus inferior muscle

Test procedure: The examiner looks at the patient's right eye.

Instruction: "Look up and to the right."

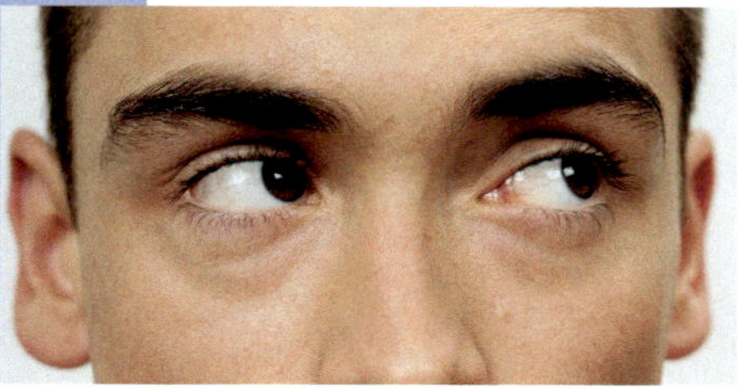

Test of the left obliquus inferior muscle

Test procedure: The examiner looks at the patient's left eye.

Instruction: "Look up and to the left."

 Clinical relevance

- A deficit in any of the individual nerves of the eye causes double vision. The experienced clinician can use the position of the double images to determine which nerve is affected. The distance between the double images increases if the patient is asked to look in the direction into which the paralyzed muscle takes the eyeball.

 Problems/comments

- The left rectus superior muscle is simultaneously involved in directing the gaze upwards and to the left.
- The right rectus superior muscle is simultaneously involved in directing the gaze upwards and to the right.

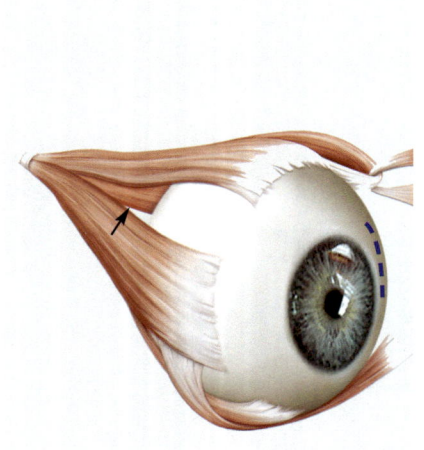

---- Insertion of the rectus medialis muscle

Rectus medialis muscle

The rectus medialis (medial straight) muscle of the eye adducts the eyeball towards the nose and, in doing so, acts solely in the horizontal plane if it contracts in isolation. The eyeball is also retracted dorsally into the orbit. Depending on the direction of gaze, all the other muscles of the eye may also be its synergists or antagonists. The only synergist listed here is the one which needs to contract in order to move the other eye in the same direction.

Origin	Common tendinous ring
Insertion	Medially at the eyeball, ventrally of its equator
Innervation	Inferior branch of the oculomotor nerve (cranial nerve III)

Functions

Synergists

Synergist on the other eye
Rectus lateralis muscle

Muscle function testing

Muscle activity

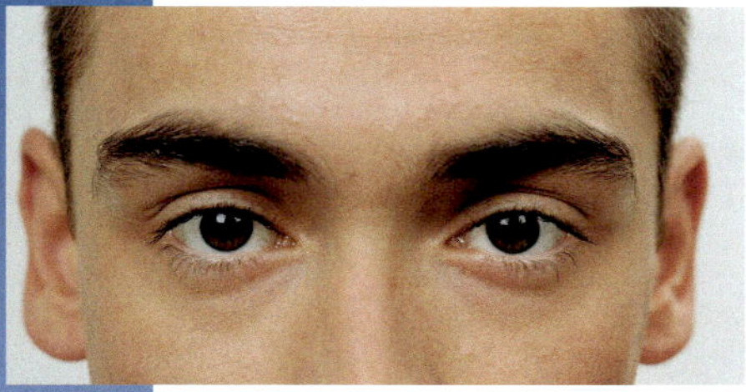

Starting position: The patient looks straight ahead with his face relaxed.

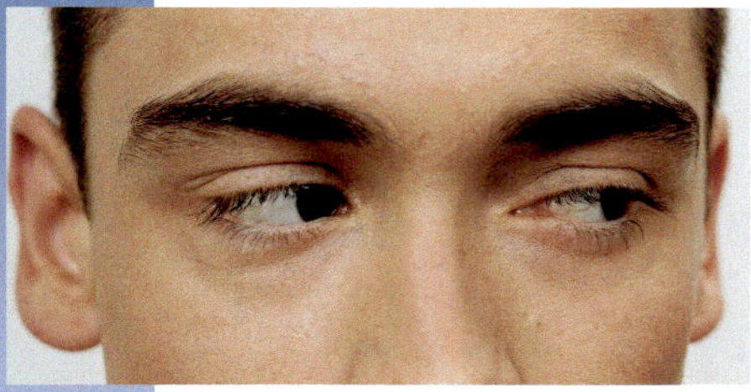

Test of the right rectus medialis muscle

Test procedure: The examiner looks at the patient's right eye.

Instruction: "Look left."

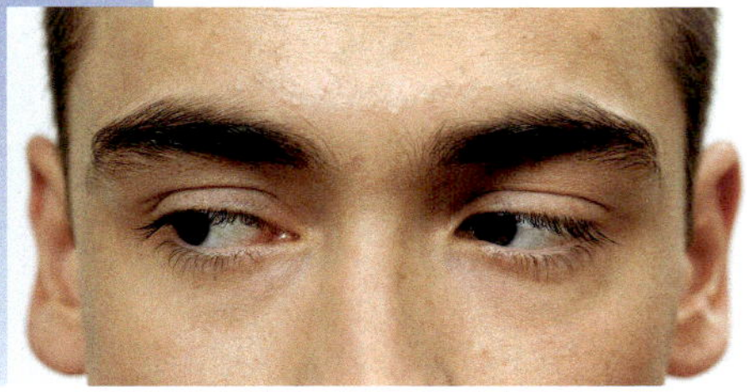

Test of the left rectus medialis muscle

Test procedure: The examiner looks at the patient's left eye.

Instruction: "Look right."

 Clinical relevance

- A deficit in any of the individual nerves of the eye causes double vision. The experienced clinician can use the position of the double images to determine which nerve is affected. The distance between the double images increases if the patient is asked to look in the direction into which the paralyzed muscle takes the eyeball.

! Problems/comments

- When the gaze is directed medially, the muscle which acts simultaneously in the eye being assessed is the rectus medialis, while in the other eye it is the rectus lateralis.

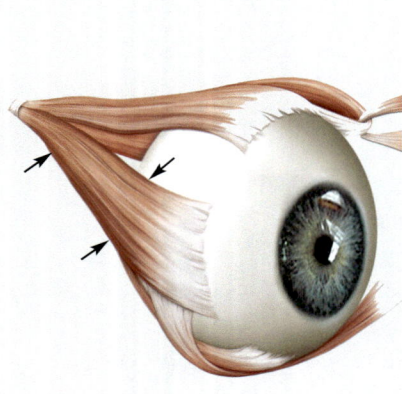

Rectus lateralis muscle

The rectus lateralis (lateral straight) muscle of the eye abducts the eyeball towards the temple and, in doing so, like its medial counterpart, acts solely in the horizontal plane if it contracts in isolation. The eyeball is also retracted dorsally into the orbit. Depending on the direction of gaze, all the other muscles of the eye may also be its synergists or antagonists. The only synergist listed here is the one which needs to contract in order to move the other eye in the same direction.

Origin	Common tendinous ring
Insertion	Laterally at the eyeball, ventrally of its equator
Innervation	Abducent nerve (cranial nerve VI)

Functions

 Synergists

Synergist on the other eye
Rectus medialis muscle

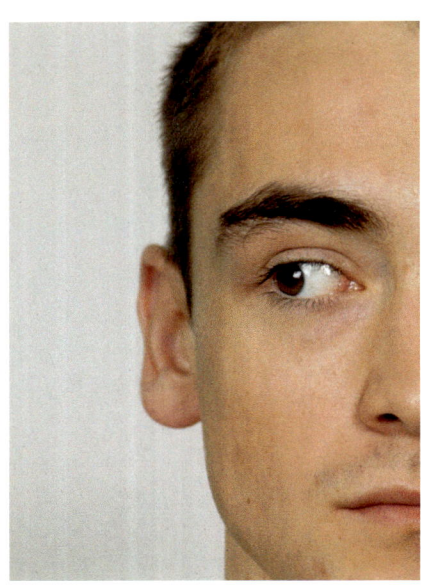

Muscle function testing

Muscle activity

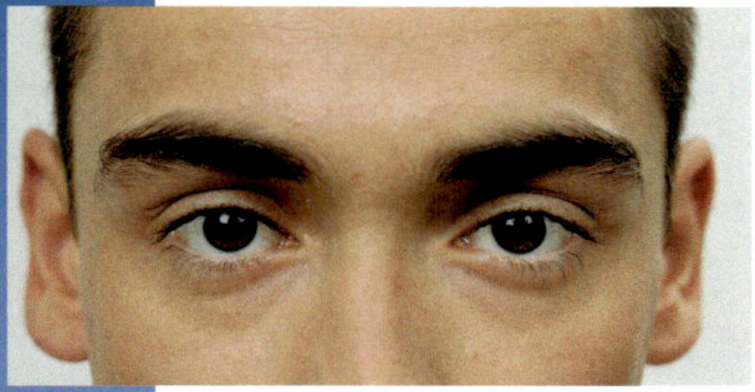

Starting position: The patient looks straight ahead with his face relaxed.

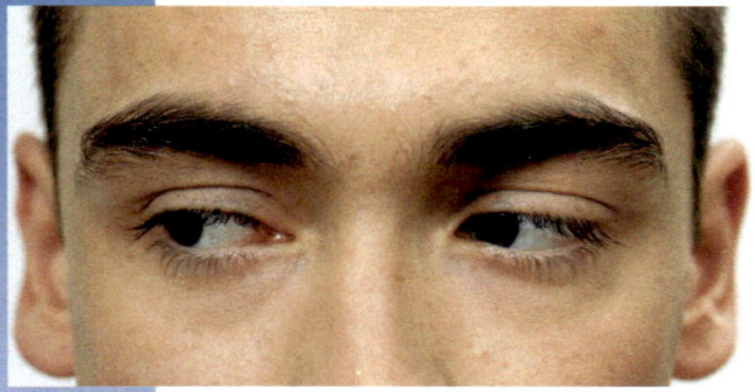

Test of the right rectus lateralis muscle
Test procedure: The examiner looks at the patient's right eye.
Instruction: "Look right."

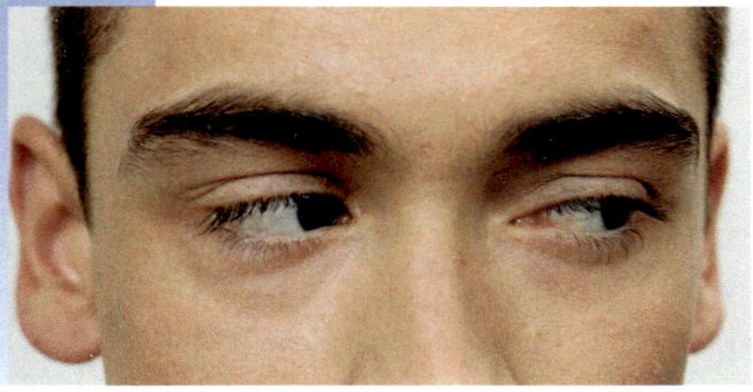

Test of the left rectus lateralis muscle
Test procedure: The examiner looks at the patient's left eye.
Instruction: "Look left."

 Clinical relevance

- Paralysis of the rectus lateralis muscle causes the eye to deviate inwards. The double images are close together and become magnified when the gaze is directed outwards.
- This is the most common form of eye muscle paralysis.

 Problems/comments

- When the gaze is directed laterally, the muscle which acts simultaneously in the eye being assessed is the rectus lateralis, while in the other eye it is the rectus medialis.

Innervation of the muscles of the head		
Nerve	**Innervated muscles**	**Alternative designation**
Oculomotor nerve		Cranial nerve III
	Levator palbebrae superioris muscle	
	Rectus superior muscle	
	Rectus inferior muscle	
	Obliquus inferior muscle	
	Rectus medialis muscle	
Trochlear nerve		Cranial nerve IV
	Obliquus superior muscle	
Trigeminal nerve		Cranial nerve V
	Temporalis muscle	
	Masseter muscle	
	Pterigoideus medialis muscle	
	Pterigoideus lateralis muscle	
Abducent nerve		Cranial nerve VI
	Rectus lateralis muscle	
Facial nerve		Cranial nerve VII
	Epicranius muscle	
	Corrugator supercilii muscle	
	Procerus muscle	
	Orbicularis oculi muscle	
	Nasalis muscle	
	Levator labii superioris alaeque nasi muscle	
	Levator labii superioris muscle	
	Zygomaticus major/minor muscle	
	Risorius muscle	
	Levator anguli oris muscle	
	Buccinator muscle	
	Orbicularis oris muscle	
	Depressor anguli oris muscle	
	Depressor labii inferioris muscle	
	Mentalis muscle	
	Platysma	
Hypoglossal nerve		Cranial nerve XII
	Intrinsic muscles of the tongue	

Appendix

Segment zones, ventral

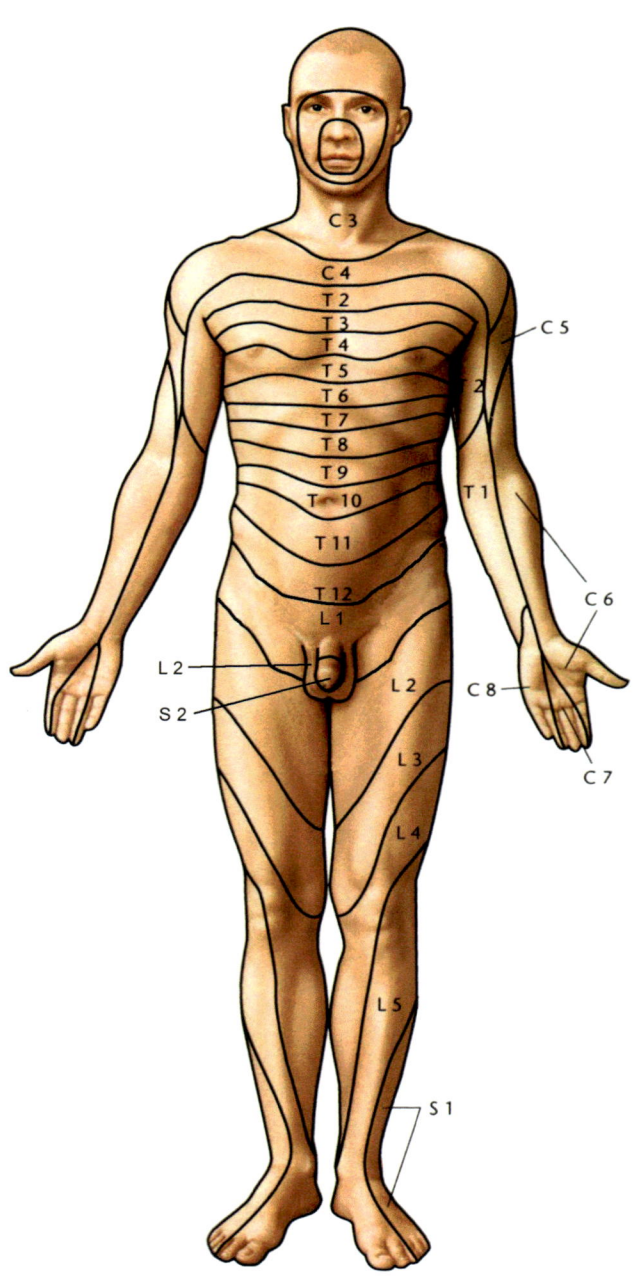

Segment zones, dorsal

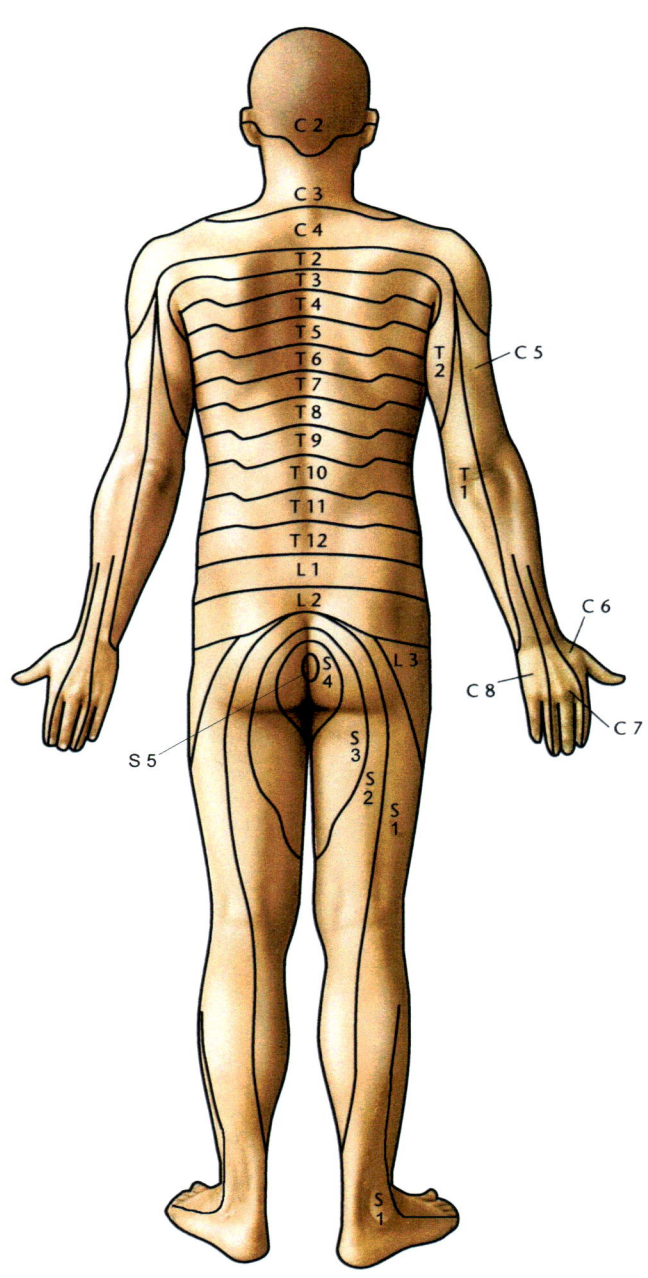

List of muscles according to innervation and innervation level

The standard indicator muscles used in the literature and in physiotherapy are highlighted in blue

C1
Longus capitis muscle
Obliquus capitis superior muscle
Omohyoideus muscle
Rectus capitis anterior muscle
Rectus capitis posterior major muscle
Rectus capitis posterior minor muscle
Semispinalis capitis muscle
Sternohyoideus muscle
Sternothyroideus muscle
Thyrohyoideus muscle

C2
Longus capitis muscle
Obliquus capitis inferior muscle
Omohyoideus muscle
Rectus capitis anterior muscle
Rectus capitis posterior major muscle
Semispinalis capitis muscle
Sternocleidomastoideus muscle
Sternohyoideus muscle
Sternothyroideus muscle
Trapezius muscle, descending part
Thyrohyoideus muscle

C3
Diaphragm
Iliocostalis cervicis muscle
Levator scapulae muscle
Longissimus capitis muscle
Longus capitis muscle
Longus colli muscle
Multifidus cervicis muscle
Omohyoideus muscle
Rotator cervicis muscle
Semispinalis capitis muscle
Splenius capitis muscle
Sternohyoideus muscle
Sternothyroideus muscle
Trapezius muscle, descending part

C4
Diaphragm
Iliocostalis cervicis muscle
Levator scapulae muscle

Longissimus capitis muscle
Longissimus cervicis muscle
Longus capitis muscle
Longus colli muscle
Multifidus cervicis muscle
Omohyoideus muscle
Rhomboideus major muscle
Rhomboideus minor muscle
Rotator cervicis muscle
Scalenus medius muscle
Semispinalis cervicis muscle
Semispinalis capitis muscle
Splenius capitis muscle
Sternohyoideus muscle
Sternothyroideus muscle
Supraspinatus muscle
Trapezius muscle, descending part

C5
Diaphragm
Biceps brachii muscle
Brachialis muscle
Brachioradialis muscle
Coracobrachialis muscle
Deltoideus muscle, acromial part
Deltoideus muscle, clavicular part
Deltoideus muscle, spinal part
Iliocostalis cervicis muscle
Infraspinatus muscle
Levator scapulae muscle
Longissimus capitis muscle
Longissimus cervicis muscle
Longus colli muscle
Multifidus cervicis muscle
Pectoralis major muscle, clavicular part
Pectoralis major muscle, sternocostal part
Rhomboideus major muscle
Rotator cervicis muscle
Scalenus anterior muscle
Scalenus medius muscle
Semispinalis cervicis muscle
Semispinalis capitis muscle
Serratus anterior muscle
Splenius capitis muscle
Splenius cervicis muscle

See Index for the relevant page numbers for each muscle

Subclavius muscle
Subscapularis muscle
Supinator muscle
Supraspinatus muscle
Teres major muscle
Teres minor muscle

C6
Abductor pollicis longus muscle
Biceps brachii muscle
Brachialis muscle
Brachioradialis muscle
Coracobrachialis muscle
Deltoideus muscle, acromial part
Deltoideus muscle, clavicular part
Deltoideus muscle, spinal part
Extensor carpi radialis brevis muscle
Extensor carpi radialis longus muscle
Extensor carpi ulnaris muscle
Extensor digiti minimi muscle
Extensor digitorum muscle
Extensor indicis muscle
Extensor pollicis brevis muscle
Extensor pollicis longus muscle
Flexor carpi radialis muscle
Iliocostalis cervicis muscle
Infraspinatus muscle
Latissimus dorsi muscle
Longissimus capitis muscle
Longissimus cervicis muscle
Longus colli muscle
Multifidus cervicis muscle
Pectoralis major muscle, clavicular part
Pectoralis major muscle, sternocostal part
Pectoralis minor muscle
Pronator quadratus muscle
Pronator teres muscle
Rotator cervicis muscle
Scalenus anterior muscle
Scalenus medius muscle
Semispinalis capitis muscle
Serratus anterior muscle
Serratus posterior superior muscle
Spinalis capitis muscle
Splenius cervicis muscle
Subclavius muscle
Subscapularis muscle
Supinator muscle

Supraspinatus muscle
Teres major muscle
Teres minor muscle
Triceps brachii muscle

C7
Abductor pollicis brevis muscle
Abductor pollicis longus muscle
Anconeus muscle
Brachialis muscle
Coracobrachialis muscle
Extensor carpi radialis brevis muscle
Extensor carpi radialis longus muscle
Extensor carpi ulnaris muscle
Extensor digiti minimi muscle
Extensor digitorum muscle
Extensor indicis muscle
Extensor pollicis brevis muscle
Extensor pollicis longus muscle
Flexor carpi radialis muscle
Flexor carpi ulnaris muscle
Flexor digitorum profundus muscle
Flexor digitorum superficialis muscle
Flexor pollicis brevis muscle
Flexor pollicis longus muscle
Iliocostalis cervicis muscle
Latissimus dorsi muscle
Longissimus capitis muscle
Longissimus cervicis muscle
Multifidus cervicis muscle
Opponens pollicis muscle
Palmaris brevis muscle
Palmaris longus muscle
Pectoralis major muscle, clavicular part
Pectoralis major muscle, sternocostal part
Pectoralis minor muscle
Pronator quadratus muscle
Pronator teres muscle
Rotator cervicis muscle
Rotator thoracis muscle
Scalenus anterior muscle
Scalenus medius muscle
Scalenus posterior muscle
Semispinalis cervicis muscle
Serratus anterior muscle
Serratus posterior superior muscle
Spinalis capitis muscle
Spinalis cervicis muscle

Splenius cervicis muscle
Teres major muscle
Triceps brachii muscle

C8

Abductor digiti minimi muscle
Abductor pollicis brevis muscle
Abductor pollicis longus muscle
Adductor pollicis muscle
Anconeus muscle
Extensor carpi ulnaris muscle
Extensor digiti minimi muscle
Extensor digitorum muscle
Extensor indicis muscle
Extensor pollicis brevis muscle
Extensor pollicis longus muscle
Flexor carpi radialis muscle
Flexor carpi ulnaris muscle
Flexor digiti minimi brevis muscle
Flexor digitorum profundus muscle
Flexor digitorum superficialis muscle
Flexor pollicis brevis muscle
Flexor pollicis longus muscle
Iliocostalis cervicis muscle
Interossei muscles of the hand, dorsal
Interossei muscles of the hand, palmar
Latissimus dorsi muscle
Longissimus capitis muscle
Longissimus cervicis muscle
Lumbricales muscles of the hand
Multifidus cervicis muscle
Opponens digiti minimi muscle
Opponens pollicis muscle
Palmaris brevis muscle
Palmaris longus muscle
Pectoralis major muscle, abdominal part
Pectoralis major muscle, clavicular part
Pectoralis major muscle, sternocostal part
Pectoralis minor muscle
Pronator quadratus muscle
Rotatores cervicis muscle
Rotatores thoracis muscle
Scalenus anterior muscle
Scalenus medius muscle
Scalenus posterior muscle
Semispinalis cervicis muscle
Serratus posterior superior muscle
Spinalis capitis muscle
Spinalis cervicis muscle
Triceps brachii muscle

T1

Abductor digiti minimi muscle
Abductor pollicis brevis muscle
Adductor pollicis muscle
Flexor carpi ulnaris muscle
Flexor digiti minimi brevis muscle
Flexor digitorum profundus muscle
Flexor digitorum superficialis muscle
Flexor pollicis brevis muscle
Flexor pollicis longus muscle
Iliocostalis cervicis muscle
Iliocostalis thoracis muscle
Intercostal muscles, external
Intercostal muscles, internal
Interossei muscles of the hand, dorsal
Interossei muscles of the hand, palmar
Longissimus capitis muscle
Longissimus cervicis muscle
Longissimus thoracis muscle
Lumbricales muscles of the hand
Multifidus thoracis muscle
Opponens digiti minimi muscle
Opponens pollicis muscle
Palmaris brevis muscle
Palmaris longus muscle
Pectoralis major muscle, abdominal part
Pectoralis major muscle, sternocostal part
Pronator quadratus muscle
Rotatores thoracis muscles
Serratus posterior superior muscle
Spinalis capitis muscle
Spinalis cervicis muscle

T2

Iliocostalis thoracis muscle
Iliocostalis cervicis muscle
Intercostal muscles, external
Intercostal muscles, internal
Longissimus capitis muscle
Longissimus cervicis muscle
Longissimus thoracis muscle
Multifidus thoracis muscle
Rotatores thoracis muscles
Serratus posterior superior muscle
Spinalis capitis muscle
Spinalis cervicis muscle
Spinalis thoracis muscle

See Index for relevant page numbers for each muscle

T3
Iliocostalis cervicis muscle
Iliocostalis thoracis muscle
Intercostal muscles, external
Intercostal muscles, internal
Longissimus capitis muscle
Longissimus cervicis muscle
Longissimus thoracis muscle
Multifidus thoracis muscle
Rotatores thoracis muscles
Spinalis capitis muscle
Spinalis cervicis muscle
Spinalis thoracis muscle

T4
Iliocostalis cervicis muscle
Iliocostalis thoracis muscle
Intercostal muscles, external
Intercostal muscles, internal
Longissimus cervicis muscle
Longissimus thoracis muscle
Multifidus thoracis muscle
Rotatores thoracis muscles
Semispinalis capitis muscle
Semispinalis thoracis muscle
Spinalis thoracis muscle

T5
Iliocostalis cervicis muscle
Iliocostalis thoracis muscle
Intercostal muscles, external
Intercostal muscles, internal
Longissimus cervicis muscle
Longissimus thoracis muscle
Multifidus thoracis muscle
Obliquus externus abdominis muscle
Obliquus internus abdominis muscle
Rectus abdominis muscle
Rotatores thoracis muscles
Semispinalis capitis muscle
Semispinalis thoracis muscle
Spinalis thoracis muscle
Transversus abdominis muscle

T6
Iliocostalis cervicis muscle
Iliocostalis thoracis muscle
Intercostal muscles, external
Intercostal muscles, internal
Longissimus cervicis muscle
Longissimus thoracis muscle

Multifidus thoracis muscle
Obliquus externus abdominis muscle
Obliquus internus abdominis muscle
Rectus abdominis muscle
Rotatores thoracis muscles
Semispinalis capitis muscle
Semispinalis thoracis muscle
Spinalis thoracis muscle
Transversus abdominis muscle

T7
Iliocostalis cervicis muscle
Iliocostalis lumborum muscle
Iliocostalis thoracis muscle
Intercostal muscles, external
Intercostal muscles, internal
Longissimus thoracis muscle
Multifidus thoracis muscle
Obliquus externus abdominis muscle
Obliquus internus abdominis muscle
Rectus abdominis muscle
Rotatores thoracis muscles
Spinalis thoracis muscle
Transversus abdominis muscle

T8
Iliocostalis lumborum muscle
Iliocostalis thoracis muscle
Intercostal muscles, external
Intercostal muscles, internal
Longissimus thoracis muscle
Multifidus thoracis muscle
Obliquus externus abdominis muscle
Obliquus internus abdominis muscle
Rectus abdominis muscle
Rotatores thoracis muscles
Spinalis thoracis muscle
Transversus abdominis muscle

T9
Iliocostalis lumborum muscle
Iliocostalis thoracis muscle
Intercostal muscles, external
Intercostal muscles, internal
Longissimus thoracis muscle
Multifidus thoracis muscle
Obliquus externus abdominis muscle
Obliquus internus abdominis muscle
Rectus abdominis muscle
Rotatores thoracis muscles
Transversus abdominis muscle

T10

Iliocostalis lumborum muscle
Iliocostalis thoracis muscle
Intercostal muscles, external
Intercostal muscles, internal
Longissimus thoracis muscle
Multifidus thoracis muscle
Obliquus externus abdominis muscle
Obliquus internus abdominis muscle
Rectus abdominis muscle
Rotatores thoracis muscles
Spinalis thoracis muscle
Transversus abdominis muscle

T11

Iliocostalis lumborum muscle
Iliocostalis thoracis muscle
Intercostal muscles, external
Intercostal muscles, internal
Longissimus thoracis muscle
Multifidus thoracis muscle
Obliquus externus abdominis muscle
Obliquus internus abdominis muscle
Rectus abdominis muscle
Rotatores thoracis muscles
Semispinalis thoracis muscle
Serratus posterior inferior muscle
Spinalis thoracis muscle
Transversus abdominis muscle

T12

Iliocostalis lumborum muscle
Iliocostalis thoracis muscle
Intertransversarii muscles, lateral lumbar
Longissimus thoracis muscle
Multifidus thoracis muscle
Obliquus externus abdominis muscle
Obliquus internus abdominis muscle
Quadratus lumborum muscle
Rectus abdominis muscle
Rotatores thoracis muscles
Semispinalis thoracis muscle
Serratus posterior inferior muscle
Spinalis thoracis muscle
Transversus abdominis muscle

L1

Cremaster muscle
Iliocostalis lumborum muscle
Iliocostalis thoracis muscle
Intertransversarii muscles, lateral lumbar

Intertransversarii muscles, medial lumbar
Longissimus thoracis muscle
Multifidus lumborum muscle
Obliquus externus abdominis muscle
Obliquus internus abdominis muscle
Quadratus lumborum muscle
Rectus abdominis muscle
Rotatores lumborum muscles
Serratus posterior inferior muscle
Spinalis thoracis muscle
Transversus abdominis muscle

L2

Adductor brevis muscle
Adductor longus muscle
Adductor magnus muscle, ventral part
Cremaster muscle
Gracilis muscle
Iliocostalis lumborum muscle
Iliopsoas muscle
Intertransversarii muscles, lateral lumbar
Intertransversarii muscles, medial lumbar
Longissimus thoracis muscle
Multifidus lumborum muscle
Obturatorius externus muscle
Pectineus muscle
Quadriceps femoris muscle
Rectus femoris muscle
Rotatores lumborum muscles
Sartorius muscle
Serratus posterior inferior muscle
Transversus abdominis muscle
Vastus intermedius muscle
Vastus lateralis muscle
Vastus medialis muscle

L3

Adductor brevis muscle
Adductor longus muscle
Adductor magnus muscle, ventral part
Gracilis muscle
Iliocostalis lumborum muscle
Iliopsoas muscle
Intertransversarii muscles, lateral lumbar
Intertransversarii muscles, medial lumbar
Longissimus thoracis muscle
Multifidus lumborum muscle
Obturatorius externus muscle
Pectineus muscle
Quadriceps femoris muscle
Rectus femoris muscle

See Index for relevant page numbers for each muscle

Rotatores lumborum muscles
Sartorius muscle
Vastus intermedius muscle
Vastus lateralis muscle
Vastus medialis muscle

L4

Adductor brevis muscle
Adductor longus muscle
Adductor magnus muscle, dorsal part
Gluteus medius muscle
Gluteus minimus muscle
Gracilis muscle
Iliocostalis lumborum muscle
Iliopsoas muscle
Intertransversarii muscles, lateral lumbar
Intertransversarii muscles, medial lumbar
Longissimus thoracis muscle
Multifidus lumborum muscle
Obturatorius externus muscle
Quadriceps femoris muscle
Rectus femoris muscle
Rotatores lumborum muscles
Tensor fasciae latae muscle
Tibialis anterior muscle
Vastus intermedius muscle
Vastus lateralis muscle
Vastus medialis muscle

L5

Biceps femoris muscle
Extensor digitorum brevis muscle
Extensor digitorum longus muscle
Extensor hallucis brevis muscle
Extensor hallucis longus muscle
Flexor digitorum longus muscle
Flexor hallucis longus muscle
Gluteus medius muscle
Gluteus maximus muscle
Gluteus minimus muscle
Iliocostalis lumborum muscle
Intertransversarii muscles, lateral lumbar
Intertransversarii muscles, medial lumbar
Longissimus thoracis muscle
Multifidus lumborum muscle
Obturatorius internus muscle
Peroneus longus muscle
Peroneus tertius muscle
Piriformis muscle
Popliteus muscle
Rotatores lumborum muscles

Semimembranosus muscle
Semitendinosus muscle
Tensor fasciae latae muscle
Tibialis anterior muscle
Tibialis posterior muscle

S1

Abductor hallucis muscle
Biceps femoris muscle
Extensor digitorum brevis muscle
Extensor digitorum longus muscle
Extensor hallucis brevis muscle
Extensor hallucis longus muscle
Flexor digitorum brevis muscle
Flexor digitorum longus muscle
Flexor hallucis brevis muscle
Flexor hallucis longus muscle
Gastrocnemius muscle
Gluteus maximus muscle
Gluteus medius muscle
Gluteus minimus muscle
Lumbricales muscles of the foot
Multifidus lumborum muscle
Obturatorius internus muscle
Peroneus brevis muscle
Peroneus longus muscle
Peroneus tertius muscle
Piriformis muscle
Plantaris muscle
Popliteus muscle
Semimembranosus muscle
Semitendinosus muscle
Soleus muscle
Tibialis posterior muscle

S2

Abductor digiti minimi muscle
Abductor hallucis muscle
Adductor hallucis muscle
Biceps femoris muscle
Bulbospongiosus muscle
Flexor digiti minimi brevis muscle
Flexor digitorum brevis muscle
Flexor digitorum longus muscle
Flexor hallucis brevis muscle
Flexor hallucis longus muscle
Gastrocnemius muscle
Gluteus maximus muscle
Interossei muscles of the foot, dorsal
Interossei muscles of the foot, plantar
Ischiocavernosus muscle

Lumbricales muscles of the foot
Obturatorius internus muscle
Piriformis muscle
Plantaris muscle
Quadratus plantae muscle
Semimembranosus muscle
Semitendinosus muscle
Soleus muscle
Sphincter ani externus muscle
Tibialis posterior muscle
Transversus perinei profundus muscle
Transversus perinei superficialis muscle

S3
Abductor digiti minimi muscle
Adductor hallucis muscle
Bulbospongiosus muscle
Flexor digiti minimi brevis muscle
Flexor hallucis brevis muscle
Iliococcygeus muscle
Interossei muscles of the foot, dorsal
Interossei muscles of the foot, plantar
Ischiocavernosus muscle
Ischiococcygeus muscle
Levator ani muscle
Lumbricales muscles of the foot
Pubococcygeus muscle
Puboprostaticus muscle
Puborectalis muscle
Pubovaginalis muscle
Quadratus plantae muscle
Sphincter ani externus muscle
Transversus perinei profundus muscle
Transversus perinei superficialis muscle

S4
Bulbospongiosus muscle
Iliococcygeus muscle
Ischiocavernosus muscle
Ischiococcygeus muscle
Levator ani muscle
Pubococcygeus muscle
Puboprostaticus muscle
Puborectalis muscle
Pubovaginalis muscle
Sphincter ani externus muscle
Transversus perinei profundus muscle
Transversus perinei superficialis muscle

See Index for relevant page numbers for each muscle

Principal muscles used in individual movements

Joints of the upper and lower extremities

Upper extremity

Shoulder

Flexion/anteversion
Pectoralis major muscle, sternocostal part   50
Deltoideus muscle, clavicular part   30
Biceps brachii muscle, long head   58
Coracobrachialis muscle   54

Extension/retroversion
Latissimus dorsi muscle   44
Triceps brachii muscle, long head   64
Teres major muscle   46
Deltoideus muscle, spinal part   32

Abduction
Deltoideus muscle, acromial part   34
Biceps brachii muscle, long head   58
Deltoideus muscle, clavicular part   30

Adduction
Pectoralis major muscle   48, 50, 52
Latissimus dorsi muscle   44
Teres major muscle   46
Coracobrachialis muscle   54

Internal rotation
Subscapularis muscle   42
Pectoralis major muscle, clavicular part   52
Deltoideus muscle, clavicular part   30
Latissimus dorsi muscle   44
Teres major muscle   46

External rotation
Infraspinatus muscle   38
Teres minor muscle   40
Deltoideus muscle, spinal part   32

Elbow

Flexion
Brachialis muscle   60
Biceps brachii muscle   58
Brachioradialis muscle   62
Pronator teres muscle   70

Extension
Triceps brachii muscle   64

Forearm

Supination
Supinator muscle   68
Biceps brachii muscle   58
Brachioradialis muscle   62

Pronation
Pronator quadratus muscle   72
Pronator teres muscle   70
Brachioradialis muscle   62
Flexor carpi radialis muscle   82

Wrist

Extension
Extensor digitorum muscle   90
Extensor carpi radialis longus muscle   76
Extensor carpi radialis brevis muscle   78
Extensor carpi ulnaris muscle   80
Extensor indicis muscle   92

Flexion
Flexor digitorum superficialis muscle   102
Flexor digitorum profundus muscle   104
Flexor carpi ulnaris muscle   86
Flexor carpi radialis muscle   82
Flexor pollicis longus muscle   110

Ulnar abduction
Flexor carpi ulnaris muscle   86
Extensor carpi ulnaris muscle   80
Flexor digitorum profundus   104

Radial abduction
Flexor carpi radialis muscle   82
Extensor carpi radialis longus muscle   76
Extensor carpi radialis brevis muscle   78
Flexor pollicis longus muscle   110
Abductor pollicis longus muscle   112

Joints of the metacarpus, fingers and thumb
Flexion

Flexor digitorum superficialis muscle   102
Flexor digitorum profundus muscle   104
Palmar interossei muscles   120
Dorsal interossei muscles of the hand   118
Flexor pollicis brevis muscle   108
Abductor digiti minimi muscle of the hand   116
Flexor digiti minimi brevis muscle of the hand   106

Extension

Extensor digitorum muscle   90
Extensor indicis muscle   92
Extensor digiti minimi muscle   94

Joints of the fingers and thumb
Flexion

Flexor digitorum profundus muscle   104
Flexor digitorum superficialis muscle   102

Extension

Extensor digitorum muscle   90
Extensor indicis muscle 92
Extensor digiti minimi muscle   94
Lumbricales muscles of the hand   100
Dorsal interossei muscles of the hand   118
Palmar interossei muscles   120
Extensor pollicis longus muscle   98

Carpometacarpal joint of the thumb
Flexion

Flexor pollicis longus muscle   110
Flexor pollicis brevis muscle   108
Adductor pollicis muscle   122

Extension

Extensor pollicis longus muscle   98
Extensor pollicis brevis muscle   96
Abductor pollicis brevis muscle   114

Abduction

Abductor pollicis brevis muscle   114
Abductor pollicis longus muscle   112
Extensor pollicis brevis muscle   96

Adduction

Dorsal interosseus muscle of the hand   118
Adductor pollicis muscle   122

Opposition

Opponens pollicis muscle   124
Flexor pollicis longus muscle   110
Flexor pollicis brevis muscle   108
Adductor pollicis muscle   122

Flexion

Flexor pollicis longus muscle   110
Flexor pollicis brevis muscle   108
Adductor pollicis muscle   122

Extension

Extensor pollicis longus muscle   98
Extensor pollicis brevis muscle   96
Abductor pollicis longus muscle   114

Lower extremity

Hip
Flexion

Iliopsoas muscle   136
Rectus femoris muscle   168
Tensor fasciae latae muscle   144
Gluteus medius muscle   140
Sartorius muscle   138

Extension

Gluteus maximus muscle   134
Semimembranosus muscle   178
Semitendinosus muscle   180
Biceps femoris muscle, long head   176
Gluteus medius muscle   140

Abduction

Gluteus medius muscle   140
Gluteus minimus muscle   142
Piriformis muscle   156
Tensor fasciae latae muscle   144
Gluteus maximus muscle   134

Adduction

Adductor longus muscle   148
Adductor brevis muscle   150
Adductor magnus muscle   154
Pectineus muscle   146
Gracilis muscle   152

Classification of muscles in the myofascial system

Upper extremity

Muscles of the pectoral girdle

2–3	Trapezius muscle, ascending part
1–2	Trapezius muscle, transverse part
1–2	Trapezius muscle, descending part
3	Levator scapulae muscle
2–3	Rhomboideus major muscle
2–3	Rhomboideus minor muscle
2–3	Serratus anterior muscle
1–2	Pectoralis minor muscle
n.c.	Subclavius muscle

Muscles of the shoulder

2	Deltoideus muscle, clavicular part
2	Deltoideus muscle, spinal part
2	Deltoideus muscle, acromial part
1	Supraspinatus muscle
1	Infraspinatus muscle
1	Teres minor muscle
1	Subscapularis muscle
2–3	Latissimus dorsi muscle
2	Teres major muscle
2–3	Pectoralis major muscle, abdominal part
2–3	Pectoralis major muscle, sternocostal part
2–3	Pectoralis major muscle, clavicular part
2	Coracobrachialis muscle

Muscles of the elbow

2–3	Biceps brachii muscle
2	Brachialis muscle
2	Brachioradialis muscle
2–3	Triceps brachii muscle
1–2	Anconeus muscle
1–2	Supinator muscle
1–2	Pronator teres muscle
1–2	Pronator quadratus muscle

Muscles of the wrist

3	Extensor carpi radialis longus muscle
2	Extensor carpi radialis brevis muscle
2	Extensor carpi ulnaris muscle
2	Flexor carpi radialis muscle
2	Palmaris longus muscle
2	Flexor carpi ulnaris muscle

Muscles of the fingers

2	Extensor digitorum muscle
2	Extensor indicis muscle
2	Extensor digiti minimi muscle
1–2	Extensor pollicis brevis muscle
2–3	Extensor pollicis longus muscle
1	Lumbricales muscles of the hand I to IV
3	Flexor digitorum superficialis muscle
2	Flexor digitorum profundus muscle
1–2	Flexor digiti minimi brevis muscle
2	Flexor pollicis brevis muscle
3	Flexor pollicis longus muscle
2	Abductor pollicis longus muscle
1–2	Abductor pollicis brevis muscle
1–2	Abductor digiti minimi muscle
1–2	Dorsal interossei muscles of the hand
1–2	Palmar interossei muscles
1–2	Adductor pollicis muscle
1–2	Opponens pollicis muscle
1–2	Opponens digiti minimi muscle
1	Palmaris brevis muscle

Lower extremity

Muscles of the hip

2–3	Gluteus maximus muscle
1–2, 2–3	Iliopsoas muscle
3	Sartorius muscle
1–2	Gluteus medius muscle
2	Gluteus minimus muscle
3	Tensor fasciae latae muscle
1	Pectineus muscle
3	Adductor longus muscle
2	Adductor brevis muscle
3	Gracilis muscle
2	Adductor magnus muscle
1–2	Piriformis muscle
1–2	Gemellus superior muscle
1–2	Obturatorius internus muscle
1	Gemellus inferior muscle
1	Obturatorius externus muscle
1	Quadratus femoris muscle

1: local   2: global, single joint   3: global, multi-joint   n.c.: not classifiable

Muscles of the knee

1–2–3	Quadriceps femoris muscle
3	Quadriceps femoris muscle
1–2	Quadriceps femoris muscle
2	Vastus intermedius muscle
2	Vastus lateralis muscle
2–3	Biceps femoris muscle
3	Semimembranosus muscle
3	Semitendinosus muscle
1	Popliteus muscle

Muscles of the ankle

3	Gastrocnemius muscle
3	Plantaris muscle
1–2	Soleus muscle
1–2	Tibialis posterior muscle
2–3	Tibialis anterior muscle
2	Peroneus longus muscle
1–2	Peroneus brevis muscle
1–2	Peroneus tertius muscle

Muscles of the toes

1–2	Extensor hallucis brevis muscle
2–3	Extensor hallucis longus muscle
1–2	Extensor digitorum brevis muscle
2–3	Extensor digitorum longus muscle
2–3	Flexor hallucis brevis muscle
3	Flexor hallucis longus muscle
2–3	Flexor digitorum brevis muscle
3	Flexor digitorum longus muscle
1–2	Quadratus plantae muscle
1–2	Flexor digiti minimi brevis muscle
1–2	Dorsal interossei muscles of the foot
2–3	Abductor hallucis muscle
2–3	Abductor digiti minimi muscle
1–2	Adductor hallucis muscle
1–2	Plantar interossei muscles
1–2	Lumbricales muscles of the foot

Trunk

Deep musculature of the back, lumbar

2–3	Iliocostalis lumborum muscle
1–2	Lateral lumbar intertransversarii muscles
1–2	Medial lumbar intertransversarii muscles
1	Rotatores lumborum muscles
1–2	Multifidus lumborum muscle

Deep musculature of the back, thoracic

3	Iliocostalis thoracis muscle
3	Longissimus thoracis muscle
2–3	Spinalis thoracis muscle
1–2	Rotatores thoracis muscles
1	Multifidus thoracis muscle
1–2	Semispinalis thoracis muscle

Deep musculature of the back, cervical

3	Iliocostalis cervicis muscle
1–2	Longissimus capitis muscle
1–2	Longissimus cervicis muscle
2–3	Splenius cervicis muscle
2–3	Splenius capitis muscle
1–2	Spinalis cervicis muscle
1–2	Spinalis capitis muscle
1–2	Rotatores cervicis muscles
1–2	Multifidus cervicis muscle
2	Semispinalis cervicis muscle
2	Semispinalis capitis muscle
1	Rectus capitis posterior minor muscle
1	Rectus capitis posterior major muscle
1	Obliquus capitis superior muscle
1	Obliquus capitis inferior muscle

Ventral musculature, abdominal

3	Rectus abdominis muscle
2–3	Obliquus externus abdominis muscle
1	Obliquus internus abdominis muscle
n.c.	Cremaster muscle
1–2	Transversus abdominis muscle
1–2	Quadratus lumborum muscle

Ventral musculature, thoracic

1-2	External intercostal muscles
1–2	Serratus posterior superior muscle
1–2	Internal intercostal muscles
1–2	Serratus posterior inferior muscle
1	Diaphragm

Pelvic floor muscles

1	Levator ani muscle
1	Pubovaginalis muscle
1	Puboprostaticus muscle
1	Puborectalis muscle
1	Pubococcygeus muscle
1	Iliococcygeus muscle
1	Ischiococcygeus muscle
1–2	Sphincter ani externus muscle
1–2	Transversus perinei profundus muscle
1–2	Transversus perinei superficialis muscle

1	Ischiocavernosus muscle
1	Bulbospongiosus muscle

Neck

Ventral musculature

3	Sternocleidomastoideus muscle
1–2	Longus capitis muscle
1	Rectus capitis anterior muscle
1–2	Longus colli muscle
2–3	Scalenus anterior muscle
2–3	Scalenus medius muscle
2–3	Scalenus posterior muscle
2	Sternohyoideus muscle
2–3	Omohyoideus muscle
2–3	Sternothyroideus muscle
2	Thyrohyoideus muscle
2–3	Digastricus muscle
2–3	Stylohyoideus muscle
2	Mylohyoideus muscle
2	Geniohyoideus muscle

Head

Muscles of facial expression

n.c.	Epicranius muscle
n.c.	Corrugator supercilii muscle
n.c.	Procerus muscle
n.c.	Orbicularis oculi muscle
n.c.	Levator palpebrae superioris muscle
n.c.	Nasalis muscle
n.c.	Levator labii superioris alaeque nasi muscle
n.c.	Levator labii superioris muscle
n.c.	Zygomaticus major muscle
n.c.	Zygomaticus minor muscle
n.c.	Risorius muscle
n.c.	Levator anguli oris muscle
n.c.	Buccinator muscle
n.c.	Orbicularis oris muscle
n.c.	Depressor anguli oris muscle
n.c.	Depressor labii inferioris muscle
2–3	Platysma

Muscles of mastication

2–3	Temporalis muscle
1–2	Masseter muscle
1–2	Pterygoideus lateralis muscle
1–2	Pterygoideus medialis muscle

Muscles of the tongue

n.c.	Intrinsic tongue muscles
n.c.	Extrinsic tongue muscles

Muscles of the eye

n.c.	Rectus superior muscle
n.c.	Rectus inferior muscle
n.c.	Obliquus superior muscle
n.c.	Obliquus inferior muscle
n.c.	Rectus medialis muscle
n.c.	Rectus lateralis muscle

1: local   2: global, single joint   3: global, multi-joint   n.c.: not classifiable

References

BAJEK, S., D. BOBINAC ET AL. (2000): Muscle fiber type distribution in multifidus muscle in cases of lumbar disc herniation. Acta Med Okayama 54(6): 235–241.

BELAVY, D. L., J. K. NG ET AL. (2010): Influence of prolonged bedrest on spectral and temporal electromyographic motor control characteristics of the superficial lumbo-pelvic musculature. J Electromyogr Kinesiol 20(1): 170–179.

BELAVY, D. L., C. A. RICHARDSON ET AL. (2007): Tonic-to-phasic shift of lumbo-pelvic muscle activity during 8 weeks of bed rest and 6-months follow up. J Appl Physiol 103(1): 48–54.

BERGMARK, A. (1989): Stability of the lumbar spine. Acta Orthopedia Scandavica 60(supplement 230): 1–54.

CHOLEWICKI, J. and S. M. MCGILL (1996): Mechanical stability of the in vivo lumbar spine: implications for injury and chronic low back pain. Clinical Biomechanics 11(1): 1–15.

GANDEVIA, S. C., J. E. BUTLER ET AL. (2002): Balancing acts: respiratory sensations, motor control and human posture. Clin Exp Pharmacol Physiol 29(1–2): 118–121.

GOLDBY, L. J., A. P. MOORE ET AL. (2006): A randomized controlled trial investigating the efficiency of musculoskeletal physiotherapy on chronic low back disorder. Spine (Phila Pa 1976) 31(10): 1083–1093.

HIDES, J. A., G. A. JULL ET AL. (2001): Long-term effects of specific stabilizing exercises for first-episode low back pain. Spine (Phila Pa 1976) 26(11): 243–248.

HIDES, J. A., M. J. STOKES ET AL. (1994): Evidence of lumbar multifidus muscle wasting ipsilateral to symptoms in patients with acute/subacute low back pain. Spine (Phila Pa 1976) 19(2): 165–172.

HODGES, P. W. AND S. C. GANDEVIA (2000): Activation of the human diaphragm during a repetitive postural task. J Physiol 1(522): 1165–1175.

HODGES, P. W., I. HEIJNEN ET AL. (2001): Postural activity of the diaphragm is reduced in humans when respiratory demand increases. J Physiol 537(Pt 3): 999–1008.

HODGES, P., A. KAIGLE HOLM ET AL. (2003): Intervertebral stiffness of the spine is increased by evoked contraction of transversus abdominis and the diaphragm: in vivo porcine studies. Spine (Phila Pa 1976) 28(23): 2594-2601.

HODGES, P. W. and C. A. RICHARDSON (1996): Inefficient muscular stabilization of the lumbar spine associated with low back pain. A motor control evaluation of transversus abdominis. Spine (Phila Pa 1976) 21(22): 2640–2650.

HODGES, P. W. and C. A. RICHARDSON (1997): Feedforward contraction of transversus abdominis is not influenced by the direction of arm movement. Exp Brain Res 114(2): 362–370.

HODGES, P. W., R. SAPSFORD ET AL. (2007): Postural and respiratory functions of the pelvic floor muscles. Neurourol Urodyn 26(3): 362–371.

HOFFER, J. and S. ANDREASSEN (1981): Regulation of soleus muscle stiffness in premammillary cats. Journal of Neurophysiology 45(2): 267–285.

HOGAN, N. (1990): Mechanical impedance of the single- and multi-articular systems. In: J. M. Winters and S. L.-W. Woo, editors. Multiple Muscle Systems: Biomechanics and Movement Organization. Springer, New York: 149–164.

JANDA, V. (1996): Evaluation of muscular dysbalance. Williams & Wilkins, Baltimore.

JOHANSSON, H., P. SJOLANDER ET AL. (1991): A sensory role for the cruciate ligaments. Clinical Orthopaedic and Related Research 268: 161–178.

JULL, G., P. TROTT ET AL. (2002): A randomized controlled trial of exercise and manipulative therapy for cervicogenic headache. Spine (Phila Pa 1976) 27(17):1835–1843; discussion 1843.

KJAER, P., T. BENDIX ET AL. (2007), Are Mri defined fat infiltrations in the multfidus muscles associated with low back pain? BMC Medicine 5: 2.

KLEIN-VOGELBACH, S. (1991): Therapeutic Exercises in Functional Kinetics. Springer, Berlin – Heidelberg.

LLOYD, D. G. (2001): Rationale for training programs to reduce anterior cruciate ligament injuries in Australian football. J Orthop Sports Phys Ther 31(11): 645–654.

MACINTOSH, J. E., N. BOGDUK ET AL. (1993): The effects of flexion on the geometry and actions of the lumbar spine erector spinae. Spine (Phila Pa 1976) 18(7): 884–893.

MANDELL, P., E. WEITZ ET AL. (1993): Isokinetic trunk strength and lifting strength measures: Differences and similarities between low-back-injured and noninjured workers. Spine (Phila Pa 1976) 18(16): 2491–2501.

MANNION, A. F., A. JUNGE ET AL. (2001): Active therapy for chronic low back pain: part 3. Factors influencing self-rated disability and its change following therapy." Spine (Phila Pa 1976) 26(8):920–929.

MOSELEY, G. L., M. K. NICHOLAS ET AL. (2004): A randomized controlled trial of intensive neurophysiology education in chronic low back pain. Clin J Pain 20(5): 324–330.

NADLER, S. F., G. A. MALANGA ET AL. (2002): Hip muscle imbalance and low back pain in athletes: influence of core strengthening. Med Sci Sports Exerc 34(1): 9–16.

NG, J. K., M. PARNIANPOUR ET AL. (2001): Functional roles of abdominal and back muscles during isometric axial rotation of the trunk. J Orthop Res 19(3): 463–471.

NG, J. K. F., C. A. RICHARDSON ET AL. (2002): Fatigue-related changes in torque output and electromyographic parameters of trunk muscles during isometric axial rotation exertion: an investigation in patients with back pain and in healthy subjects. Spine (Phila Pa 1976) 27(6): 637–646.

O'Sullivan, P. B., G. D. Phyty et al. (1997): Evaluation of specific stabilizing exercise in the treatment of chronic low back pain with radiologic diagnosis of spondylolysis or spondylolisthesis. Spine (Phila Pa 1976) 22(24): 2959–2967.

Richardson, C., P. Hodges et al. (2004): Therapeutic exercise for lumbopelvic stabilization: A motor control approach for the treatment and prevention of low back pain. 2nd edition. Churchill Livingstone, London.

Richardson, C. A. and M. I. Bullock (1986): Changes in muscle activity during fast, alternating flexion-extension exercises of the knee. Scandinavian Journal of Rehabilitative Medicine 18: 51–58.

Sahrmann, S. (1990): Diagnosis and treatment of muscle imbalances associated with regional pain syndromes. Washington University School of Medicine, Seatle.

Sterling, M., G. Jull et al. (2004): Characterization of acute whiplash-associated disorders. Spine (Phila Pa 1976) 29(2): 182–188.

Stuge, B., I. Holm et al. (2006): To treat or not to treat postpartum pelvic girdle pain with stabilizing exercises? Man Ther 11(4): 337–343.

Tsao, H. and P. W. Hodges (2007): Immediate changes in feedforward postural adjustments following voluntary motor training. Exp Brain Res 3: 3.

Twomey, L. T. and J. R. Taylor (1979): A description of two new instruments for measuring the ranges of sagittal and horizontal plane motions in the lumbar region. Australian Journal of Physiotherapy 25(5): 201–204.

van den Berg, F. (1999): Angewandte Physiologie 1. Das Bindegewebe des Bewegungsapparates verstehen und beeinflussen. Thieme, Stuttgart: 181–196.

Wang, H. K. and T. Cochrane (2001): Mobility impairment, muscle imbalance, muscle weakness, scapular asymmetry and shoulder injury in elite volleyball athletes. J Sports Med Phys Fitness 41(3): 403–410.

White, A. A. and M. M. Panjabi (1990): Clinical Biomechanics of the Spine. 2nd edition, JB Lippincott, Philadelphia.

Zhao, W. P., Y. Kawaguchi et al. (2000): Histochemistry and morphology of the multifidus muscle in lumbar disc herniation: comparative study between diseased and normal sides. Spine (Phila Pa 1976) 25(17): 2191–2199.

Further reading

Benninghoff, A. (2008): Anatomie 1: Makroskopische Anatomie, Histologie, Embryologie, Zellbiologie. Vol. 1, Urban & Fischer, München.

Brügger, A. (1986): Die Erkrankungen des Bewegungsapparates und seines Nervensystems. Gustav Fischer, New York

Cody, J. (1991): Visualizing muscles: a new ecorche approach to surface anatomy. University Press of Kansas, Lawrence.

Cutter, N. C. and C. G. Kevorkian (1999): Handbook of manual muscle testing. McGraw Hill, New York.

Field, D. (2006): Field's anatomy, palpation and surface markings. Butterworth Heinemann, Oxford.

Hahn von Dorsche, H. and Dittel, R. (2005): Anatomie des Bewegungssystems. Neuromedizin Verlag, Bad Hersfeld.

Henne-Bruns, D. et al. (2001): Chirurgie, Thieme, Stuttgart.

Hislop, H. J. and J. Montgomery (2000): Daniels' und Worthinghams Muskeltests. Urban & Fischer, München.

Kendall, F. P. et al. (2001): Muskeln. Urban & Fischer, München.

Leonhardt, H., B. Tillmann et al. (1998): Rauber/Kopsch, Anatomie des Menschen. Thieme, New York.

Palastanga, N., D. Field et al. (2002): Anatomy and human movement. 4th edn., Butterworth-Heinemann, Oxford.

Rössler H. and W. Rüther (2000): Orthopädie. Urban & Fischer, München.

Sobotta, J. (2007): Atlas der Anatomie des Menschen: Allgemeine Anatomie – Bewegungsapparat – Innere Organe – Neuroanatomie. Urban & Fischer, München.

Terminologia Anatomica (1998): Thieme, Stuttgart.

Trepel, M. (2008): Neuroanatomie. Urban & Fischer, München.

Wayne, A. W., L. R. Mitchell et al. (2007): Gray's Anatomie für Studenten. Urban & Fischer in Elsevier, München.

Index